Eco-Landscape Design

Eco-Landscape Design

Editor

Carlos Schweig

Eco-Landscape Design
Edited by **Carlos Schweig**

ISBN: 978-1-68117-245-3
Library of Congress Control Number: 2016934786

© 2017 by
SCITUS Academics LLC,
www.scitusacademics.com
Box No. 4766, 616 Corporate Way,
Suite 2, Valley Cottage,
NY 10989

Preface

Landscape design is an independent profession and a design and art tradition, practised by landscape designers, combining nature and culture. In contemporary practice, landscape design bridges between landscape architecture and garden design.Eco-Landscape Design demonstrates that an intelligent and thoughtful approach to landscape design can not only ensure survival, it can reap compound benefits and rewards far in excess of those originally envisaged.The ability to adapt to a changing environment has ensured the continued survival of the human race into the 21st century. The effects of drought, melting polar ice and increased incidences of extreme weather events will impact on the diverse landscapes of the earth and a human population predicted to be 9 billion by the middle of the 21st century, a three-fold increase in less than one hundred years.Landscape design focuses on both the integrated master landscape planning of a property and the specific garden design of landscape elements and plants within it. The practical, aesthetic, horticultural, and environmental sustainability components merit Landscape design inclusion. It is often divided into hardscape design and softscape design. Landscape designers often collaborate with related disciplines such as architecture and geography, soils and civil engineering, surveying, landscape contracting, botany, and artisan specialties. Design project focus can tend towards: in landscape design - artistic composition and artisanship, horticultural finesse and expertise, and a detailed site involvement emphasis from concepts through construction; whereas in landscape architecture - focus of urban planning, city and regional parks, civic and corporate landscapes, large scale interdisciplinary projects, and delegation to contractors after completing designs. The precious resources of water and the air that we breathe are no longer taken for granted; rivers flowing through the world's mega-cities are now being cleaned, restored and given pride of place in the landscapes they flow through. Conservation projects provide evidence that even fragile island and desert landscapes can be protected from the negative impacts of population. This book, Eco-Landscape Design, provides a valuable comprehension into landscaping activity worldwide by those tasked with housing, feeding and nurturing all species that share the planet.

Table of Contents

CHAPTER 1

Ecological Landscape Design

Filiz Çelik

¹ Selçuk University, Faculty of Agriculture, Department of Landscape Architecture, Konya, Turkey

1. INTRODUCTION

The most critical changes in the world over the last century have been derived from the variety of environmental problems. Growing environmental problems now affect entire the world. The majority of environmental problems originates in human greed and interference.

It is well known that planet Earth is experiencing a so-called environmental crisis (ecological crisis). This crisis is characterized by three major themes:

- Rapid growth of the human population and its associated economic activity,
- The depletion of both non-renewable and renewable resources, and
- Extensive and intensive damage caused to ecosystems and biodiversity.

The environmental crisis is a predicament of inappropriate design-it is a consequence of how cities have been developed, industrialization undertaken, and ecoscapes used. Fundamentally, the problem has been one of inadequate integration of ecological concerns into planning (Shu-Yang et al., 2004).

In many ways, the environmental crisis is a design crisis. It is clear that design has not been given a rich enough context. Design is a hinge that inevitably connects culture and nature through exchanges of materials, flow of energy, and choices of land use. The every world of buildings, artifacts, and domesticated landscape is a design world, one shaped by human (Van Der Ryn and Cowan, 1996).

Some environmental problems have arisen from design problems. Design can have a crucial impact upon the environment in many different ways. This is because every design decision is an environmental decision. Design is a consequence of how things are made, and the world has been shaped by the designers. The present forms of everything in the world have been derived from design. It is clear that design has been previously used only to meet human needs. Unfortunately, in many past situations environmental effects were ignored during the design stage. Design has not been taught in the context of its ecological impact. Many practices in the design field have been done with unsustainable design principles. The environmental problems have boosted the sustainable explorations necessary for protecting ecological system in order to address and find solutions to the problems. Scientists, planners and designers have questioned the effectiveness of design and have suggested incentives as alternatives. At the end of 20th century, the power of design for to solve the problem and the potential of design for sustainability have been noticed; an integration that goes from ecological processes and functions to design has started. Design and its potential have been regarded a creative problem solving activity. While ecological sciences provide the knowledge and guidance, design provides creative solutions for the environmental problems.

In a world facing a future characterized both by expanding metropolitan regions and by ecological crisis, it is imperative that we re-think the relationship of urban dwellers to the natural environment. The 21st century is expected to be the first in history in which a majority of humanity lives in cities, and if present trends continue, it may also be the one in which those urban populations inflict irreversible damage on the earth's living systems (Eisensten, 2001).

Designers and design critics are increasingly emphasizing the actual or, potentially, radical nature of an ecological approach to design which implies a new critique-a recognition of the fact that to adopt an ecological approach to design is, by definition, to question and oppose the status quo (Madge, 1997). In this context design has a crucial role to play in achieving sustainability and to provide solutions for environmental problems. In parts of the world dominated by humans, landscape design can have significant and positive environmental effects (Helfand et al., 2006).

Ecological design explicitly addresses the design dimension of the environmental crisis. It is not a style. It is a form of engagement and partnership with nature that is not bound to a particular design profession (Van Der Ryn and Cowan, 1996).

In recent years ecological design has been applied to an increasingly diverse range of technologies and innovative solutions for the management of resources. Ecological technologies have been created for the food sector, waste conversion industries, architecture and landscape design, and to the field of environmental protection and restoration (Todd et al., 2003).

As environmental problems escalate, ecological design in landscape architecture has increasing in academia and practice. Ecological design is an integrative ecologically responsible design discipline. Ecological design has been emerged as a means to model ecological processes and functions, and therefore as a model for sustainability. Today's ecological landscape design movement tends to address design problems.

2. THE RELATIONSHIP BETWEEN ECOLOGY, SUSTAINABILITY AND DESIGN

Ecology, sustainability and design are different fields, but they have been merged together in recent years. This is because human lifestyle is having an increasingly negative impact on the surrounding environments.

Ecology, in the 100 years since its inception, has increasingly provided the scientific foundation for understanding natural processes, managing environmental resources and achieving sustainable development. By the 1960s, ecology's association with the environmental movement popularized the science and introduced it to the design professions (e.g. landscape architecture, urban design and architecture) (Makhzoumi, 2000).

"Ecology" in the profession of landscape architecture and planning can't be understood solely as meaning the relationship between nonhuman life forms and their environment. The term ecology is traditionally used as shorthand for the sum of the biophysical forces that have shaped and continue to shape the physical world. Thus there are other dimensions to be recognized if we are to understand the key nature of ecology: that of process, integration, and humanity (Ahern et al., 2001).

The relationship between design and ecology is a very close one, and makes for some unexpected complexities (Papanek, 1995). Ecology explains how the natural world is and how it behaves, and design is also the key intervention point for making sustainability in ecology (Figure 1.). The knowledge gained from ecology can influence landscape design.

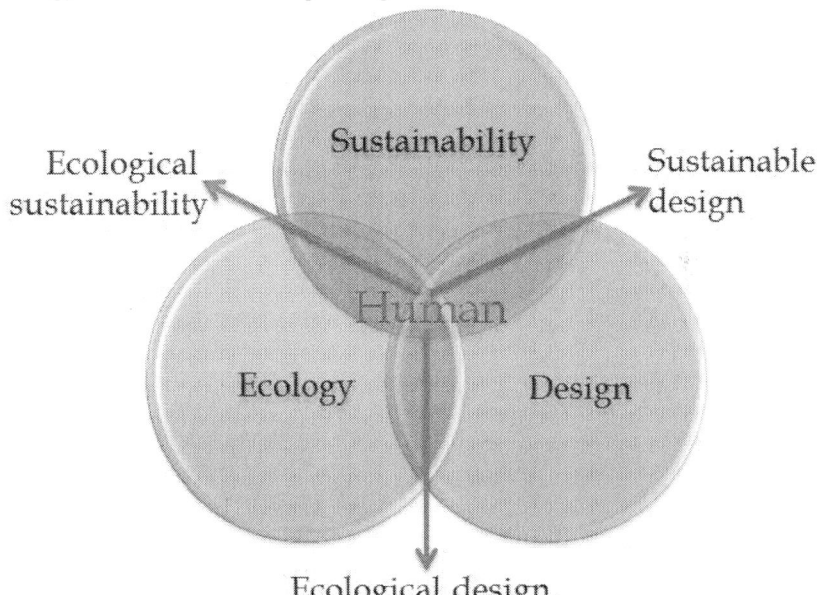

Figure 1. The relationship between ecology, sustainability and design

In landscape architecture ecology's emphasis on natural processes and the interrelatedness of landscape components influenced outlook and method and prompted an ecological approach to design (Makhzuomi and Pungetti, 1999). The ecological component is crucial in landscape design according to the principles of sustainability.

The typical relationship of designer and scientist presumes that most of what can be known is known. The designer is the creative partner; the scientist is an interactive book. Since the scientific base for ecological design is nascent, the nature of this relationship is flawed. Science and design are complementary ways to generate knowledge (and therefore both are creative endeavors). Scientists solve problems inductively, forming generalized principles from specific observations (Figure 2.). Designers use general principles to solve specific problems deductively. The knowledge available for ecological design would greatly increase if designed landscapes were used as ecological research sites. Designed landscapes that are typical of the surrounding region, with one to a few clear themes and repeated patterns (replication), are potential ecological research sites (Galatowitsch, 1998).

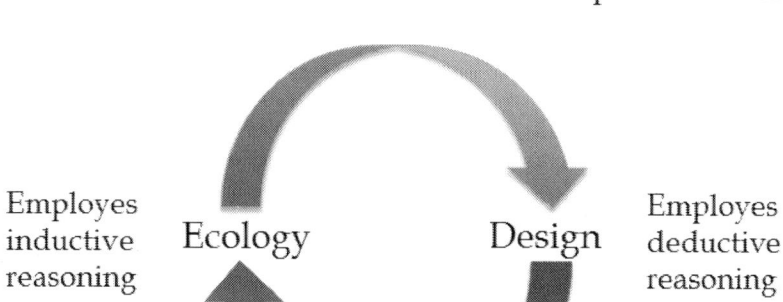

Applying general principles to a specific situation.

Employes inductive reasoning

Ecology

Design

Employes deductive reasoning

Making observation from specific situations and generalizing.

Figure 2. Design and ecology are complementary problem-solving techniques (Galatowitsch, 1998)

3. ECOLOGICAL SUSTAINABILITY

Sustainability is not a single movement or approach. It is varied as the communities and interests currently grappling with the issues it raises. One the one hand, sustainability is the province of global policy makers and environmental experts. One the one hand, sustainability is also the domain of grassroots environmental and social groups, indigenous peoples preserving traditional practices, and people committed to changing their own communities. The environmental educator David W. Orr calls these two approaches technological sustainability and ecological sustainability. While both are coherent responses to the environmental crisis, they are far apart in their specifics. Technological sustainability, which seems to get most of the airtime, may be characterized this way: "every problem has either a technological answer or a market solution. There are no dilemmas to be avoided, no domains where angels fear to tread." Ecological sustainability is the task of finding alternatives to the practices that got us into trouble in the first place; it is necessary to rethink agriculture, shelter, energy use, urban design, transportation, economics, community pattern, resource use, forestry, the importance of wilderness, and our central values. While the two approaches have important points of contact, including a shared awareness of the extent of the global environmental crisis, they embody two very different visions of a sustainable society (Van Der Ryn and Cowan, 1996).

A goal of ecological design is to help meet this vision of ecological sustainability, by finding ways of manufacturing goods, constructing buildings, and planning more complex enterprises, such as business and industrial parks, while reducing resource consumption and avoiding ecological damage to the degree possible (Shu-Yang et al., 2004).

Ecological design strives to achieve an increasing reliance on renewable sources of energy and materials, while maintaining standards of quality of goods and services and reducing overall resource consumption, waste generation, and ecological damage through efficiencies of use, re-use, and recycling.

Ecological design provides a framework for uniting conventional perspectives on design and management with environmental ones, by incorporating the consideration of ecological concerns at relevant spatial and temporal scales. If the principles of ecological design are rigorously applied, important progress will be made towards ecological sustainability (Shu-Yang et al., 2004).

Landscape design mostly depends on natural resources, so ecological sustainability is very important. Landscape design contributes to the ecological sustainability.

4. SUSTAINABLE DESIGN

There is no verifiable starting point for the current sustainable design movement. It seems to have converged from several different broad ideas concerning our relationship with the natural world. Some of the key figures who have contributed to the discussion include Frederick Law Olmsted and Calvert Vaux, John Muir, Theodore Roosevelt and Gifford Pinchot, Aldo Leopold, Rachel Carson, and Ian McHarg (Cook and VanDerZanden, 2011).

Sustainability is an ecological term that has been used since the early 1970s to mean: "the capacity of a system to maintain a continuous flow of whatever each part of that system needs for a healthy existence," and when applied to ecosystems containing human beings refers to the limitations imposed by the ability of the biosphere to absorb the effects of human activities. The term sustainable development was first used in the early '80s, but was popularized by the Brundtland Report of 1987. "Sustainable" has become the buzzword of the '90s in the same way "green" was in the '80s, and is equally open to different interpretations and misuse. The Brundtland Report adopted a global perspective on the consumption of energy and resources, and emphasized the imbalance between rich and poor parts of the world, arguing that: "Sustainable development requires that those who are more affluent adopt lifestyles within the planet's ecological means." However, because the report also argued that economic growth or development is still possible as long as it is green growth, this has been interpreted by many to endorse a "business as usual" approach, with just a nod in the direction of environmental protection. This ignores the real meaning of sustainable development, which is enshrined in the widely quoted concept of "futurity":..."meeting the needs of the present without compromising the ability of future generations to meet their own needs."

When applied to design, this not only introduces or reintroduces the ideas of ethical and social responsibility, but also the notion of time and timescale. Thinking about the life cycle of products through time, and considerations about design for recycling, have led to the concept of DfD (Design for Disassembly) followed by the idea of going Beyond recycling towards the design of long-life, durable products. These two concepts are not as contradictory as they sound, as Victor Papanek has recently remarked: "To design durable goods for eventual disassembly may sound like an oxymoron, yet it is profoundly important in a sustainable world. The term "sustainable design" has begun to be used in the last 15 years or so to refer to a broader, longer-term vision of ecological design. At the Centre for Sustainable Design, established at the Surrey Institute of Art and Design in July 1995, sustainable design means "analyzing and changing the 'systems' in which we make, use, and dispose of products," as opposed to more limited, short-term DFE. The ECO2 group makes a similar distinction between "green design, project-based, single issue and relatively short-term; and sustainable' design, which is system-based, long-term" ethical design. Emma Dewberry and Phillip Goggin have also explored the distinctions between ecological design and sustainable design; arguing that, whereas ecological design can be applied to all products and used as a suitable guide for designing at product level: "The concept of sustainable design, however, is much more complex and moves the interface of design outwards toward societal conditions, development, and ethics.... This suggests changes in design and the role of design, including an inevitable move from a product to a systems-based approach, from hardware to software, from ownership to service, and will involve concepts such as dematerialization and "a general shift from physiological to psychological needs." Finally, they emphasize the extent to which consumption patterns must change, and refer to the inequality between developed and developing nations, the fact that 20 percent of the world's population consumes %80 of the world's resources and conclude that ecological design does fit into a global move toward sustainability, but has many limitations in this context. This is the point made by Gui Bonsiepe, who has expressed the fear that ecological design will remain the luxury of the affluent countries while "the cost of environmental standards would be shifted onto the shoulders of the Third World." (Madge, 1997).

Sustainability can be viewed as the long-term outcome of maintaining landscape integrity. Designing for sustainable landscapes necessitates a holistic and integrative outlook that is based on ecological understanding and awareness of the potentialities and limitations of a given landscape. Such understanding ensures that in accommodating future uses their impact on existing ecosystems and essential ecological processes and biological and landscape diversity is anticipated. This will allow for healthy ecosystems and long-term ecological stability (Makhzuomi and Pungetti, 1999).

Designs that promote sustainable landscapes should be simultaneously aware of local values and resources as well as regional and national ones, as sustainability is the domain of both. Further, achieving landscape sustainability requires patience, humility and a design approach that attends to scale, community, self-reliance, traditional knowledge and the wisdom of nature's own (Van der Ryn and Cowan, 1996).

Whereas maintaining landscape integrity and designing for sustainability can be seen as the practical objectives of ecological landscape design, the design of creative and meaningful places addresses aesthetic concerns.

The following is a palette of terms that in some way define or refer to sustainable design:

- Design for environment,
- Ecological design (ecodesign/eco-design),
- Environmental design,
- Environmentally oriented design,
- Ecologically oriented design,
- Environmentally responsible design,
- Socially responsible design,
- Environmentally sensitive product design,
- Sustainable product development,
- Green design,
- Life-cycle design,
- Dematerialization,
- Eco-efficiency design,
- Energy efficient design, and
- Biodesign (Deniz, 2002).

5. THE ROLE OF TECHNOLOGY IN ECOLOGICAL DESIGN

Environmental problems become an increasingly important aspect of the designer's work to minimize the risks and to solve the problems. Because of the rapid technological development, environmental problems increase day by day. On the other hand, new technologies often tend to be less dangerous than what they replace, and hence designers may find themselves in the forefront of identifying problems which must be addressed by technology. Sometimes, existing technologies may not be able to provide the solution, and the designer may have to influence the development of a new technological approach. Designers must also follow technological developments in order to be sure of incorporating the most environmentally advanced technologies (Deniz, 2002).

Technology has been the principal method by which we intervene on the land and modify the ecosystems to ensure our existence, yet its various manifestations are most often ignored in discussions of the designed landscape. In fact, much of the rationale for this exhibit might be based upon the obfuscation of ecological clarity by technology and the subsequent employment of more benign and expressive techniques for bringing back such clarity. In the ordinary landscape, the instances in which intentional land design aims at a higher, symbolic meaning in some decipherable form are few when compared with the countless millions of ordinary landscape structured by the dominant, operative, contemporary technological paradigms. In one sense, we have covered up our ecosystems with our technologies; we have obscured a degree of innate clarity of the former with the vast complexities of the latter. While science and technology have made it possible to comprehend deeper levels of ecosystem knowledge, they have also enabled the physical cover-up and subsequent concealment of dimensions of the landscape once readily accessible to more primal peoples. With technological hegemony, our ecosystems have gained little and lost a lot (Thayer Jr., 1998).

This raises the whole issue of the relationship between design and the "Appropriate Technology" (AT) movement in the last twenty to thirty years. Schumacher (1973) coined the term "intermediate technology" to signify "technology of production by the masses, making use of the best of modern knowledge and experience, conducive to decentralization, compatible with the laws of ecology, gentle in its use of scarce resources, and designed to serve the human person instead of making him the servant of machines". The central tenet of appropriate technology is that a technology should be designed to be compatible with its local setting. Examples of current projects that are generally classified as appropriate technology include passive solar design, active solar collectors for heating and cooling, small windmills to provide electricity, roof-top gardens and hydroponic greenhouses, permaculture, and worker-managed craft industries. There is general agreement, however, that the main goal of the appropriate technology movement is to enhance the self reliance of people on local level. Characteristics of self reliant communities that appropriate technology can help facilitate include: low resource usage coupled with the extensive recycling; preference for renewable over nonrenewable resources; emphasis on environmental harmony; emphasis on small-scale industries; and a high degree of social cohesion and sense of community (Roseland, 1997).

6. EMERGE OF ECOLOGICAL LANDSCAPE DESIGN

Landscape architecture is a multi-disciplinary field, incorporating aspects of; botany, horticulture, the fine arts, architecture, industrial design, geology and the earth sciences, environmental psychology, geography, and ecology.

Landscape architecture has ecological thinking at the core of its legacy (Mozingo, 1997). As a result of a trend favoring ecological perspectives in design, significant changes have occurred in the landscape architecture profession in recent decades through the move to integrate ecological perspectives (Hooper et al., 2008).

Thinking ecologically about design is certainly not a "new" idea. Since ancient times "designers" looked to nature for "solutions" to their common problems; they saw nature as the perfect model to follow. Even though, in recent times, an increase in ecological education and environmental awareness is apparent among design professionals, there is still the need to better understand the expression of ecology through design (Lomba-Ortiz, 2003). In the face of the environmental problems new approaches to reconciling the divide between ecology and design have been explored in landscape architecture.

Since the 1960s, ecology has increasingly influenced the design professions, providing for a holistic and dynamic outlook on nature, environment and landscape. The different dimensions of ecology have come to imply the ability to think broadly, to search for patterns that connect and to observe nature with insight. Alternatively, ecological knowledge allows a comprehensive understanding of landscape as the outcome of interacting natural and cultural evolutionary processes which account for pattern, diversity, sustainability and stability (Makhzuomi and Pungetti, 1999).

To date, however, ecological design has been principally concerned with the realistic emulation of ecological form, function, and, where possible, process. As an outgrowth of, and to some degree, a fusion between landscape architecture, ecology, environmental planning, and the building science aspects of architecture, there is a distinctive functional emphasis in the discipline. Ironically, artistic elements and visual aesthetics have not been a priority in a discipline that bears the label of "design." I would attribute this principally to the dominance of landscape architecture in influencing ecological design, itself (until recently) a discipline characterized by a schism between garden design and horticulture in one domain, and technical ecologists concerned with ecological restoration and reconstruction in the other. This remediative, reactive "applied ecology" practice of landscape architecture along with related environmental professions have understandably been the progenitors of the new discipline of ecological design, largely (and understandably) as a response to global environmental crises (Lister, 2005).

Motivated by environmental values, landscape architects became increasingly knowledgeable about ecological principles and systems (Meyer, 2000). Ecology, the study of interactions between organisms and their environments, has long been a compelling theme for faculty, practitioners, and students of landscape design and planning. Frederick Law Olmsted's visionary public designs, Jens Jensen's native plantings, May Watt's observations of vernacular landscapes, and Ian McHarg's book, Design with Nature, are all milestones of ecological thinking in landscape design and planning (Johnson and Hill, 2001). McHarg (1969), Spirn (1984) and Hough (1995) played seminal roles in applying theories and principles of ecological landscape design to urban areas (Özgüner et al., 2007). Ian McHarg who, perhaps more than any other, popularized ecology in landscape architecture. Patrick Geddes is the initiator of an ecological approach in design and planning and because he offered an integrative view of the environment that embraced urban design, landscape design and planning. John Tillman Lyle offers a comprehensive approach embracing theory, practice and method (Makhzuomi and Pungetti, 1999).

In the late 1860's Frederick Law Olmstead supported the idea that landscape architects were stewards of the land. Olmstead's designed landscapes borrowed aesthetically from the picturesque but he was overtly conscious of ecological processes playing a critical role in the function and design of landscape spaces (Ware, 2004).

The early influence of ecology can be traced to the work of late nineteenth century visionary biologist Patrick Geddes, the conceptual initiator of an ecological approach to urban and landscape design and landscape planning. Patrick Geddes had a clear, overall conceptual strategy for improving the manmade environment and for advocating a sympathetic coexistence with the natural environment. In his 'biological principles of economics' he came closest to the present day concept of sustainability (Makhzuomi and Pungetti, 1999).

Ecological thinking was only resumed with the publication of Ian McHarg's (1969) 'Design with Nature'. The significance of McHarg's work, however, lies elsewhere, namely in introducing ecological understanding to the profession. McHarg believed that ecology had the potential to emancipate landscape architects from the static scenic images of ornamental horticulture by steering them away from arbitrary and capricious designs (Makhzuomi and Pungetti, 1999). Ian McHarg's work fore grounded much of the early sustainable design discussions of the 1970's and into the 1980's. Carl Steinz's, Fred Steiner's, and Rob Thayer's earliest work was a critique of McHarg's methods (Ware, 2004).

John Tillman Lyle's (1985) 'Design for Human Ecosystems' is a comprehensive integration of ecological concepts and landscape design. The term human ecosystems is proposed by Lyle to signify the totality of the landscape at hand as a warning against a strongly visual notion of landscape assessment and as a reminder that the landscape needs to be evaluated as the outcome of natural and cultural processes. Lyle argues the necessity of making full use of ecological understanding in the process of designing ecosystems; only then can "we shape ecosystems that manage to fulfill all their inherent potentials for contributing to human purposes, that are sustainable, and that support nonhuman communities as well".

Three aspects of Lyle's (1985) work are of direct relevance in establishing the conceptual foundation for ecological design. The first is that he attempts to tackle the complexity of design method and offers a critical investigation of the design process in the context of ecosystem, its function, structure and ecological (rather than economic) rationality. The second is that he includes 'management' as an integral part of ecosystem design, arguing that ecosystems like any organic entity have a variable future and as such, their design should be probabilistic; it is difficult to predict the changes that will take place. The implication here is that design is an ongoing process and that the final product of design is only one stage in this process; it should not be the objective. It also implies that design is interactive because it takes into account future change resulting from the designed system's interaction with its environment. A third aspect of Lyle's work is that he breaches the professional categorization of landscape architecture and landscape planning. The terms 'landscape design' and 'landscape planning' are often used interchangeably, however, uses 'design' as giving form to physical phenomena 'to represent such activity at every scale'. In this he follows others (Steinitz, 1979 and McHarg, 1969) who refer to the regional planning scale while using 'design'. Lyle viewed landscape planning's focus on the rational as inevitably excluding the intuitive (Makhzuomi and Pungetti, 1999).

More recently, designers such as Le Corbusier and Frank Lloyd Wright, among many others, have attempted, with some degree of success, to address ecological issues through their designs. "Green Architecture," "Alternative Architecture," "Sustainable Design," and "Ecological Design," are some of the terms commonly used today to describe a special expression of design that takes as its primary driving force nature's processes. Van Der Ryn and Cowan (Ecological Design, 1996) defined this form of expression as "any form of design that minimizes environmentally destructive impacts by integrating itself with living processes." A "new" movement among design professionals has been developing for some time now with many of its principles synthesized by the current "green" movement in design (Lomba-Ortiz, 2003).

Ecological design is an emerging interdisciplinary field of study and practice. In fact, many would argue that it is a transdisciplinary field, concerned with the creation of entirely new applications that may emerge from its progenitor disciplines or arise from a synthesis of several. Influenced principally by ecology, the environmental sciences, environmental planning, architecture, and landscape studies, ecological design is one of several rapidly evolving (theoretical and practical) approaches to more sustainable, humane, and environmentally responsible development. As such, it may also be considered a critical approach to navigating the interface between culture and nature. In the broadest sense, ecological design emerges from the interdependent and dynamic relationship between ecology and decision making.

Van Der Ryn and Cowan (1996) described ecological design as a hinge that connects culture and nature, allowing humans to adapt and integrate nature's processes with human creations. In modern industrialized societies, human culture and nature are perceived and treated as separate realms, yet their interface offers fertile ground for the creation of new, hybridized natural/cultural ecologies and the rehabilitation and re(dis)covery of others. Ecological design is inspired by the nexus of these worlds and the urgent need to blur the boundary between them; it seizes on the creative tensions between them and, as such, may offer opportunities for and insights to a re(dis)covered place of "living lightly" with the land (Lister, 2007).

By the beginning of the 21st century, ecological design had emerged as an expression of a sustainability world-view, which seeks to integrate the human enterprise with a sustainable harvest of resources, while ensuring that stresses caused to natural ecosystems are within the bounds of viability. If this can be achieved, the integrity of both the human economy and of natural ecosystems can be maintained. As such, ecological design is an all-encompassing concept, as it deals with the sustainability of:

- The enterprises of families, neighborhoods, and cities;
- The construction of buildings in a manner that decreases resource use and environmental damage to the degree possible;
- The manufacturing of certifiably green products;
- The organic production of foods and other renewable resources;

- The integration of these various activities within ecologically planned mutualisms, such as industrial and business parks, which are designed to maintain high production while reducing the use of resources and minimizing waste; and
- The maintenance of indigenous biodiversity (Shu-Yang et al., 2004).

Landscape architects continue to speculate how we can design with the materials of nature and not have the result be confused for nature itself (Ware, 2004). Beth Meyer asserts, 'to some it might seem odd that landscape architects looked toward art and design theory and practice when seeking direction about folding ecological principles and environmental values into their creative processes. But this simultaneous look to art as well as science and to theories of site specificity and phenomenology as well as ecology was critical to the successful integration of environmentalism into landscape architectural design.' (Meyer, 2002).

7. ECOLOGICAL LANDSCAPE DESIGN

Ecological landscape design is based on an ecological understanding of landscape which ensures a holistic, dynamic, responsive and intuitive approach (Figure 3.). It is holistic because it simultaneously considers past and present as well as local and regional landscape patterns and processes. It is responsive because it develops from a realization of the constraints and opportunities of context whether natural, cultural or a combination of both. Ecological landscape design is guided by three fundamental, mutually inclusive objectives: the maintenance of landscape integrity; promoting landscape sustainability; and reinforcing the natural and cultural spirit of place. Ecological landscape design engages the designer's rational, intellectual, emotional and creative capabilities (Makhzoumi and Pungetti, 1999).

Ecological design develops out of two areas of inquiry. On the one hand, it is the outcome of ecology's interface with the environmental design professions. Despite the differing perspectives and focus of interest, a number of common concepts have been outlined. On the other hand, ecological landscape design also utilizes fundamental ecological. Input from these two areas of inquiry forms the foundation for ecological landscape design which is here seen as integrating four overlapping attributes (Figure 3.).

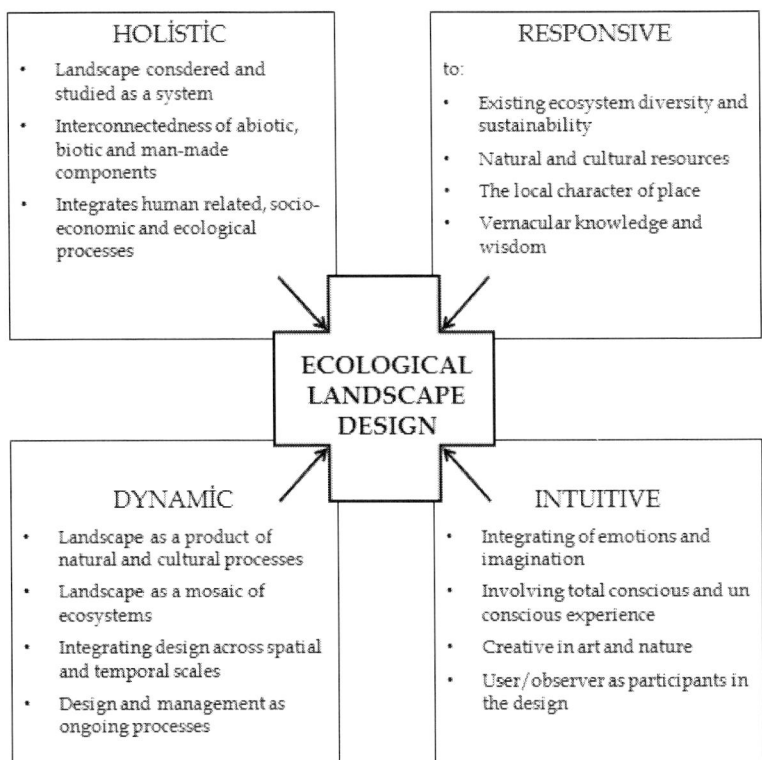

Figure 3. Framework for ecological landscape design, drawing on concepts from ecology (left) and ecological design (right)

The first is a holistic approach to landscape understanding, integrating abiotic, biotic and cultural landscape components. The second is a dynamic approach in which landscape is investigated along two continuums: a spatial one, i.e. movement between a larger scale and a local one; and a temporal one representing the evolutionary historical development of the landscape. The third is ecological landscape design's responsiveness to the constraints and opportunities of context whether natural, cultural or a combination of both. Responsiveness also dictates an anticipatory approach that considers the impact of the design on existing ecosystems and resources. Finally, ecological landscape design is intuitive, encompassing not only the rationality of the outer world but also the neglected 'intangible relationships' of the inner world. This intuitive approach embraces a new definition of creativity that departs from the formal, i.e. object-centered, appearance-oriented aesthetics to a phenomenological participatory aesthetics where the emphasis is on the totality of human experience of the object (Makhzuomi and Pungetti, 1999).

Reviewing ecology's interaction with the environmental design professions reveals a wide range of concepts, solutions and approaches (Figure 4.). The contributions in architecture and the urban landscape design include practical strategies (e.g. energy conservation, ecological networks) and design solutions to specific problems (e.g. earth-sheltered architecture and bioclimatic design). The interaction of ecology and landscape architecture has been more extensive, leading to a holistic approach to landscape design. All the contributions, however, find inspiration in nature and aim to shape man's environment sustainably and 'beautifully'.

Ecological landscape design integrates input from landscape ecology and design, both of which are seen as providing parallel and complementary, albeit different methodological approaches. The analytic and descriptive nature of landscape ecology, the science, provides for a holistic understanding of existing landscapes, while the intuitive and creative problem-solving capabilities of design prescribe alternative courses for future landscape development (Makhzoumi, 2000).

In the different steps of the design process a lot of information has been needed to analyze and evaluate ecological processes and functions. Thus ecological design has been interdisciplinary field of study and practice.

Over the past 20 years landscape architecture has re-invested in ecologically driven design. Ware (2004) investigates the following typologies:

- Interpretation and Environmental Education
- Environmental Remediation/Re-vegetation
- Re-Use/Re-programming
- Eco-Revelatory Design
- The Art of Landscape Function
- Intertwining Ecologies
- Constructed Ecologies
- Simulated 'natural' Attractions

The typological framework aims to illustrate and differentiate current methods of approaching ecological design in landscape architecture. The eight categories include a critical reflection as to how the work itself may not be addressing much of the dynamic, ecological processes that the projects are predicated upon (Ware, 2004).

7.1. Principles of ecological landscape design

The main ecological principles concerning cities are that:

- Cities are ecosystems;
- Cities are spatially heterogeneous;
- Cities are dynamic;
- Human and natural processes interact in cities; and
- Ecological processes are still at work and are important in cities.

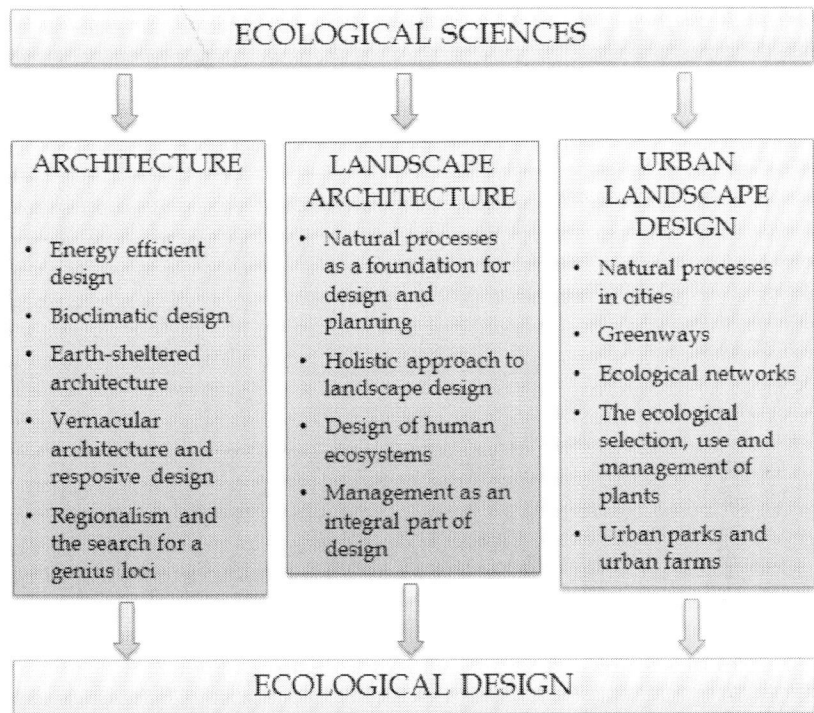

Figure 4. The interface of ecology with architecture, landscape architecture and urban landscape design (Makhzuomi and Pungetti, 1999)

The first three principles address the structure of cities and the change in structure through time. The remaining two principles focus on ecological processes in cities (Table 1.).

The first principle suggests that landscape design theory and management practice must address all the components of such systems. Urban ecosystems include four broad kinds of components (organisms, a physical setting and conditions, social structures, and the built environment) all interacting with one another. Landscape designs and management strategies that are aimed at one or two of these components or interactions, in reality have the potential to affect them all. Landscape designs that acknowledge and work with the connections between the social, biological, physical, and built components of the system are much less likely to produce unintended negative consequences, and are more likely to contribute to ecological sustainability. Furthermore, enhanced quality of urban life depends on all components of the urban ecosystem, not just some of them (Cadenasso and Pickett, 2008).

Table 1. A brief summary of the general implications of each of the five principles of urban ecology for ecologically motivated landscape design and management

Principles	Summary of Implication for Landscape Design
Cities are ecosystems	Design affects all four components of human ecosystems.
Cities are heterogeneous	Design should enhance heterogeneity, and its ecological functions.
Cities are dynamic	Design must accommodate internal and external changes projects can experience.
Human and natural processes interact in cities	Design should recognize and plan for feedbacks between social and natural processes.
Ecological processes remain important in cities	Remnant ecological processes yielding ecological services should be maintained or restored.

The second principle suggests that interactions and transfers among patches within the urban matrix are affected by landscape design and management. Urban landscape design should carefully consider the heterogeneity and its role in maintaining desirable functions such as biodiversity, storm water retention, microclimate mitigation, and carbon sequestration. The interaction between a particular landscape project and adjacent patches of similar or contrasting landscape structure can enhance the function and value of individual projects. This may mean paying particular attention to the boundaries between contrasts within or between projects to enhance or protect from exchanges.

The third principle means that landscape designs should accommodate change. Natural disturbances, extreme climate events, shifting economic investment or disinvestment, the maturation of households, and the aging of or renovation of infrastructure are but some of the examples of the kinds of dynamism that landscape designs and management will have to respond to. Persistent equilibrium in cities is unlikely. Designs that plan for successional changes in vegetation have redundancies in the face of disturbance, or that encourage use by different age groups may be more resilient in changing cities.

The fourth principle suggests that both of these major categories must be addressed as landscape design goals. A design that satisfies only obvious social criteria, such as recreation or efficiency of commerce, misses an opportunity to contribute to ecosystem services that may ultimately have great social value. All landscape designs and management schemes should be judged for their ability to contribute to both social and ecological goods and services, and to reduce both social and ecological risks and vulnerabilities.

The fifth principle means that landscape designs and management practices have the opportunity to preserve and promote those basic biological processes upon which human health and well-being depend. It will be important to provide for these functions even in areas beyond the large green parcels usually targeted for this kind of benefit. The control of water flow and infiltration, the retention of limiting and hence potentially polluting nutrients, the sequestration of carbon dioxide, the neutralization of toxics, the maintenance of soil respiration, the production of biomass, the amelioration of climate extremes, the mitigation of natural disturbance, and the preservation of biodiversity, are but some of the processes that can exist in various places in designed systems. Landscape designs and management protocols can be purposefully planned so as to maintain, or in some cases restore, as many of these kinds of natural processes as possible throughout the urban matrix. As such, landscape design and management can provide creative new ways to insinuate ecological processes in cities (Cadenasso and Pickett, 2008).

Ecologically designed urban landscapes are ones that can use both ecological processes and human values as form-giving elements. In addition to their many environmental benefits, these landscapes -which include systems such as energy efficient buildings, storm water infiltration, sewage treatment wetlands, and urban forests- can also contribute to local cultures of sustainability that, like all cultures, both shape and are shaped by the built and designed environment. If they are to do so, however, their designers must think clearly about the experience of the users of the urban landscape, and particularly about the meanings and lessons that they derive from their surroundings. The ways that people learn from and respond to the urban environment are critical to the prospects for sustainability, if for no other reason than that for most of us, it is the landscape of the city that helps to shape our view of nature and our relation to it (Eisenstein, 2001).

Ecological landscape designs fall into four categories:

1. Preservation of existing, functioning ecological systems;
2. Enhancement or re-establishment of degraded ecological systems;
3. Intensification of ecological processes to mitigate potential or existing ecological degradation; and
4. Environmental interventions which reduce nonrenewable resource consumption (Mozingo, 1997).

Van Der Ryn and Cowan (1996) have pointed out principles of ecological design Table 2.. The first principle grounds the design in the details of place. In the words of Wendell Berry, we need to ask, "What is here? What will nature permit us to do here? What will nature help us to do here?" The The second principle provides criteria for evaluating the ecological impacts of a given design. The third principle suggests that these impacts can be minimized by working in partnership with nature. The fourth principle implies that ecological design is the work not just of experts, but of entire communities. The fifth principle tells us that effective design transforms awareness by providing ongoing possibilities for learning and participation. Taken together, these five principles help us to think about the integration of ecology and design.

Table 2. Principles of ecological design

Principles	Summary of Implication for Landscape Design
Solutions grow from place	Ecological design grows from an intimate, detailed knowledge of the place and its nuances.
Make nature visible	Make sure natural cycles and processes are visible to bring the designed environment back to life.
Design with nature	Nature's living processes offer opportunities to design using natural cycles, natural waste, and regeneration as part of the total design.
Ecological accounting informs design	By tracing the environmental impacts of a design, we can discover the more ecologically sound options.
Everyone is a designer	Listen to every voice in the design process.

7.2. Examples of ecological landscape design

7.2.1. The West Davis Pond

The West Davis Pond in Davis, California, is exemplary of the new landscape space of ecological design. The subdivision of a single family and low-rise apartment neighborhood required capacity improvement of an existing storm-water-treatment settling pond. This prosaic infrastructure requirement innovatively integrates a constructed habitat for numerous over-wintering migratory birds and resident wildlife whose wetland habitats have largely been destroyed in the Central Valley.

The pond had pre-existing development on three sides: an arterial roadway edged by backyard fences, a long edge of directly adjacent backyard fences, and warehouse commercial uses. On the fourth side, the project developers and their team of engineers, environmental scientists, and landscape architects conceived of the pond as integral to the open space of the new development. In lieu of a more typical suburban park, between the housing and the pond, a bike path, part of a famous city-wide system, incorporates two pond overlooks and a constructed arroyo channel as a children's play area. Between the manicured, exotic landscape of the housing and the habitat planting of the pond, transitional "native planting" envelopes the bike path, overlooks, and play area (Figure 5.). Most of the species are not native to this part of California, and many are unhealthy or dying.

As one of the first storm-water-treatment wildlife ponds in the Central Valley, and one of the first wetland restoration projects within an urban context, the project is laudable in ways-it is based on sound ecological science, it achieves its clearly stated ecological goals, it is innovative, and it manifests strong community support. The project was done with conscience, care, and the considerable risk that precedents always entail (Mozingo, 1997).

Bike path   Birding walks

How it looked in 2007   How it looked in 2010

Early fall at the pond

Figure 5. The West Davis Pond (Anonymous, 2012)

The West Davis Pond is a new kind of ecologically integrated project, with measurable ecological benefits that we want to increasingly infiltrate into the landscape. The Pond is an enhanced wetland wildlife habitat, while its primary purpose is to retain storm water runoff and help prevent flooding. In the dry months, water is provided by a supplementary well. The Pond is enclosed by a security fence and is designated a "Wildlife Preserve" and "Sensitive Habitat Area" by the City of Davis. Native trees and shrubs grow on the slopes around the Pond and provide habitat for a diversity of wildlife (Anonymous, 2012).

7.2.2. The Glenn W. Daniel King Estate Park

The Glenn W. Daniel King Estate Park encompasses eighty acres of a north-south ridge overlooking the East Bay Hills and an expansive panorama of the San Francisco Bay. The park is the largest open space and only urban wild land west of Interstate 580, the city's social and physical divide. The Glenn W. Daniel King Estate Park is not blueprint for a park constructed as a single project. Rather, it is a guide for a sustained effort to bring to fruition a park that is ecologically healthy and well integrated into the social life of its community (Figure 6.). The park lies within a home owning, middle-class, primarily African American neighborhood considerably integrated with European American, Latino and Asian American residents (Mozingo et al., 1998).

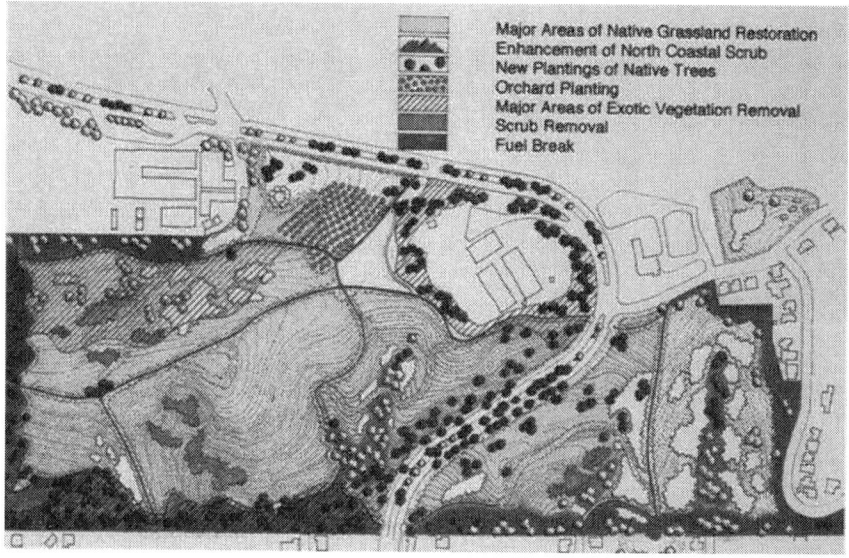

Figure 6. The Glenn W. Daniel King Estate Park Master plan (Mozingo et al., 1998)

A partnership ethic respects both cultural diversity and biodiversity. In the hills above Oakland, California, a culturally diverse middle-class neighborhood consisting of a majority of African Americana along with many European, Asian, and Latin Americans worked in partnership with the each other and with landscape architect Louise Mozingo of the University of California, Berkeley. The goal was to restore biodiversity to the oak groves from which the city derived its name and ecological heritage. Together they devised a plan to develop the neighborhood's The Glenn W. Daniel King Estate Park to benefit from the diversity of perennials grasses, oak savannahs, and brushy chaparral indigenous to the area. At the same time, they revamped hiking trails, added a recreation center, and increased security. The resulting master plan provided "a template for how communities can become active partners in the fulfillment of their own environmental visions" (Merchant, 2004).

7.2.3. Village of Yorkville Park

The idea of this urban park dates back to the late 1950s when a block of Victorian-era row houses was demolished along Cumberland Street to allow for the construction of the Bloor Danforth subway line. The park sits at the cusp of two neighborhoods: the small-scale old Yorkville neighborhood with its late 19th and early 20th century row houses, and the high-rise commercial core that has built up along the Bloor Street corridor since the subway opened. For years, this highly visible site remained a parking lot. Activist neighbors fought to build a public place to bring the neighborhood together rather than to divide it. Finally, in 1991, the City of Toronto Department of Parks, Forestry and Recreation announced an international design competition (Figure 7.).

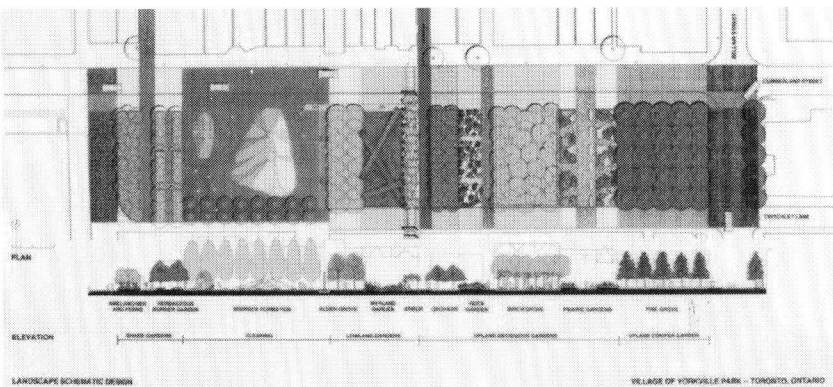

Figure 7. Village of Yorkville Park landscape schematic design (Anonymous, 2012a)

The community wanted a park that reflected the scale and context of the neighborhood, incorporated the native ecology of the surrounding region, and made connections with the circulation of local streets and a system of midblock passageways. The design strategy for the competition was to design the park to express the Victorian style of collecting. In this case, "collecting" landscapes of Ontario -pine groves, prairies, marshes, orchards, alder woods, rock outcroppings and so on -and arranging them in the pattern of the nineteenth century row houses.

The park design creates a series of linear subdivisions with contextual alignments to the building lot lines across the street and connections to mid-block passageways in the adjoining blocks. Each linear park segment is distinct in character but related to the next, creating a park of diversity and unity. To anchor this space with an element of regional glacial geology, a large 700-ton bedrock outcrop of native Muskoka granite was taken apart along natural crevices, moved 150 miles south, and reconstructed on site. Immense yet inviting, the outcrop has a wonderful tactile surface for sitting and absorbs warmth on cool sunny days. Moveable tables and chairs next to the boulder offer a nice contrast of permanence and flexibility (Figure 8.).

Figure 8. Village of Yorkville Park

Figure 9. Downsview Park (Anonymous, 2012b)

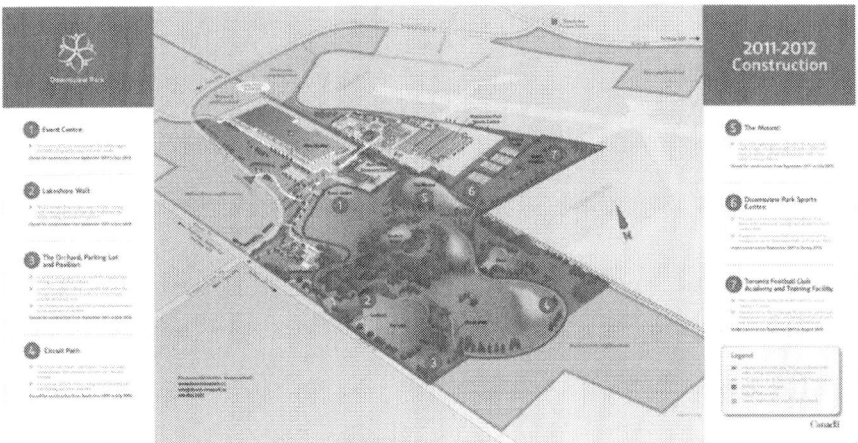

Figure 10. Recreational, educational and cultural amenities in the Downsview Park (Anonymous, 2012d)

The park has become a local landmark. While small in size, Yorkville's park has played an important role in the revitalization of the neighborhood since its completion in 1994. The neighborhood has continued its redevelopment with several new high-rise buildings rising along the edge or near the park. Recently, the park underwent some restoration work, but its original design integrity as a distillation of regional ecology, along with its role as a neighborhood connection point, remain as strong as ever. The park is owned and maintained by the City of Toronto Department of Parks, Forestry and Recreation. The Bloor-Yorkville Business Improvement Area takes an active role in the management and programming of the park (Anonymous, 2012a).

7.2.4. Downsview Park
In 1999, the Parc Downsview Park announced an International Design Competition in attempt to turn Downsview Park into an urban park, and potentially one of the largest ones in the world, in which Bruce Mau Design, Rem Koolhaas, Oleson Worland, and Petra Blaisse submitted the winning design scheme, known as "Tree City." Parc Downsview Park has since come up with a new plan to construct commercial and residential developments instead (Anonymous, 2012b). This 320-acre federal park will provide natural and formal garden environments, offering both passive and active recreation while promoting such themes as environmental sustainability, new ecologies, and the rich heritage of the site. Contributors to this volume analyze the entries of the competition finalists and consider a range of issues raised by the competition, including landscape architecture, geography, landscape ecology, and contemporary urbanism (Czerniac, 2002).

Downsview Park is designed to support environmental, social and economic sustainability. The vision for the park is the creation of a recreational space incorporating expansive open space areas, as well as the repurposing of an inventory of historic aviation-related buildings to create a year-round setting

(Figure 9.). Downsview Park is a model development demonstrating sustainable practices in its design, construction, operation and maintenance. It is intended to be a recreational, educational and cultural amenity for all Canadians (Figure 10.); a diverse, healthy and livable community for its occupants, visitors and neighbors; and an educational demonstration project of international significance. In addition to creating a unique park on the majority of the lands, portions of the property will be developed to facilitate creating and maintaining Downsview Park. More than $20 million has been spent to date on construction, improvements to infrastructure and renovations of older buildings. The investment that Downsview Park is making in the public realm will have a significant impact well beyond its 231.5 hectares (572 acres) -job creation, increased real estate values, social and cultural engagement and numerous environmental benefits are all a direct result of the work being performed in the creation of the Park (Anonymous, 2012c).

REFERENCES

1. J Ahern, ., R France, ., M Hough, ., J Burley, ., W Turner, ., S Schmidt, ., D Hulse, ., J Badenhope, . and G Jones, . 2001. Integration Ecology "across" the Curriculum of Landscape Architecture. B. R. Johnson and K. Hill (eds). Ecology and Design Frameworks for Learning, Island Press, 1-55963-813-3

2. Anonymous, 2012. http://daviswiki.org/Friends_of_West_Pond (accessed 15 October 2012).

3. Anonymous, . http://www.asla.org/2012awards/034.html (accessed 15 October 2012).

4. Anonymous, 2012b. http://www.cicadadesign.ca/portfolio/downsviewpark.html (accessed 22 November 2012

5. Anonymous, . http://en.wikipedia.org/wiki/Downsview_Park (accessed 22 November 2012).

6. d Anonymous, . http://downsviewtownhomes.com/wp-content/uploads/2011/12/constructionmap_nov2011_EN.jpg (accessed 22 November 2012).

7. M. L. Cadenasso, and S. T. A. Pickett, 2008Urban Principles for Ecological Landscape Design and Management: Scientific Fundamentals. Cities and the Environment, 116

8. J. Czerniak, 2002Downsview Park TorontoPrestel, USA.

9. T. W. Cook, and VanDerZanden, A. M., 2011Sustainable Landscape Management: Design, Construction, and Maintenance,John Wiley & Sons, Inc., USA.

10. D. Deniz, 2002Sustainability and Environmental Issues in Industrial Product DesignIzmir Institute of Technology. Master Thesis on Industrial Design, Turkey.

11. W Eisenstein, . 2001. Ecological Design, Urban Places, and the Culture of Sustainability. San Francisco Planning and Urban Research Association, http://www.spur.org/documents/pdf/010901_article_02.pdf (accessed 02 October 2012).

12. S. M. Galatowitsch, 1998Ecological Design for Environmental Problem Solving. Landscape Journal, Fall 98, 17299107

13. G. E. Helfand, J. S. Park, J. I. Nassauer, and S. Kosek, 2006The Economics of Native Plants in Residential Landscape Designs. Landscape and Urban Planning, 78229240

14. V. H. Hooper, J. Endter-wada, and C. W. Johnson, 2008Theory and Practice Related to Native Plants: A Case Study of Utah Landscape ProfessionalsLandscape Journal, 27127141

15. B. R Johnson, . and K Hill, . 2001. Ecology and Design Frameworks for Learning, Island Press, 1-55963-813-3

16. N. M. Lister, 2007Sustainable Large Parks: Ecological Design or Designer Ecology?. In: J. Czerniak & G. Hargreaves (eds.), Large Parks. Princeton NJ: Princeton Architectural Press, 3151

17. N. M. Lister, 2005Industrial Ecology as Ecological Design: Opportunities for Re(dis)covery, In R. Coté, J. Tansey and A. Dale (eds). Linking Industry and Ecology: A Question of Design. Vancouver: UBC Press, 1528

18. E. A Lomba-Ortiz, ., 2003. Questioning Ecological Design: A Deep Ecology Perspective. Ecotecture, http://www.ecotecture.com/library_eco/appropriate_tech/lomba-ortiz_ questioningeco.html (accessed 04 December 2012).

19. P. Madge, 1997Ecological Design: A New CritiqueDesign Issues132A Critical Condition: Design and Its Criticism (Summer, 1997), 4454

20. J. M. Makhzoumi, 2000Landscape Ecology as a Foundation for Landscape Architecture: Application in MaltaLandscape and Urban Planning50167177

21. J. M. Makhzuomi, and G. Pungetti, 1999Ecological Landscape Design and Planning, Taylor & Francis, ISBN-139780419232506USA.

22. I. L. Mcharg, 1969Design with NatureNatural History Press, New York.

23. C. Merchant, 2004Reinventing Eden: the Fate of Naturein Western Culture, Taylor & Franchis Books, Inc., 238USA.

24. E. K. Meyer, 2000The Post-Earth Day Conundrum: Translating Environmental Values into Landscape Design, Environmentalism in Landscape Architecture (Edt: Michel Conan), Dumbarton Oaks Research Library and Collection, 22187244USA.

25. L. A. Mozingo, 1997The Aesthetics of Ecological Design: Seeing Science as CultureLandscape JournalSpring 97, 14659

26.   L. A. Mozingo, A. Baker, J. London, N. Ancel, I. Cheng, and M. Dohi, 1998The Glenn W. Daniel King Estate Park Master PlanLandscape Journal171214

27.   H. Özgüner, A. D. Kandle, and R. J. Bisgrove, 2007Attitudes of Landscape Professionals Towards Naturalistic Versus Formal Urban Landscapes in the UKLandscape and Urban Planning813445

28.   V. Papanek, 1995The Green Imperative, Ecology and Ethics in Design and Architecture. Thames and Hudson, 256London.

29.   M Roseland, ., 1997. Dimensions of the Eco-city. Cities, 14 4 1972002.

30.   F Shu-Yang, ., B Freedman, . and R Cote, . (2004). Principles and Practice of Ecological Design. Environmental Review, 12 97112 , doi: 10.1139/A04-005, Canada.

31.   Thayer JrR. L. 1998Landscape as an Ecologically Revealing LanguageLandscape Journal17isuue: 2, 118129

32.   J Todd, ., E. J. G Brown, . and E Wells, ., 2003. Ecological Design Applied. Ecological Engineering, 20 421440 .

33.   S Van Der Ryn, . and S Cowan, . 1996. Ecological Design. 1-55963-389-1

34.   S. Ware, 2004The Nature of Design, 2004 AILA Nationla Conference (200 MILE CITY), Landscape Architecture Online: an online magazin published by the Australian Institute of Landscape Architects, Australia.

CHAPTER 2

Eco-Revelatory Design

Nurgül Konaklı Arısoy

¹ Selçuk University, Agricultural Faculty, Department of Landscape Architecture, Turkey

1. INTRODUCTION

Built environments ignore people's need and their potential for learning. The negative effects related to the ignorance of natural systems in human development are evident. Making natural cycles and processes visible bring the designed environment back to life. Effective design helps inform us of our place within nature. Landscape architects have developed theories and methodologies which represent a new, ecologically oriented approach to design.

Eco-revelatory design (ERD) is an ecological design concept in the field of landscape architecture means "a design strategy that attempts to enhance site ecosystems as well as engage users by revealing ecological and cultural phenomena, processes and relationships affecting a site " [1]. Landscape architects reveal nature through their form, materials and formation, and they also reveal the nature of the person who designed them.ERD is a new approach to landscape architecture, one where ecological processes and the environment is a fundamental determinant of the design.

ERD is a different exposition and interpretation –updated version- of design. It is an integrative and ecologically responsible design. It is a partnership between people and nature. ERD attempts to enhance site ecosystems and engage users by revealing ecological and cultural phenomena, processes, and relationships affecting a site. It aspires to reveal endemic ecological process and affords a more direct connection between fundamental ecological process and the phenomenological experience of landscape. It is important to involve using knowledge about how interact with environment to form objects and spaces with skill and artistry [4]. This approach to design should be applied not only to urban sides, but also to non-urban sides such as wetlands, arboretums. The theory has received heavy criticism about its ability to absorb an audience in ecological understanding or improve site conditions [2,3].

Another criticism about ERD is some designs make tangible improvements in local ecological health while others are symbolic gestures [2]. It is adequate for people to develop ecological perception by means of the visibility of ecological process in design. The theory has received heavy criticism about its ability to absorb an audience in ecological understanding or improve site conditions.

It aspires to reveal endemic ecological process and affords a more direct connection between fundamental ecological process and the phenomenological experience of landscape. It is important to involve using knowledge about how interact with environment to form objects and spaces with skill and artistry (Ndubisi, 2002). This approach to design should be applied not only to urban sides, but also to non-urban sides such as wetlands, arboretums.

They are typically designed landscapes that elucidate natural phenomenon such as the cleansing action of wetlands. Ecorevelatory landscapes have also been referred to as "educational and enlightening". They are reference sites for what we understand about our environment and its workings. Designs can convey knowledge through direct experience as well as by interpretation. Interpretation seeks to create connections between the resources that are being interpreted and relevant everyday knowledge that everyone has by engaging emotions, creating experiences, and entertaining ideas through engagement [5]. By highlighting the particular ecological relationships at any given site, such design can punctuate and enliven our environment and sensitize us to what is known about its interlocking complexities" [1]. These sites often still use traditional interpretive media such as wayside exhibits and publications to fully convey their meaning and function, but eco-revelatory design can bring to light processes that usually remain unseen and forgotten.

2. HISTORY

This study explores descriptive theory, principals, techniques and practice of eco-revelatory design, can help to plan a sustainable development, which uses and reveals natural systems to reconciling human systems and its effects on the surrounding environment. The goal of this study, therefore, is to generate ideas and to begin a discussion about how the design within an eco-revelatory framework will be.

Most of the "eco" prefixes such as eco-city, eco-technique, eco-efficient was began to used nineties; also ERD emerged in 1998 as a new theory within the field of landscape architecture.

The term and practice of ERD was coined from an exhibit Nature Constructed/Nature Revealed sponsored by the University of Illinois at Urbana-Champaign. Two predominant schools of thought, one insistently cultural and the other assertively ecological, reigned over the conceptual and theoretical dialog in landscape design and planning. In 1998 a group of practitioners and landscape scholars published a special issue of Landscape Journal as a catalogue and record of the exhibition Eco-Revelatory Design: Nature Constructed/Nature Revealed. Brenda Brown, Terry Harkness, and Douglas Johnston chaired the exhibition and served as guest editors of the journal (Figure 1). The exhibit opened at the University of Illinois in 1998 and closed at the Washington DC's National Building Museum in 2000.

Brown identifies three areas of investigation on landscapes as/of sound: listening gardens, sound, or listening trails, and sound designs [3]. She is particularly interested in what she calls "the reciprocal revelations of landscapes and sounds, how sounds can reveal landscapes and how landscapes can reveal sounds." She is therefore also concerned with how people perceive, understand and engage with landscapes.

ERD roots are based on ecologically minded landscape architects such as Ian McHarg and Frederick Law Olmsted but they didn't use the term as ERD. These landscape architects created works to foster function in natural systems and processes aesthetically. Ian McHarg was accepted as the first person who to apply his ecological knowledge into design. He began advocating the use of ecology as a basis for design in the early 1960s. He accomplished his goal of merging design with ecology. He gave importance of ecological principles in design.

Thirty years ago, in *Design with Nature*, McHarg proposed a system of ecological inventories to help explain the way natural processes may influence regional and urban planning and design [6]. Also, it makes the case for an ecological approach to design. McHarg gave a new dimension to the historical goal of 'imitating nature' (mimesis). He was concerned both with the practicalities of 'design with nature' and with the aesthetic results of a naturalist approach to landscape and garden design. It was demanded that local landscapes should follow ecological principles that implied that landscape could not represent some particular claim of social identity.

Figure 1. Eco-revelatory design exhibit

Figure 2. Central Park, NYC

In the second half of the 19th century, Frederick Law Olmsted completed a series of parks which continue to have a huge influence on the practices of Landscape Architecture today. Among these were Central Park in New York City (Figure 2), Prospect Park in Brooklyn (Figure3), New York and Boston's Emerald Necklace park system (Figure 4).

Figure 3. Prospect Park in Brooklyn

Figure 4. Boston's Emerald Necklace park system

The Emerald Necklace consists of a 1,100-acre (4.5 km^2) chain of parks linked by parkways and waterways in Boston and Brookline, Massachusetts.

Several components of the Emerald Necklace pre-date the plan to unite them. Some links of the Emerald Necklace not only offer an opportunity for recreation in a wooded environment, but are also ecologically important urban wilds that provide nesting places for migratory birds, fishes and other animals and improve the air quality of the city. The Emerald Necklace Project successfully ties together conservation, land restoration, sewage treatment, solid waste disposal, recreation, transportation, and water and visual quality [7]. The park system provides opportunities for people to learn about natural systems by enjoying, observing, and appreciating these systems.

Today, landscape architects recognize the effects of ignorance on natural systems in human development. So they coined a new design approaches.

Arcata Marsh is one of the ERD examples of today design approach. The Arcata Marsh and Wildlife Sanctuary was constructed in 1981 (Figure 5). The City of Arcata incorporated wastewater treatment to the system in 1986. The City of Arcata's unique wastewater treatment facility, marsh, and wildlife sanctuary attracts approximately 150,000 visitors per year [8]. Arcata's wastewater treatment plant is an example of a community involvement in environmental politics, innovative uses of land, and applications of appropriate technology in a small urban community. The Arcata Wastewater Treatment Plant combined with the Arcata Marsh and Wildlife Sanctuary has multiple uses, including wastewater treatment, recreation, wildlife habitat, education, and research. The residents of Arcata who stroll by the wetland can, for instance; see that wastewater treatment wetlands can be important habitat for fish and birds, as well as an energy-efficient, biologically based method of controlling water pollution. The experience of the Arcata wetland shows that ecological processes can be brought into a constructive partnership with human settlements.

Figure 5. The Arcata Marsh and Wildlife Sanctuary (source: http://www.humboldt.edu/engineering/graduate/facilities)

Students are centrally involved in the original design and development of the Marsh's constructed wetland water treatment system, and they continue to play a key role through projects and research geared towards the continued optimization of the system.

The second example for ERD is the Gateway Business Center - post-industrial area- is a 25-acre site consisting of 10 industrial tilt-up buildings located at the San Gabriel Mountain foothills on a great alluvial fan that crosses the San Gabriel Valley on its way to the Pacific Ocean. The site concept transformed an industrial building environment into a sustainable and nature-inspiring experience. The layout of cairns act as way finding and directional markers and add to the overall landscape display and experience of bioswales, infiltration zones, rain harvesting, solar canopies and an array of recycled materials.

The site landscape is designed to mimic the natural feel of the local native landscape character and capture it into the renovated industrial complex. The goal is to create "biophilia" (an instinctive bond between man and nature). The bioswale has taken the place of the previous concrete swale, while the concrete catch basin remains to collect any additional runoff in a deluge, and as a reminder of the previous form of site storm water drainage. More than 95 percent of the runoff water in this parking lot is captured.

The storm garden is irrigated by the adjacent rain harvesting tanks. A natural wash picks up excess rainwater and roof water not captured by the parking lot bioswale. [9] (Figure 6)

Figure 6. The Storm Garden of Gateway Business Center

3. DESIGN AND ECOLOGY RELATIONSHIP

Design is the intention that redefines how we relate to each other and our environment. The process of designing places and artefacts are opportunities to reimagine a new relationship with our environment, especially when it seems remote or difficult to create. Design that reveals hidden systems, whether ecological or economic, is a powerful way to meet the challenges of ecologically benign communities. Landscape design is a discipline which transfer the knowledge developed in landscape ecology to application [10].

The term ecology, like the term landscape, has multiple meanings. Ecology has historically focused on pristine, natural environments, however, by the 1970's many ecologists began to turn their interest towards the ecological interactions taking place in, and caused by urban environments. Urban ecology is recognized as a diverse and complex concept which differs in application between North America and Europe. The European concept of urban ecology examines the biota of urban areas while to the North American concept which has traditionally examined the social sciences of the urban landscape as well as the ecosystem fluxes and processes.

Environmental knowledge have been part of the intent of design in landscape architecture and various conceptions and methodologies have been improved for measuring the environmental consequences of design.

Nassauer and Opdam define design as intentional change of landscape pattern, for the purpose of sustainably providing ecosystem services while recognizably meeting societal needs and respecting societal values [11]. Design is both a product, landscape pattern changed by intention, and the activity of deciding what that pattern could be.

Ecology plays a big role in ERD. Disciplines in applied ecology such as urban ecology, landscape ecology etc. helps planning and design practice in landscape architecture. These disciplines help put theories into practice. The most important thing is to find how to use ecological thinking in design.

4. LANDSCAPE ARCHITECTURE AND LANDSCAPE ECOLOGY

Landscape architecture is informed by scientific knowledge and aspires to provide aesthetic expressions in landscapes across a range of spatial scales. Landscape ecology has been defined as the study of the effect of landscape pattern on process, in heterogeneous landscapes, across a range of spatial and temporal scales [12]. The logical reasons for integrating these two fields are clear and compelling, with a great potential to support sustainable landscapes through ecologically based planning and design.

Proponents of ERD recognized that landscape architecture alters and directs both cultural and ecological systems. Furthermore, they acknowledged landscape architects' capacity to direct human experience and reveal, through design, aspects of ecology and culture. This integrated approach provides opportunity for people to place themselves in and as part of an interconnected socio-ecologic world, reinforcing the relationships between humans and the bio-geosphere.

Landscape ecology is defined as a problem-oriented science [13]. It has developed from the growing awareness of environmental problems since the nineteen seventies. Spatial planning and landscape design are disciplines which transfer the knowledge developed in landscape ecology to application. To optimize this process of knowledge transfer, landscape ecology must co-evolve with spatial planning [14]. The development of ecologically sustainable landscapes requires that patterns of future landscapes sustain the necessary ecological processes in the landscape. Therefore, we must know how landscape patterns relate to these processes.

Humans are the driving force behind urban ecology and influence the environment in a variety of ways, such as modifying and altering land surfaces and waterways, introducing foreign species. In this context, changes between human-environment affect the styles of landscape design directly.

However, the present concept of ecological design and its interpretation in this sense does not refer to information charts or written explanations, which people encounter at places like zoological gardens or arboretums.

Generally, "ecological designs blend with their contexts and results in a diffuse visual pattern" [15]. Consequently, this perceptual subtlety can make ecological landscapes difficult for inhabitants to recognize and care about. ERD make ecological considerations perceivably a "visible part of landscape experience". It achieve this is by benefiting from the contrast between cultural and ecological domains. One way to achieve this is by exploiting the power of contrast, particularly the contrast between cultural and ecological domains. Ecological design has tended to diffuse edges to provide transition.

We can use basic principles of landscape ecology in landscape design. Such as linear parks, neighbourhood parks, playgrounds as for patch, greenways for corridor.

Every human directly or indirectly contributes towards enriching as well as degrading the quality and experience of cities. In order to create a successful ecological design it is important to recognize and interpret the historic and cultural significance of the landscape. In other words, "eco-revelatory design expands by hitching human habitat and their inevitable cultural determinants in to an environmentally inclusive vision" [16].

Farina observed that landscape design is an important component in practical landscape ecology as it expresses the relationship of spatial patterns and processes in a practical manner [17]. It provides in depth understanding on wildlife habitats and movements and biological interdependency within a region. The two disciplines should be complementary [17].

People use land for its scenic and recreational uses. Thus, the aesthetic use is important for them. However, natural systems are the important part of the designing decisions.

It is important to understand the social and cultural backgrounds and expectations of the society who use the land are determinant. It depends on two factors [18], in how people experience and use their landscapes, and their understanding of ecological processes.

It is important to organize a linkage between cultural expectations and ecological process. Man made land uses affect the ecological system activities such as wildlife crossing and subsurface water movement. Lyle has explained the following six basic ecological processes that are vital for operational integrity of natural systems [19].

In order to participate creatively in natural processes and to do so with reasonable hope of success, we need to include as subject of design the inner workings of the landscape, the systems that motivate and maintain it, and reveal them through creative, imaginative, and visible form of the landscape [19].

5. THEORY

A design theory is a procedure for how to set about a design project. The classical twentieth century approach to landscape design has been Survey-Analysis- Design (SAD). It has been elevated to the status of a design methodology and cruelly overworked. The resultant places lacked clarity of intention. The intention of ERD is to "connect people with the natural environment". A dynamic balance between natural environment and society, intended to reveal and interpret (resolve and educate the relationship between user and the designed area) and finally to provide awareness on ecological understanding.

Most of the landscape designers' style ignores natural systems and have only one purpose, primarily focused on aesthetics. ERD has to combine aesthetics and science. It is one of the endeavours that contribute to rediscovering aesthetics. It is hard to measure aesthetics outcome statistically or collectively. But can measure the environmental effects in numbers. So you can measure the sustainability by its effects on environment not by aesthetics.

6. METHOD

It is important to choose a method that is most compatible design strategy. It depends on the design intent, the place and the designer. How methods fit into design process should be determined clearly.

Generally two methods are used at ERD as follows;
a. Mimic the nature
b. Use the nature
c. We need to design a natural area to show the ecological processes and relationships. Artificial waterfalls can perfectly mimic the nature a way that highlight the system of the real one (Figure 7). Native trees and shrubs mimic the adjacent natural areas. The site landscape is designed to mimic the natural feel of the local native landscape character.
d. ERD utilizes the natural capabilities of the environment. In an ecorevelatory approach the pre-existing natural, structural and functional characteristics of the place are integrated into design and made them observable and understandable.

There are a lot of methods for ERD that landscape architecture can use such as environmental method, man centred method, evaluation method, interpretive analogy method. The design method is up to designer's idea and knowledge.

Through our senses, we form concrete relationships, we reconciliate, with the world. With this understanding, phenomenology can be used as a method for eco-revelatory design. Suitability analysis is also one of the approaches for ERD.

Figure 7. Artificial Waterfalls

7. DESIGN PROCESS

When we examine the literature about ERD we can't see any knowledge about the design process. Landscape ecology principles integrate to design process. How can we adopt the ecology approach and principles into design process? We can use basic principles of landscape ecology in landscape design such as linear parks, neighbourhood parks, playgrounds as for patch, greenways for corridor. Landscape ecology principles influence every stage of landscape design process such as site planning.

One of the design processes is traditional design process that landscape architects undertake. It typically starts with the selection of a site based on a set of criteria. Once a site is selected, the typical design process will move through a series of phase including site inventory, site analysis, conceptual design, design development, construction documentation and finally implementation [20].

Landscape designers need to understand how natural and human systems work and design as an integral part of a nature and establish relationship with nature. They can achieve this by combining those with new technology to meet changing cultural and ecological needs.

8. PRINCIPLES (DESIGN STRATEGY)

The hosts of the exhibit, Barbara Brown, Terry Harkness and Doug Johnston, take great care in describing what deserves to be called eco-revelatory design, and stressed stringent and ambitious goals for the competition. Those chosen for the exhibit represent rigorous application of eco-revelatory design, and utilize some or all of the following strategies [1]:

- Abstraction and simulation of natural processes
- New uses of landscapes producing deeper caring for life and ecological processes
- Signifying features that speak for natural/cultural processes that might otherwise remain invisible
- Expose infrastructure and process
- Reclaim landscapes so that the past is remembered
- Change perspectives by structuring how we interact with the Landscape.

Interpretation of ecological processes refers to the ability of the design to reveal ecological processes at work. This process of revealing can only be successful if the environments created are visible, observable, legible, and have the ability to raise curiosity in visitors to explore and understand the complexity of the landscape. Principles for ERD are described generally as follows:

9. VISIBILITY

Visibility is a cardinal point for ERD. Many designers and planners have become concerned in recent years with "revealing" ecological processes in their designs so that the users of the environment may experience, learn about, and appreciate those processes. In practice, "revelation" of ecological process has meant everything from capturing stormwater on the surface of the land before it drains away to the storm sewers. In addition, the ecological processes that are revealed may themselves be truly "natural," in the sense that they could continue to exist without the management of humans, or they may be highly artificial, engineered systems that need constant supervision if they are to persist in an urbanized context.

"Most of the time, natural systems themselves are not visible and readily engaging. What are visible are the surface manifestations and the material conclusions of these natural systems, for example layers of rocks are not ecological process, but the result of it" [21]. Thus the most important challenge for designers is to recognize which ecological processes can actually be made visible and how they can interpret these dynamic processes or their material conclusions to form and inform landscapes.

The Arcata Marsh and Wildlife Sanctuary, a real-world example of eco-revelatory design, highlights some of the positive potential of using visibility as a design strategy (Figure 8).

Some ERDs, for example, have sought to bring ecological processes (such as water flows) into the open, but then blend them in with the surrounding landscape as much as possible. Many proposals to capture rainfall in grassy areas and infiltrate it into the soil before it runs off into the storm sewers would use either parks or front lawns for this purpose. Although this strategy reveals an ecological process occurring during and just after a rainfall, the rest of the time these spaces would simply look like what they have always looked like--large grassy expanses--and would forfeit an opportunity to communicate a clear, consistent, and meaningful landscape message. Making natural cycles and processes visible bring the designed environment back to life.

In reference [22] the quantitative interpretation challenges aesthetics by rendering it negative, segregated and unstable. Under this "hegemony" various endeavours in which eco-revelatory design played a part involved in. It is important for people to "read" or "experienced" the ecological knowledge tried to given. This attempt is achieved by visibility.

Figure 8. An Oxidation Pond at the Arcata Wastewater Treatment Plant

10. SUSTAINABILITY

The second concern for ERD is to provide sustainability. The term sustainability was first used in 1980 in IUCN's World Conservation Strategy. The sustainability concept is arguably relevant to systems from the global to the local scale. Sustainable ecologically-based approaches to design are desirable but their application is not widely seen.

Sustainability reshapes environmental ethics, available technologies, planning techniques, and assessment criteria, which in turn influence environmental design disciplines. This approach suggests a need to contemplate spiritual aspects of sustainable design. A set of criteria is developed for sustainability to evaluate the environmental performance. In this context, sustainability becomes the key consideration in design.

If we propose to use ecological design of urban places to promote cultural change in the human relationship to the environment, then, we should be thinking about how to create physical settings with cues for sustainable behaviour.

The design of sustainable systems is consistent with ecological principles, which integrate human society with its natural environment for the benefit of both. There are many compelling reasons why environmental and resource problems should be placed in a dynamic perspective. It is important to provide sustainability by using moderate and efficient resource use. ERD strives for moderation and efficiency in resource use.

11. THE USE OF NATIVE PLANTS (NATURALNESS)

Naturalness provide to "sustainability" and "knowledge" about native plants. It also engages people and nature as it used to be. We can see the usage of native plants example on stormwater garden (Figure 9). It gives opportunity to people to have information about their native plants. Native trees and shrubs mimic the adjacent natural riparian areas, native plant communities and wildlife habitat. Natural water sources and planting zones will be considered when choosing plant material to minimize the need for irrigation.

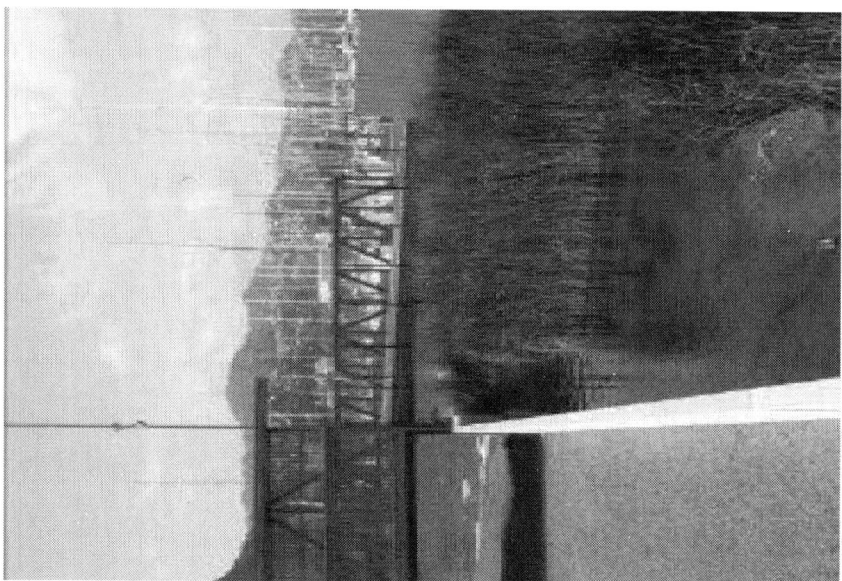

Figure 9. Stormwater management- the use of native plants

You can help stop the exotic plant invasion by using and nurturing native plants around your home and on your property. Native plants generally grow well and require less care than exotic species when grown on the proper soils under the right environmental conditions.

- Protective cover for most animals.
- Seeds, nuts, and fruits for squirrelsand other mammals.
- Seeds, fruits, and insects for birds.add beauty to the landscape and preserve our natural heritage provide food and habitat for native wildlife
- serve as an important genetic resource for future food crops or other plant-derived products

If you enjoy observing nature, are concerned about the environment, or wish to make a long-term contribution to your community's ecosystem, then using native plants is a responsible, money-saving, long-term, positive investment to both your property and your community.

Landscaping with native plants improves the environment. Native plants are hardy because they have adapted to the local conditions. The native plants increase our connection to nature, help educate our neighbours, and provide a beautiful, peaceful place to relax.

The interest in the preservation and restoration of native plant communities increases as the public becomes more concerned about the environment. Native plants are valued for their economic, ecological, genetic, and aesthetic benefits in addition to the growing societal belief in their intrinsic value as living species.

12. OBSERVABILITY

It is relevant with visibility. People observe the design by hearing, listening and seeing. It makes awareness about the nature.

When ecological design incorporates "visibility" [23] and "observability" [24] it reveals ecological phenomena and processes and can be referred to as ERD.

13. MULTIFUNCTIONALITY

Multifunctionality is the most important characteristics of this design strategy. It serves people multifunctional activities such as recreation, protection and education together. For instance, The Arcata Wastewater Treatment Plant combined with the Arcata Marsh and Wildlife Sanctuary has multiple uses, including wastewater treatment, recreation, wildlife habitat, education, and research.

Some people come to learn about the innovative wastewater treatment that enhances the community. Other people come to see the more than 270 species of birds that make use of the habitat provided by the Marsh [8]. Even more people come to exercise on the trails while enjoying the natural experience.

Multifunctionality is fundamental for sustainability. Multi-functionality is generally desirable, as it encourages efficient use of land, delivers wider public benefit and builds partnerships of user groups, leading to better stewardship. [25].

A designed land has many layers for visitors such as place for leisure activities, resting areas, conservation areas, fish and wildlife areas, sports activities and educational areas together. People gain social, environmental, economical advantage from multifunctional places.

14. LEGIBILITY

The design gives its mission to the visitor clearly. ERD purpose is "educate" and illumine". In this context, the messages have to be understandable. Effective design helps inform us about our place within nature.

Legibility was interpreted as the understanding of how the landscape worked from its manifestation based on coherence between climate, soil, water and human occupation; or "understanding relationships between process and material, form and space" [26].

The spatial organization of design was utilized to clarify the ecological and cultural phenomena for people. People accompany the space by their senses. It gives people to create a spiritual connection with their environment and to have an educational consciousness.

15. ABILITY TO RAISE CURIOSITY

If the design does not take attention it can not to carry out its mission. To make people curious about the design shows its success. The perception is the most important element of design that its comprehensibleness and arouse curiosity.

It is important for people to "read" or "experienced" the ecological knowledge tried to given. This attempt is achieved by rising curiosity. If people wonder, they can learn the ecological awareness. Designers should enhance visitor's experience by encouraging interaction to interpret the ecological processes.

Effective design solutions or distinctive features are chosen as the general publics' ability to gain ecological information from their environment is limited. By using and revealing natural systems to spread consciousness and earn attention and care for our environment.

ERD overlap cultural needs and ecological processes to raise curiosity in visitors to explore and understand the complexity of the landscape.

16. CONCLUSION

ERD is a new approach in subdiscipline of design for landscape architecture, one where ecological process, the environment and the cultural awareness of people is a fundamental determinant of the design. Landscape architects should enhance visitor's experience by encouraging interaction to interpret the ecological processes. In this context, the ecological knowledge of them should be enough for interpretation of ecological processes on their designs. Also, it is important to be understood that the ecological phenomena helps people understand their environment and its cycles by their senses.

The present concept of ERD enriches the landscapes by incorporating visibility and observability. In this context, it reveals ecological phenomena and process. This process of revealing can only be successful if the environments created are visible, observable, legible, and have the ability to raise curiosity in visitors to explore and understand the complexity of the landscape.

In order to create a successful ecological design it is important to recognize and interpret the historic and cultural significance of the landscape. In other words, "ERD expands by hitching human habitat and their inevitable cultural determinants in to an environmentally inclusive vision" [16].

The *Eco-Revelatory Design Exhibit*, summarized in the 1998 special issue of *Landscape Journal* [1], clearly articulates an open challenge for landscape architects to work as environmental educators and to help heal the relationship between society and natural systems.

Explaining this new philosophy as distinctly different and complementary to ecological design, a more technical approach for designing with natural processes, ERD reveals the significant ecological aspects of a site and helps visitors build meaning and connection between the landscape and their own lives. In this sense, eco-revelatory designs are educative landscapes, pushing their visitors to think, gain perspective, and internalize new information [1].

ERD serves as a lens for reading the landscape's story. This design idea was resulted from environmental and ecological degradation, and the erosion of spiritual connections with the land.

Most of the projects have only one purpose: they are planned for recreation that ignores natural system usually or planned for protection. However, ERD is a multifunctional design that reveals natural systems and meets the cultural and ecological needs of human systems. By revealing natural systems earn people consciousness attention, and care for their environment. The basis for ERD is to create a landscape that is ecologically as well as culturally sustainable. It provides people educational and recreational opportunities.

The technology and the conditions of life break people's connections with nature. Landscape architecture can play a major role in the mission to bring together again nature and people and reveal the ecological processes by design.

REFERENCES

1. Brown B Harkness T Johnston DEco-Revelatory Design: Nature Constructed/Nature Revealed: Guest Editors' Introduction. Landscape Journal. 1998Special Issue: x-xi.

2. W Eisenstein, . Ecological design, urban places and the culture of sustainability: Can city-building foster a culture of sustainability? Retrieved 8/23/2009, from SPUR. 2001. http://www.spur.org/publications/library/article/ecologicaldesign09012 001

3. B. Brown, Holding moving landscapes. Landscape Journal 1998Special 53-685368

4. F. Ndubisi, Ecological planning: a historical and comparative synthesis.Johns Hopkins University Press, Baltimore; 2002

5.   Cox Von Ins R EDesıgnıng for Interpretatıon: Nanny's Mountaın Park, Master Of Landscape Archıtecture. Maureen Grasso Dean of the Graduate School The University of Georgia, Athens; 2006

6.   I L. Mcharg, Design with Nature. Doubleday/Natural History Press, Garden City, NJ; 1969

7.   R T. Forman, The Missing Catalyst: Design and Planning with Ecology Roots. Washington Covelo London:Island Press; 2002

8.   The Ashford Borough Council2008

9.   http://wwwlandscapeonline.com/research/article/15739

10. Opdam P Verboom J and Reijnen RLandscape cohesion assessment: determining the conservation potential of landscapes for biodiversity (submitted); 2002

11. J Nassauer, I P Opdam, .. Design in science: extending the landscape ecology paradigm, Landscape Ecol, Springer 2008: 23:633-644

12. M G. Turner, Landscape ecology: The effect of pattern on process. Annu. RevEcol. Syst 1989

13. IALE Executive Committee1998IALE Mission Statement. IALE Bulletin 16:1

14. J. Ahern, Integration of landscape ecology and landscape design: An evolutionary process. Edited by J. A. Wiens and M. R. Moss. International Association for Landscape Ecology, Guelph, Ontario, Canada.Landscape Ecology 19991999119123

15. J T. Lyle, Regenerative Design for Sustainable DevelopmentNew York: John Wiley & Sons;1994

16. L A Mozingo, . The Aesthetics of Ecological Design: Seeing Science as Culture Landscape Journa,l Spring, 1997 1997 16 146

17. A. Farina, Principles and methods in landscape ecology. London: Chapman & Hall; 1998

18. A M. Deshpande, Design Process to Integrate Natural and Human SystemsLandscape Architecture, State University, Ms Thesis, Blacksburg, VA.; 2003

19. J T. Lyle, Design for human ecosystems. Landscape, land use, and natural resources.New York: Van Nostrand Reinhold Company Inc ; 1985

20. LaGro J ASite analysis. Linking program and concept in land planning and design.New York: John Wiley and Sons: 2001

21. J I. Nassauer, The appearance of ecological systems as a matter of policyLandscape Ecology 1992

22. B. Zhang, Technicalization of Environmental Aesthetics and a Resolution of Spirituality, Architecture, Culture, and Spirituality Symposium. JUne29July 1 2011Serenbe, Georgia, USA

23. M. Hough, Cities and Natural ProcessLondon: Routledge; 1995

24. R L. Thayer, The experience of sustainable landscapes. Landscape Journal 1989

25. P. Selman, Community Essay. Department of Landscape, University of Sheffield, Crookesmoor Building, Conduit Road, Sheffield S10 1FL United Kingdom, Planning for landscape multifunctionality, Sustinnability: Science, Practice &Policy 2009

26. A W. Spirn, The language of landscapeNew Haven and London, Yale University Press; 1998

27. J T. Lyle, Regenerative Design for Sustainable DevelopmentNew York: John Wiley & Sons; 1994

CHAPTER 3

Incorporating Renewable Energy Science in Regional Landscape Design: Results from a Competition in The Netherlands

Renée M. de Waal [1,*], Sven Stremke [1,†], Anton van Hoorn [2,], Ingrid Duchhart [1,†] and Adri van den Brink [1,]

[1] Landscape Architecture Group, Wageningen University, P.O. Box 47, 6700 AA Wageningen, The Netherlands
[2] PBL Netherlands Environmental Assessment Agency, P.O. Box 30314, 2500 GH Den Haag, The Netherlands

ABSTRACT

Energy transition is expected to make an important contribution to sustainable development. Although it is argued that landscape design could foster energy transition, there is scant empirical research on how practitioners approach this new challenge. The research question central to this study is: To what extent and how is renewable energy science incorporated in regional landscape design? To address this knowledge gap, a case study of a regional landscape design competition in the Netherlands, held from 2010–2012, is presented. Its focus was on integral, strategic landscape transformation with energy transition as a major theme. Content analysis of the 36 competition entries was supplemented and triangulated with a survey among the entrants, observation of the process and a study of the competition documents and website. Results indicated insufficient use of key-strategies elaborated by renewable energy science. If landscape design wants to adopt a supportive role towards energy transition, a well-informed and evidence-based approach is highly recommended. Nevertheless, promising strategies for addressing the complex process of ensuring sustainable energy transition also emerged. They include the careful cultivation of public support by developing inclusive and bottom-up processes, and balancing energy-conscious interventions with other land uses and interests.

Keywords: energy transition; sustainable energy; renewable energy; design competition; landscape architecture; planning; energy-conscious planning and design; renewable energy; strategic landscape design; evidence based design

1. INTRODUCTION

Sustainable energy transition—the shift from a fossil-fuel based energy system to one based on renewable sources—is motivated by environmental, (socio-)economic and geopolitical factors [1,2,3]. In the coming decades, the transition to renewable energy is expected to make an important contribution to the process of sustainable development [3,4].

Historically, much of the world's energy is provided by the (natural) environment and, in turn, its exploitation often had a considerable impact on the landscape [5,6,7]. Because of this reciprocal relationship, Ghosn [8], Bloemers, et al. [9], Stremke and van den Dobbelsteen [10], Ivančić [11], Radzi [12] and van Hoorn and Matthijsen [13], amongst others, argue that energy transition represents a challenge for those involved in planning and design. From the mid-1990s, landscape architects world-wide have been involved in studying the visual impact of wind parks [14] and developing strategies for siting wind turbines in the landscape [15,16]. More recently, however, a more strategic approach to energy transition has emerged that includes fostering a sustainable realization of energy transition goals from a spatial perspective [17,18,19]. To this end, academics working in the field of landscape architecture developed spatial design concepts and principles, based on insights derived from renewable energy science, thermodynamics, systems science, and ecology [20,21,22]. The process of implementing the envisaged strategic approach would begin with surveying and mapping potential energy saving and generation resources in a selected environment using, for example, Energy Potential Mapping methodologies and GIS [1,23,24]. In addition this approach to energy-conscious planning and design would involve making spatially explicit scenarios and envisioning (long-term) interventions [1,25,26,27].

The relevance of knowledge and theory as the basis for planning and design is addressed in notions such as "knowledge-based design" [28], "evidence-based practice" [29], "evidence-based landscape architecture" [30], and "evidence-based design" [31,32]. To enhance the development of landscape architecture as an academic discipline and to provide a bases for evidence-based practice Meijering, et al. [33] and Deming and Swaffield [31] emphasize the importance of a shared and focused research agenda in landscape architecture. In the context of energy transition, the departure point for evidence-based landscape design practice can be found in renewable energy science, and the studies referred to above that translate fundamental insights into spatial design concepts, principles and procedures for energy-conscious design. That evidence-based approaches are appropriate in the context of energy transition assignments was illustrated by Twidell and Weir [3] (p. 2) who concluded that "Failure to understand the distinctive scientific principles will almost certainly lead to poor engineering and uneconomic operation". Although landscape design operates at larger levels of scale than individual technical installations, it can be argued that insights from renewable energy science will enhance effective energy-conscious planning and designing. Given the importance and availability of insights, it is surprising that there is a lack of empirical research into whether and how practitioners in landscape design take up the challenge of energy transition. Our research question, therefore, is: To what extent and how is renewable energy science incorporated in regional landscape design? To answer this question we studied the results of the Ninth Eo Wijers Regional Landscape Design Competition (The Ninth Eo Wijers Competition), because it focused on integral, strategic landscape transformations and energy transition as a major theme. Other landscape design competitions that referred to energy transition, tended to focus on smaller levels of scale and/or land art (for instance [34,35,36]). It was decided to focus on a design competition instead of (implemented) design projects for two reasons. First, studying the products of an ideas competition allowed to focus on the designer's intentions given that designs, compared to implemented projects, are less influenced by practical, financial and political factors [37]. Moreover, studying competition entries made it possible to compare designs, because each team was working on the same assignment set by the competition and subject to the same social context and time frame [38].

The structure of this paper is as follows. Section 2 contains a description of The Ninth Eo Wijers Competition. In Section 3, key-strategies from renewable energy science crucial for energy-conscious planning and design are elaborated within a theoretical framework. Research materials and methods are described in Section 4. In Section 5 there is a detailed discussion of results of this study. The conclusions are summarized in Section 6.

2. THE NINTH EO WIJERS REGIONAL LANDSCAPE DESIGN COMPETITION

Since 1985, the Eo Wijers Foundation has been promoting Dutch regional landscape design, mainly by organizing a prestigious competition every two to three years [39]. After the seventh competition in 2004, the structure of the competition was changed to some extent. A preparatory phase designed to encourage joint learning and debate among potential competition regions was added. At the end of this phase, the competition region is selected. An implementation phase was also added to the ideas competition and the Foundation committed itself to supporting the implementation of prize-winning ideas [40].

At the beginning of the Ninth competition, the Eo Wijers Foundation identified the themes of energy transition, population decline and spatial quality as being of urgent national relevance [41]. The Veenkoloniën ("Peat Colonies") region was selected as competition region because the themes of energy transition and population decline were most apparent there.

The Veenkoloniën in the North of the Netherlands covers some 800 km^2 and is an area where peat used to be extracted (Figure 1). The Foundation developed the competition brief in cooperation with a body of regional representatives, known as the "Agenda voor de Veenkoloniën". This is a partnership between the

Provinces of Drenthe and Groningen, eight municipalities and two water boards that aims to increase the socio-economic capacity of the region. In the final competition brief, the Foundation's initial themes were merged with input from the region and included four issues: "population decline", "energy", "agriculture" and "water management" [41]. The Foundation's spatial quality theme was regarded an overarching theme and was, therefore, taken up as one of the seven criteria to be considered by the jury. The instruction detailing the competition requirements with regard to energy is presented in Box 1.

Figure 1. Location of the competition region Veenkoloniën within The Netherlands.

Box 1. Specific competition instruction relating to the theme energy for the Ninth Eo Wijers Competition ([1] (p. 21); translated from Dutch).
"Alternative energy sources can reduce dependence on fossil fuels. Bio-, solar, geothermal and wind energy have potential. The national government wants to realize 400 MW extra wind energy capacity in Northeast Netherland.

Also the regional authorities want to encourage reliable and affordable energy provision with low emissions of greenhouse gases. Therefore, make the most of the opportunities for renewable energy generation and distribution, inter alia by providing adequate spatial possibilities. Also saving energy, the careful use of subterranean resources for energy provision, the storage of CO_2, green gas, natural gas and energy infrastructure are important. What do these mean for regional spatial development? The central notion in the 'Grounds for Change' philosophy is that our society must adjust to contemporary landscapes, which emerge through the use of, for example, wind power, and a more intensive use of subterranean resources. This process meets resistance in society and therefore demands careful selection and development of landscape sites and the involvement of the population."

In general, the competition entries should contain plans and designs relevant to local, regional and supra-regional levels. Because public support was highly valued, the regional representatives facilitated interaction between those taking part in the competition and local stakeholders. Two informative meetings were held and competitors were strongly recommended to collect local field data. Moreover, competitors were challenged not only to make spatially explicit designs, but also to rethink current planning processes by developing process-oriented proposals. Because of the complex and integrative nature of the assignment, the Eo Wijers Foundation suggested competitors form multi-disciplinary teams that would consist of landscape designers and planners, experts from the social and natural sciences as well as local experts [41].

The competition phase was launched in June 2011. The submission deadline was 6 January 2012 [41]. By then, 36 contributions had been submitted by 204 entrants. By 22 March 2012, the winners were announced during a special award ceremony that received considerable national attention.

3. THEORETICAL FRAMEWORK

In this section, we present the theoretical framework that was developed on the basis of the literature. This framework provided the basis for a coding scheme and this will be referred to later in the section dealing with methods. This framework was central to the subsequent analysis of competitors' entries and focuses on key-strategies for energy transition as identified by renewable energy experts. These were strategies that were relevant for energy-conscious landscape planning and design and that were readily available at the time of the competition. Four key-strategies are addressed in the following subsections: reductions in energy demand (3.1), diversity of energy supply (3.2), reduction of fossil fuel emissions (3.3), and consideration of the energy system components (3.4).

3.1. Reductions in Energy Demand

The first strategy in energy transition aims at increasing energy efficiency [3,42,43]. This is because determining the demand for energy is the starting point for organizing energy supply; when demand falls, less energy has to be provided. Energy efficiency—providing the same services with a reduced amount of energy—can be achieved by energy saving practices and by ensuring a better match between energy supply and demand. Most measures are beyond the influence of the spatial domain, for example, stimulating energy use at times of excess supply by offering reduced prices. Yet, efficiency can also be realized in the built environment. For example by improving the insulation and heat recovery capacity of buildings. This is already being applied and has considerable potential. On a larger scale, the spatial organization of the built environment can also be adapted to facilitate residual heat exchange between industry and housing, for example [22,44].

3.2. Diversity of Supply

The second strategy in energy transition focuses on increasing the use of renewable energy sources in meeting energy needs [42]. (Regional) energy supply is made up of electricity, heat and (transport) fuels, all of which are—in theory—interchangeable. At present, when focusing on the Dutch situation, a limited variety of conventional fuels such as crude oil, natural gas, coal and nuclear energy meet almost the whole energy demand [45]. However, when moving towards renewable energy sources, a more diverse mix of sources and conversion technologies will be needed. This is because, for example, there is a limit to the availability of sources of renewable energy in space and time, and their potential for generation, conversion, distribution and storage [2,3]. Moving towards renewable energy sources will affect the amount of space needed to satisfy energy demand, and conversion technologies such as wind turbines, which are bound to windy locations, will change the landscape [1,7]. This is why landscape designers have been involved in the siting and design of renewable energy technologies. Following a more strategic approach, landscape designers could support the mapping of renewable and residual energy potentials, that could well be the starting point for organizing renewable energy provision in a given region [3,23,24].

3.3. Reduction of Fossil Fuel Emissions

A third strategy in energy transition focuses on achieving a reduction in fossil fuel emissions [3,42]. The transition to other forms of energy in the past has taken time and the current transition to sustainable energy should be seen in this context [45]. In the meantime, fossil fuels will not be abandoned overnight. When fossil fuel use is unavoidable, negative effects on the environment should be reduced as far as possible, for example through Carbon Capture and Storage (CCS) techniques [43,45].

3.4. Consideration of the Energy System Components

In this paper, the term energy system refers to the interconnected whole of energy sources (areas where local energy supply exceeds the consumption) and sinks (where local energy consumption exceeds the supply) as well as the technologies associated with converting, distributing and storing energy [3]. Energy distribution is about getting energy to the right place; energy storage is about ensuring there is sufficient energy available for anticipated needs [3].

Since energy transition requires a changed approach to energy sources and conversion technologies, the infrastructure for distribution and storage must be adapted accordingly [2,3,46]. This will affect land use and landscape image [3,13,47], given that implementation will have a direct impact on the landscape. The utilization of geothermal energy, for example, involves specific pumps, pipes and associated installations for energy distribution. Moreover, due to fluctuations in supply of renewable sources, such as wind, solar and tidal energy, storage capacity may have to be increased. As with renewable energy sources and conversion technologies, surveying and environmental mapping can guide the location of potential energy storage and distribution infrastructure. In this way, for example, empty salt caverns and artificial lakes in the Netherlands were identified as potential options for storing biogas and kinetic energy [23].

4. MATERIALS AND METHODS

The research presented in this paper followed the general guidelines for case study research outlined by Yin [48]; The Ninth Eo Wijers Competition was analyzed in context and a variety of data and methods were used. The study has a single-case design, in which the competition entries are considered embedded units of analysis [48] (see Figure 2).

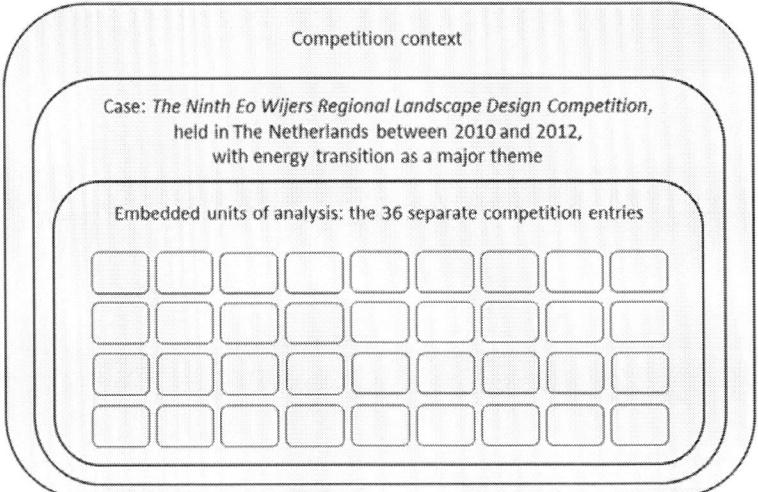

Figure 2. The single-case design with 36 embedded units of analysis (based on Yin [48] (p. 46)).

4.1. Content Analysis of the 36 Competition Entries

The 36 competition entries submitted form the empirical basis of this study. The Foundation made the original competition entries available to us digitally and in hard copy. Each entry consisted of at least three A0-sized posters and an essay of about 1500 words. An example of the posters submitted by the competitors is presented in Figure 3. All entries can be accessed online, at http://www.veenkolonien.nl/83-eo-wijers.html (in Dutch).

Evaluating designs is a reflective activity, and involves several interpretive steps including sorting out, analysis and comparison which lead to a deeper understanding of the diverse aspects of the design process and its results [38]. In analyzing the competition entries we adopted a qualitative content analysis procedure using a combination of predetermined and emerging codes [49]. Coding was executed manually. Results were gathered and clustered in an Excel spreadsheet for (comparative) analysis. The text and the visual data from the entries—plans and designs, schemes, graphs, sections, 'bird's eye' perspectives, and artist impressions—were analyzed using the same procedure and coding scheme. This was possible because the focus was on exploring the content of the data, and not on how words and images were used to solicit a particular effect (see also [50]).

The predetermined codes were drawn from renewable energy science and the competition brief itself. The codes derived from renewable energy science addressed the four key-strategies for realizing energy transition (see the theoretical framework discussed in Section 3). Dimensions and indicators were defined in order to operationalize these strategies. For the diversity of supply strategy, amongst others, the following dimensions were defined:

1) Catering for the regional energy demand using at least two options derived from electricity, heat, and (transport) fuels;
2) Making use of at least one renewable energy source and/or conversion technology;
3) Making use of more than one kind of renewable energy source and/or conversion technology and in this way acknowledging the need for a diversified energy mix.

Indicators were used to establish the presence of these dimensions and the presence of key-strategies for energy transition in the competition entries. These refer to the codes in the coding scheme, for example "electricity", "heat" and "(transport) fuels" for the first dimension referred to above, and various kinds of renewable energy sources and technologies for the second and the third dimension.

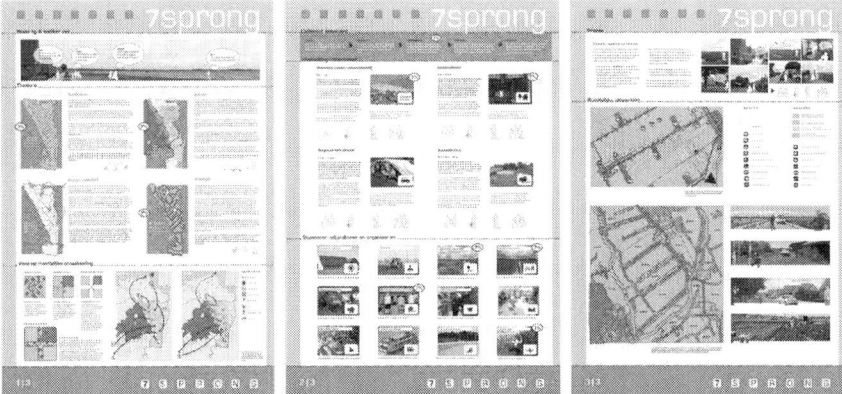

Figure 3. The three A-0 sized posters that accompanied the entry "7Sprong". Reproduced with permission of the authors: Frank Stroeken, Jan Maurits van Linge, Sander van den Helm, Jannemarie de Jonge, Rianne Knoot and Ruth Dobbelsteen (from [51] published by Eo Wijers-stichting, 2012).

The competition brief, as has been explained in Section 2, suggested entrants create spatially explicit designs at three levels of scale but that they should also rethink the planning procedures currently being used in the area. In addition, the integrative nature of the assignment and the suggestion to work in multidisciplinary teams played an important role in this competition. Therefore, we also chose to incorporate these strategies in this study. In order to analyze the nature of the proposals, we coded whether an entry focused on process-oriented proposals and/or spatially explicit designs. To examine the integration of energy-conscious interventions with other (competition) themes and interests, we coded indicators for population decline, agriculture and water management and spatial quality. A number of codes also emerged here that referred to land use, for example. Table 1 in Section 5.1 provides an overview of the strategies and the dimensions we used.

4.2. Study of the Competition Context and Triangulation of Data and Methods

In addition to the content analysis of competition entries, three other sources of data were used. This was done to enable the triangulation of results and to establish the context in which the competition entries were created and judged. First, the competition process was observed by the lead author of this paper, who attended the meetings organized by the Foundation, and followed the judging process. The third author of this paper was in fact a member of the professional jury. In this way, the outcomes of the content analysis and the opinion of the jury could be compared. Second, a survey was held to gain insight into the background of those who had entered the competition, their opinion about the theme of energy transition in relation to landscape design, and the sources and reference projects they had used in developing their entry. The survey was posted online directly after the award ceremony and remained online for three weeks. The 242 people who had initially signed up for the competition were invited to take part in the survey. In total some 126 people responded, 100 of whom had actually participated in the competition. This represented 33 out of the total of 36 participating teams. Prior to being submitted, the survey had been tested by four colleague researchers. Third, in order to build up a complete image of the competition process as a whole, the Foundation's written records and online resources including its website (www.eowijers.nl) the competition brief [41] and the jury report [51], were also taken into account.

4.3. Research Quality and Limitations

Although analyzing an individual case has limitations as far as the generalization of results is concerned, we argue that this study—because of its uniqueness and its empirical character—provides new insights into the (potential) contribution that landscape design can make to energy transition. In addition to the triangulation of data sources and methods such as those described above, the content analysis has been cross-checked by the authors themselves and reviewed with two members of the Foundation. Even so the limitations of single case study methods should be borne in mind when assessing the implications of the conclusion.

5. RESULTS AND DISCUSSION

In this section, the findings of the study are presented and discussed. In Sub-section 5.1, the results of the content analysis are explained and in Sub-section 5.2 the judging process and outcomes are described. Sub-section 5.3 deals with the results of the survey.

5.1. The Competition Entries

We now present the results of a content analysis of the 36 competition entries. For an overview, see Table 1. In Sub-subsection 5.1.1, Sub-subsection 5.1.2, Sub-subsection 5.1.3 and Sub-subsection 5.1.4 the results are described of analyzing the extent to which key-strategies derived from renewable energy science were present in the entries. Sub-subsection 5.1.5 shows the extent to which the entries contained spatially explicit designs and/or process-oriented proposals and 5.1.6 deals with the extent to which energy-conscious interventions were integrated with other (competition) themes and objectives.

5.1.1. Reductions in Energy Demand
In 31% of the entries energy efficiency measures were proposed and these ranged from car-pooling to installing insulation (see Table 1). Next an assessment was made as to whether the entrants had defined the goals for energy efficiency and/or renewable energy generation in specific, quantitative terms that they wanted to achieve with their proposals. Only 8% of the teams had done

so (see Table 1). Two of these teams specified the targets for regional energy self-sufficiency for the years 2025 and 2040 respectively. A third team quantified the regional energy demand and expressed the goal of regional energy self-sufficiency, but did not specify a specific time period.

Table 1. Overview of results of a content analysis of competition entries.

Origin	Strategy	Dimension Present in the Entries	% of the Entries
Renewable energy science	Reductions in energy demand	Proposals for improving energy efficiency, e.g., by energy saving measures	31%
		Reference to goals for energy efficiency and/or renewable energy generation in specific, quantitative terms	8%
	Diversity of supply	Catering for the regional energy demand using at least two options derived from electricity, heat, and (transport) fuels	78%
		Making use of *at least one* renewable energy source and/or conversion technology	97%
		Making use of *more than one* kind of renewable energy source and/or conversion technology and in this way acknowledging the need for a diversified energy mix	86%
		Calculations indicating the contribution of proposed energy-conscious interventions	14%
	Reduction of fossil fuel emissions	Proposals for the use of CCS (Carbon Capture and Storage) and/or alternative solutions for reducing fossil fuel emissions	19%
	Consideration of the energy system components	Proposals for at least two of the following: energy generation, energy distribution, and energy storage	58%
		Proposals for energy generation, energy distribution and energy storage that acknowledging the fact that energy-conscious interventions should be seen as components of a larger energy system	25%
The competition brief	The nature of proposals	Process-oriented proposals for realizing energy-conscious designs and plans	89%
		Spatially explicit designs for locating, planning and/or designing energy-conscious interventions	58%
	Integration with other (competition) themes and interests	Combinations of energy-conscious interventions and other (competition) themes that show energy transition from an integrative landscape design perspective	89%
		Extent to which entries document the motivations for energy-conscious interventions	97%

Taken together, these results indicated that the strategy of reducing the energy demand was only applied by one third of the teams, although a diverse palette of interventions was proposed. Hardly any team mentioned explicit targets for energy efficiency and/or renewable energy generation as a starting point for purposefully matching energy demand with supply. Strikingly, none of the entrants took the 400 MW of wind energy that the national government wants to realize in the North of the Netherlands into their proposed design, even though this was part of the competition brief.

5.1.2. Diversity of Supply

Regarding the strategy to diversify energy supplies, 78% of the entries proposed the provision of multiple forms of energy, namely electricity, heat and fuels (see Table 1). Almost all entries, 97%, proposed at least one renewable energy technology to meet regional energy demand (see Table 1). Eighty-six percent of the entries (see Table 1) proposed more than one renewable energy technology. Table 2 provides an overview of renewable energy sources and technologies proposed. Biomass (from landscape maintenance and/or energy crops; excluding biomass waste streams) was proposed most frequently, in 75% of the entries (see Table 2) and the specific rural and agricultural character of the region was often given as the reason for proposing this source. Onshore wind energy came in at second place and was proposed in 64% of the entries (see Table 2). The fierce opposition to onshore wind turbines was explicitly mentioned by entrants as one of the reasons for looking into alternative renewable sources and technologies. On average, between three and four different renewable sources and technologies were proposed per entry. Therefore, although biomass and onshore wind turbines were proposed relatively frequently a number of other sources and technologies capable of meeting regional energy demand were also taken into account. In Figure 4 it can be seen how one of the entrants envisioned a landscape in which a variety of renewable energy sources and technologies were integrated. Only 14% of the teams explained their energy-conscious interventions by quantifying the expected contribution of their proposals in, for example, joule or kWh (see Table 1).

Case results show that it can be concluded that the entrants were well aware of the concept of diversity of supply and the technologies available for renewable energy generation. It could also be concluded that they showed an understanding of the fact that different forms of energy (electricity, heat and fuels) needed to be catered for [2,3,43]. In this particular competition, the regional landscape and local social conditions clearly played a role in the choice of renewable energy resources and technologies. Some teams explained their proposals regarding energy transition in quantitative terms, but this did not lead to a purposeful and efficient matching of energy demand and supply, as is common in energy planning [3].

Table 2. Renewable energy sources and conversion technologies proposed.

Renewable Energy Sources and Conversion Technologies	% of the Entries
Biomass (excluding biomass waste streams)	75%
Wind energy: onshore wind turbines	64%
Solar energy: photovoltaic cells	50%
Biomass waste streams	47%
Residual heat	36%
Heat and cold storage	19%
Solar energy: solar collectors	17%
Geothermal energy: heat	17%
Combined heat and power	17%
Geothermal energy: electricity	6%
Other	36%

Figure 4. "Verborgen kracht—Veenkoloniën 3.0": An example of a design visualizing diverse renewable energy generation via solar, wind and biomass in the Veenkoloniën. Reproduced with the permission of the author Tim Snippert (from [51] published by Eo Wijers-stichting, 2012).

5.1.3. Reduction of Fossil-Fuel Emissions

With regard to this strategy, 19% of the entries (see Table 1) proposed to reduce emissions from remaining fossil fuels. None of the teams proposed CCS techniques as they have been developed for coal-fired power plants for example. Instead, five of the entries focused on the re-use of CO_2 in greenhouses and/or for algae production. Two other entries proposed to (re)create peat lands or swamps to support CO_2 sequestration in a natural way.

With regard to this strategy, we can conclude that only a few of the teams felt the need to reduce emissions by curtailing the use of fossil fuels by applying technical or natural solutions.

5.1.4. Consideration of the Energy System Components

Two or more components of energy systems—energy generation, energy distribution and energy storage—were proposed in 58% of the entries (see Table 1). Only nine of these entries—25% of the total number of entries—addressed all three components of energy systems as defined in the literature (see Table 1). Figure 5 gives an example drawn from an entry that addressed the components of energy generation, energy distribution and energy storage in a design model that can be scaled up to regional level.

Findings in respect of this strategy indicate that understanding and optimizing the energy system was a challenge and we found this surprising. Renewable energy may be a relatively new subject but landscape designers are used to approaching the landscape as system [28] and to working on solutions by going back and forth between interrelated levels of scale [52]. For these reasons, we had expected more entries to develop systemic approaches that addressed all three components of the energy system.

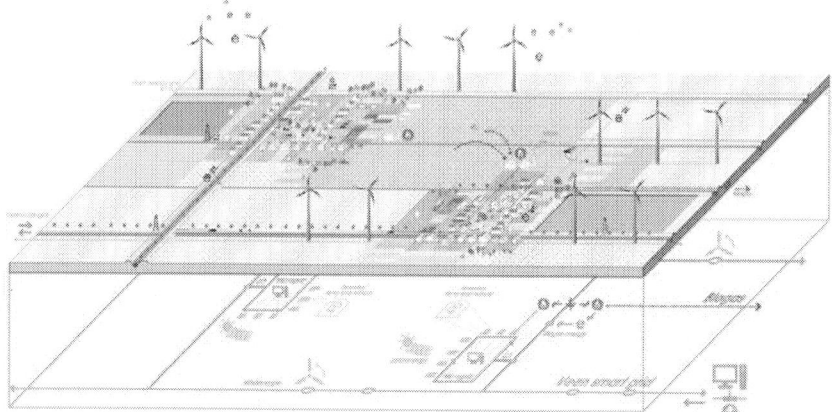

Figure 5. Veennet: an example of a schematic visualization of an energy system including the components of renewable energy generation, energy distribution and energy storage. Reproduced with permission from the authors Boris Hocks, Han Dijk, Emile Revier, Iris Wijn, Justina Muliuolyte, Dion van Dijk, Machiel Bakx and Michiel Brouwer (from [51] published by Eo Wijers-stichting, 2012.

5.1.5. The Nature of Proposals: Spatially Explicit and/or Process-Oriented

Because the competition brief stressed the need to develop alternative, bottom-up, and inclusive planning processes, the entries were studied to determine whether they contained spatially explicit designs and/or process-oriented proposals relating to energy-conscious interventions. The number of process-oriented proposals at 89% outnumbered the 58% of spatially explicit designs. However, both process orientated proposals and spatially explicit designs were included in 53% of the entries.

Spatially explicit designs indicated, for example, the location and design of renewable energy technologies and how these could be integrated into the Veenkoloniën landscape. Process proposals, for example, included ideas for involving farmers in developing a bio-based economy and stimulating local cooperative car sharing. Sixty nine percent of the entrants used inputs from the region to develop new process proposals. Innovative strategies for stimulating interaction with local people and to harvest their input for making plans and designs included the use of social media. This has been discussed in de Waal, et al. [53].

From this analysis it became clear that it was not the spatially explicit designs that were central in this competition, which one would expect in a landscape design ideas competition, but process-oriented proposals. This is an interesting finding in the light of the cyclical governance model for (energy) transitions discussed by Loorbach, et al. [54]. Here (spatially explicit) envisioning is seen as one of many activity sets in transition processes. The other sets in this model are agenda building and networking, experimenting and diffusion, and monitoring, evaluating and adaptation. Each process-oriented proposal in the competition addressed at least some of these. In the Ninth Eo Wijers Competition, the competition brief—with its focus on process-oriented proposals—and the way the competition's preparatory and implementation phases were organized, emphasized joint agenda building, networking, and the diffusion of results. It can, therefore, be concluded, that the scope of the concept of regional landscape design in relation to energy transition was significantly widened in the context of this competition.

5.1.6. Integration of Energy Transition Interventions with Other Competition Themes and Interests

As discussed in Section 2, the competition presented an integrated assignment which addressed not only the theme of energy transition, but also drew attention to factors such as population decline, agriculture and water management. We therefore decided to list which themes or land uses the entrants integrated with energy transition. In total, 89% of the teams responded to the call for integrated solutions. Table 3 specifies the combinations that were created.

Table 3. Integration of energy-conscious interventions with other competition themes and land uses.

Combinations with Other Competition Themes and Land Uses	% of the Entries
Agriculture	72%
Water	44%
Nature	22%
Habitation	22%
Industry	19%
Recreation	17%
Amenities (shopping, nightlife *etc.*)	11%
Education and Research	8%
Welfare (healthcare, childcare *etc.*)	8%
Infrastructure	6%

Interestingly, connections were also made beyond the themes outlined in the competition brief and included nature development, habitation, industries and infrastructure. Figure 6 illustrates how one competition entry visualized the integration of renewable energy generation with agricultural production, health care, and education in a new type of farmyard structure.

In addition, we listed the motivations given by the entrants for energy transition that were relevant to energy-conscious interventions. Besides 'direct' motivations for energy transition—such as improving sustainability and ensuring a secure and affordable supply—more indirect motivations were also provided. For example, stimulating the regional economy by renewable energy generation was put forward as a motivation in 55% of the entries while 47% stressed that energy-conscious interventions should be carefully integrated with landscape image and regional identity (see Table 4). Here it is shown that difficulties arise when a new type of land use—such as a renewable energy provision—has to be integrated within the existing spatial order because in pursuit of sustainability the interests involved at different levels of scale may come into conflict with each other (see also [55,56]). The entrants showed however, on the positive side, how energy transition gave rise to exploring win-win strategies, for example, by introducing energy crops as a new agricultural product, or establishing a cooperative wind turbine venture as a way of promoting a sense of community. In total, only one team failed to provide a motivation for energy-conscious interventions (see Table 4).

Figure 6. Two visualizations that show how renewable energy generation and agricultural production, habitation and welfare were combined on a new type of farmyard from the competition entry 'Wat weet een boer van saffraan'. Reproduced with permission of the authors Richard Colombijn, Claire Oude Aarninkhof, Renzo Veenstra and Arjan Boekel (from [51] published by Eo Wijers-stichting, 2012.

Table 4. Motivations energy-conscious interventions.

Motivations and Interests	% of the Entries
Stimulating regional economy	55%
Maintaining or improving landscape quality	47%
Sustainability	36%
Energy independence	19%
Raising awareness about energy transition	6%
Security of supply	3%
Affordability of supply	3%
Public acceptance of renewable energy technologies	3%
Turning the region into a testing ground	3%

From this analysis, it became clear that the teams approached the theme of energy transition in a highly integrative way, as required by the competition brief. This is important, because it was found that interventions to promote sustainability sometimes have unintended aversive effects. As Stremke [18], elaborated in his article "Sustainable energy landscape: Implementing energy transition in the physical realm" adopting an integrative perspective that goes beyond the mere implementation of renewable energy technologies, and addressing sustainable, technical, economical and socio-cultural criteria in energy-conscious landscape design, would have a definite and positive impact on sustainable development.

5.2. The Judging Process and Outcomes

On 22 March 2012, the Eo Wijers Foundation was ready to award the prizes. There was a first, second and third prize, three honorable mentions and two young professional awards. The winners were selected according to a process of blind review. The jury consisted of professional and regional representatives. The regional jury had 11 members drawn from the civil sector, for example, residents, entrepreneurs and aldermen. The professional jury consisted of eight members whose expertise was directly related to the competition's themes: population decline, energy transition, agriculture and water management [51].

The jury praised the entrants in general for the way they had approached the complex issues in the Veenkoloniën. However, it was also felt that only a third of the entrants had considered the placement of wind turbines in the area as a regional design challenge. None of the entries explicitly addressed the 400 MW of wind energy that, according to the competition brief, was needed in the North of the Netherlands. Neither did they present a clear vision of the regional energy supply when moving to renewable energy sources. During the award ceremony the chairperson explained it as follows [57]: "In the opinion of the jury, an evaluation of the landscape's capacity for holding wind turbines required detailed argumentation. If the entrants assumed this capacity to be low or reduced—an effective alternative for wind energy should have been proposed."—a conclusion that coincides with our own and which has been presented here.

5.3. Survey among the Competition Entrants

The background of competition entrants was analyzed by means of a survey. Each team consisted of an average of five to six persons. The largest team had 12 members. Only one entry came from a person working alone. The disciplinary backgrounds of the entrants varied and included, for example, landscape architecture, urban design, planning, architecture, history, energy consultancy, management, economics, industrial engineering and communication. Most teams followed the instruction of the competition brief and were—to varying degrees—multi-disciplinary in composition. The age distribution was varied. Seven percent of entrants were younger than 26 years;

34% were between 26 years and 35 years, 29% were between 36 years and 45 years. In addition 16% of entrants were between 46 years and 55 years and 13% were older than 55 years.

Moreover, because the content analysis and the jury's evaluation of results suggested that the entrants approach to the application of the key-strategies that have emerged from renewable energy science was mixed and sometimes poor, we asked entrants to give us their opinion on energy transition as this related to landscape design. Seventy-two entrants answered our questions. The results were as follows:

- The respondents were positive about the potential contribution landscape design could make to energy transition (38% responded "yes, to large extent"; 54% responded 'yes, to some extent'; seven percent responded 'no' and one percent responded "I do not know");
- More than half the respondents believed that energy transition provides an opportunity to enhance spatial quality in the Netherlands (22% fully agreed; 35% partly agreed, 25% did not agree or disagree, 10% partly disagreed, 3% percent fully disagreed and 5% percent did not know).

These results indicated that, regardless of how energy transition was dealt with in the entries, the majority of entrants agreed with the organizing Foundation that it is an important element in regional landscape design. At least half of the entrants recognized the potential for enhancing spatial quality when working on energy transition.

In addition we asked entrants to list between one to three reference projects and one to three (written) sources of information on energy transition that they used in compiling their entry. These questions were answered by the teams involved rather than by individuals and were classified according to type. From the 33 teams that answered the survey, only 17 teams listed reference projects (43 in total) and 13 teams listed information sources (23 in total). The 23 information sources referred to by the entrants showed a wide degree of variety and very little overlap (see Table 5). Only one source was mentioned by four teams. This was Energielandschappen: De 3de generatie.Over regionale kansen op het raakvlak van energie en ruimte by Noorman and de Roo from 2011—a Dutch book on energy transition from a planning and design perspective [58].

Thirty-one of the 43 reference projects referred to specific, unique projects, such as the well-known Danish energy neutral island of Samsø, or the Dutch island of Texel. The remaining 12 reference projects were rather either unspecific or not unique; for example "wind energy in Germany" and "cooperatives that were already providing solar and wind energy".

As it seemed from this part of the survey, the teams had consulted very few (scientific) documents and no standard sources on renewable energy science had been referred to. Rather, the teams had tended to focus on reference projects. When combining these findings with the analysis of competition entries and the jury's judgment, we must conclude that the underutilization of information on renewable energy science led—in the context of this regional design competition—to a less than optimal application of basic strategies for realizing energy transition.

Table 5. Types of information sources about energy transition listed by the teams.

Types of Information Sources	Counts	Different Documents
Report by engineering or design firm	6	6
Book aimed at a professional audience	6	3
Input by expert	3	3
Research report	3	3
Opinion article	1	1
Policy report	1	1
Report by NGO	1	1
Scientific literature	1	1
Symposia	1	1
Totals	23	20

6. CONCLUSIONS

The information that has been analyzed in this paper is derived from a regional landscape design competition that focused on renewable energy in the context of integral and strategic landscape transformation. Although it is argued that landscape design could foster energy transition from a spatial perspective, there is scant empirical research on how practitioners approach this new challenge. The research question addressed in this study, therefore, centres on the extent to which renewable energy science was incorporated into regional landscape design and how this was done.

Four key-strategies for energy transition were derived from renewable energy science: reductions in energy demand; diversity of energy supply; reduction of fossil fuel emissions and consideration of energy system components. By conducting a content analysis of the competition entries from the perspective of these key-strategies, we identified serious flaws in their application. All but one team, 97% of the entries, worked on renewable energy generation, and diversity of supply was addressed by 78% of the entries. Often regional landscape qualities and socio-economic conditions were the starting point for selecting renewable energy sources and technologies. However, just one third of the teams addressed strategies aimed at reducing energy demand. Only 19% of the teams addressed the problem of reducing fossil fuel emissions, and a mere 25% suggested solutions to the systemic problems of energy generation, energy distribution and energy storage. These figures suggested that, although informative literature was readily available at the time of the competition, the application of the four key-strategies that could be derived from renewable energy science was mixed and sometimes extremely poor. The competition jury had also commented negatively on the way in which energy transition had been dealt with by competition entrants. Based on these two sources, we concluded that there was a considerable gap between theory and practice.

A survey among entrants, however, showed that the respondents believed that energy transition is indeed a theme for regional landscape design, as the Eo Wijers Foundation competition brief suggested. In addition, more than half the respondents believed that energy transition provides an opportunity to enhance spatial quality in the Netherlands. Despite these opinions, the survey revealed little evidence that entrants had consulted the relevant literature on the subject. This might be the reason for the rather mixed and poor application of the key-strategies that can be derived from renewable energy science. These findings lead us to stress the importance of evidence-based approaches to landscape design, in terms of enhancing its development as a socially relevant academic discipline and one that has important implications for sustainable energy transitions.

Notwithstanding the hesitant application of renewable energy science, the competition in itself emphasized the role of landscape design in effective, sustainable energy transition. In fact, the competition brief included two strategies that were rewardingly elaborated by the entrants. These were the integration of energy-conscious interventions with other (competition) themes and interests, and the nature of proposals regarding energy transition. In terms of the latter, it appeared that spatially explicit designs, which can be seen as a more traditional product of landscape designers, were included in 58% of the entries. However, 89% of the entries directly referred to the importance of a process-oriented approach, and 53% of the entries included both. The main benefit of widening the scope of landscape design in this way was that it focused attention on the importance of public support and the development of inclusive and bottom-up processes, a development that is in line with recent insights into transition theory. A side effect was that competition entrants were confronted with the relatively new subject of energy transition, in combination with the assignment to create both spatially explicit and process-oriented proposals. We would recommend further research into the relationship between landscape design and transition theory—especially in relation to how transition goals can be realized by spatially explicit envisioning and inclusive, bottom-up planning processes. Fierce public resistance to the implementation of some renewable energy technologies at the local level and consequent delays in realizing energy transition goals, might be addressed by further research in this area.

Finally, we conclude that our study has shown that more attention should be given to an integrative approach to energy transition in regional landscape design as was emphasized by this competition. An integral approach seems to be one of the most important contributions of landscape design when energy transition is being considered in specific areas. A careful consideration of other land uses and interests including environmental impact and the implications for socio-economic and cultural aspects must be addressed in the complex task of facilitating sustainable energy transition.

ACKNOWLEDGMENTS

The authors would like to thank the Eo Wijers Foundation and competition entrants for their cooperation and making available competition materials. We are grateful to Adrie van 't Veer for his help with Figure 1, and to Edo Gies, Annet Kempenaar, Marjo van Lierop and Jurian Meijering for their help in constructing the survey. Finally, we would like to thank the anonymous referees for their useful comments on the paper.

AUTHOR CONTRIBUTIONS

Each author contributed to the research in terms of conception, research design, cross-checking data analysis and co-writing the paper. Renée M. de Waal was primarily responsible for data collection, analysis and writing of the paper. All five authors read and approved the final manuscript.

REFERENCES

1. Sijmons, D.; Hugtenburg, J.; Feddes, F.; van Hoorn, A. Landscape and Energy, Designing Transition; NAi010 Publishers: Rotterdam, The Netherlands, 2014.

2. Tester, J.W.; Drake, E.M.; Driscoll, M.J.; Golay, M.W.; Peters, W.A. Sustainable Energy: Choosing among Options; MIT Press: Cambridge, MA, USA; London, UK, 2005.

3. Twidell, J.; Weir, A.D. Renewable Energy Resources, 2nd ed.; Taylor & Francis: Oxon, UK, 2006.

4. Brundtland, G.H. Report of the World Commission on Environment and Development: "Our Common Future"; United Nations: New York, NY, USA, 1987.

5. Leenaers, H.; Camarasa, M. De Bosatlas van de Energie; Noordhoff Uitgevers: Groningen, The Netherlands, 2012.

6. Mulder, K. The technological landscape. In Landscape and Energy: Designing Transition; Sijmons, D., Hugtenburg, J., Feddes, F., van Hoorn, A., Eds.; NAi Publishers: Rotterdam, The Netherlands, 2014; pp. 368–380.

7. Pasqualetti, M.J. Reading the changing energy landscape. In Sustainable Energy Landscapes. Designing, Planning and Development; Stremke, S., van den Dobbelsteen, A., Eds.; CRC Press (Taylor & Francis Group): Boca Raton, FL, USA, 2013; pp. 11–44.

8. Ghosn, R. New Geographies 2: Landscapes of Energy; Harvard University Press: Cambridge, MA, USA, 2009.

9.   Bloemers, T.; Daniels, S.; Fairclough, G.; Pedroli, B.; Stiles, R. Landscape in a Changing World; ESF and COST: Strassbourg, France, 2010.

10. Stremke, S.; van den Dobbelsteen, A. Sustainable Energy Landscapes; Designing, Planning and Development; CRC Press (Taylor & Francis Group): Boca Raton, FL, USA, 2013.

11. Ivančić, A. Energyscapes; Editorial Gustavo Gili: Barcelona, Spain, 2010.

12. Radzi, A. 100% Renewable champions: International case studies. In 100% Renewable. Energy Autonomy in Action; Droege, P., Ed.; Routledge: New York, NY, USA, 2009; pp. 93–166.

13. Van Hoorn, A.; Matthijsen, J. De Ruimtelijke Impact van Hernieuwbare Energie: een Verkenning; Planbureau voor de Leefomgeving: Den Haag, The Netherlands, 2013; Volume 1099.

14. Mogen, E.A.H. The role of the landscape architect in the wind farm site selection process and best practices. In Conference of CELA and ISOMUL on Landscape Legacy: Landscape Architecture and Planning between Art and Science; Carsjens, G.J., Ed.; Wageningen University: Wageningen, The Netherlands, 2010.

15. Schöbel, S. Windenergie und Landschaftsästhetik: Zur Landschaftsgerechten Anordnung von Windfarmen; Jovis: Berlin, Germany, 2012.

16. Schöne, M.B. Windturbines in het Landschap. Nieuw Plaatsingsbeleid op Basis van Landschapsbeleving Gewenst voor de Jongste Generatie Windturbines; Alterra: Wageningen, The Netherlands, 2007.

17. Wächter, P.; Ornetzeder, M.; Rohracher, H.; Schreuer, A.; Knoflacher, M. Towards a sustainable spatial organization of the energy system: Backcasting experiences from Austria. Sustainability**2012**, 4, 193–209.

18. Stremke, S. Sustainable energy landscape: Implementing energy transition in the physical realm. In Encyclopedia of Environmental Management; Jørgensen, S.E., Ed.; Taylor & Francis: Abingdon, UK, 2015; pp. 1–9.

19. De Waal, R.M.; Stremke, S. Energy transition: Missed opportunities and emerging challenges for landscape planning and designing. Sustainability**2014**, 6, 4386–4415.

20. Stremke, S.; van den Dobbelsteen, A.; Koh, J. Exergy landscapes: Exploration of second-law thinking towards sustainable landscape design. Int. J. Exergy**2011**, 8, 148–174.

21. Stremke, S.; Koh, J. Ecological concepts and strategies with relevance to energy-conscious spatial planning and design. Environ. Plan. B: Plan. Des.**2010**, 37, 518–532.

22. Stremke, S.; Koh, J. Integration of ecological and thermodynamic concepts in the design of sustainable energy landscapes. Landsc. J.**2011**, 30, 2–11.

23. Van den Dobbelsteen, A.; Broersma, S.; Fremouw, M. Energy potential mapping and heat mapping: Prerequisite for energy-conscious planning and design. In Sustainable Energy Landscapes. Designing, Planning and Development; Stremke, S., van den Dobbelsteen, A., Eds.; CRC Press (Taylor & Francis Group): Boca Raton, FL, USA, 2013; pp. 71–94.

24. Grêt-Regamey, A.; Wissen Hayek, U. Multicriteria decision analysis for the planning and design of sustainable energy landscapes. In Sustainable Energy Landscapes. Designing, Planning and Development; Stremke, S., van den Dobbelsteen, A., Eds.; CRC Press (Taylor & Francis): Boca Raton, FL, USA, 2013.

25. Stremke, S.; Koh, J.; Neven, K.; Boekel, A. Integrated visions (part II): Envisioning sustainable energy landscapes. Eur. Plan. Stud.**2012**, 20, 609–626.

26. Thün, G.; Velikov, K. Conduit urbanism: Rethinking infrastructural ecologies in the great lakes megaregion, North America. In Sustainable Energy Landscapes. Designing, Planning and Development; Stremke, S., van den Dobbelsteen, A., Eds.; CRC Press (Taylor & Francis Group): Boca Raton, FL, USA, 2013; pp. 261–284.

27. Stremke, S.; van Kann, F.; Koh, J. Integrated visions (part I): Methodological framework for long-term regional design. Eur. Plan. Stud.**2012**, 20, 305–319.

28. Motloch, J.L. Introduction to Landscape Design; Wiley: New York, NY, USA, 2001.

29. Krizek, K.; Forysth, A.; Slotterback, C.S. Is there a role for evidence-based practice in urban planning and policy? Plan. Theory Pract.**2009**, 10, 459–478.

30. Brown, R.D.; Corry, R.C. Evidence-based landscape architecture: The maturing of a profession. Landsc. Urban Plan.**2011**, 100, 327–329.

31. Deming, E.M.; Swaffield, S. Landscape Architecture Research: Inquiry, Strategy, Design; John Wiley &Sons: New York, NY, USA, 2011.

32. Lenzholzer, S.; Brown, R.D. Climate-responsive landscape architecture design education. J. Clean. Prod.**2013**, 61, 89–99.

33. Meijering, J.V.; Tobi, H.; van den Brink, A.; Morris, F.; Bruns, D. Exploring research priorities in landscape architecture: An international Delphi study. Landsc. Urban Plan.**2015**, 137, 85–94.

34. Klein, C.; Monoian, E.; Ferry, R. Regenerative Infrastructures: Freshkills Park NYC; Prestel Verlag: Munich, Germany, 2013.

35.   Ozgun, K.; Weir, I.; Cushing, D. Optimal electricity distribution framework for public space: Assessing renewable energy proposals for Freshkills Park, New York City. Sustainability**2015**, 7, 3753–3773.

36.   Monoian, E.; Ferry, R. New Energies, Copenhagen; Prestel Verlag: Munich, Germany, 2014.

37.   Brinkhuijsen, M. Landscape 1:1: A Study of Designs for Leisure in the Dutch Countryside; Wageningen University: Wageningen, The Netherlands, 2008.

38.   Vroom, M.J. Lexicon van de Tuin- en Landschapsarchitectuur; Blauwdruk: Wageningen, The Netherlands, 2010.

39.   Eo Wijers-Stichting. Eo Wijers-Stichting Doelstelling. Available online: http://www.eowijers.nl/?page_id=1053 (accessed on 7 August 2014).

40.   De Jonge, J. Een Kwart Eeuw Eo Wijers-Stichting: Ontwerpprijsvraag als Katalysator voor Gebiedsontwikkeling; Habiforum: Gouda, The Netherlands, 2008.

41.   Eo Wijers-stichting. Eo Wijers-Prijsvraag 2011–2012: Nieuwe Energie voor de Veenkoloniën, op zoek naar Regionale Comfortzones. Brochure voor de Ideeënfase over Krimp, Energietransitie en Ruimtelijke Kwaliteit; Eo Wijers-stichting: Deventer, The Netherlands, 2011.

42.   Lysen, E.H. Trias Energica: Solar Energy Strategies for Developing Countries, Proceedings of the Eurosun Conference Freiburg. Freiburg, Germany, 16–19 September 1996.

43.   MacKay, D.J.C. Sustainable Energy—Without the Hot Air; UIT Cambridge Ltd.: Cambridge, UK, 2009.

44.   Tillie, N.; van den Dobbelsteen, A.; Doepel, D.; Joubert, M.; de Jager, W.; Mayenburg, D. Towards CO_2 neutral urban planning: Presenting the Rotterdam energy approach and planning (REAP). J. Green Build.**2009**, 4, 103–112.

45.   PBL; ECN. Naar een Schone Economie in 2050: Routes Verkend; Planbureau voor de Leefomgeving and Energy Research Centre of the Netherlands: Den Haag, The Netherlands, 2011.

46.   Laughton, M. Variable renewables and the grid: An overview. In Renewable Electricity and the Grid; Boyle, G., Ed.; Earthscan: London, UK, 2009; pp. 1–29.

47.   Van Hoorn, A.; Tennekes, J.; van den Wijngaart, R. Quickscan Energie en Ruimte. In Raakvlakken Tussen Energiebeleid en Ruimtelijke Ordening; Planbureau voor de Leefomgeving: The Hague, The Netherlands, 2010.

48.   Yin, R.K. Case Study: Research, Design and Methods; Sage: Thousand Oaks, CA, USA, 2009.

49. Creswell, J.W. Research Design: Qualitative, Quantitative & Mixed Methods Approaches; Sage: Thousand Oaks, CA, USA, 2013.

50. Rose, G. Visual Methodologies: An Introduction to Researching with Visual Materials; Sage: Thousand Oaks, CA, USA, 2012.

51. Eo Wijers-Stichting. Eo Wijers-Prijsvraag 2011–2012: Nieuwe Energie voor de Veenkoloniën, op zoek naar Regionale Comfortzones; Jury Report; Eo Wijers-Stichting: Deventer, The Netherlands, 2012.

52. De Zwart, B. A triptich of expertise. The design competition as an instrument to unite assignment, design and commissioner. In Designing for a Region; Meijsmans, N., Ed.; SUN Academia: Amsterdam, The Netherlands, 2010.

53. De Waal, R.; Kempenaar, A.; van Lammeren, R.; Stremke, S. Application of social media in a regional design competition: A case study in the Netherlands. In Digital Landscape Architecture 2013; Buhmann, E., Pietsch, M., Ervin, S.M., Eds.; Wichmann Verlag: Berlin, Germany, 2013; pp. 186–200.

54. Loorbach, D.; van der Brugge, R.; Taanman, M. Governance in the energy transition: Practice of transition management in the Netherlands. Int. J. Environ. Technol. Manag. **2008**, 9, 294–315.

55. Olwig, K.R. The earth is not a globe: Landscape versus the 'globalist' agenda. Landsc. Res. **2011**, 36, 401–415.

56. Van der Horst, D.; Vermeylen, S. Local rights to landscape in the global moral economy of carbon. Landsc. Res. **2011**, 36, 455–470.

57. Feddes, Y. Juryverslag Eo Wijers-Prijsvraag; Eo Wijers-Stichting: The Hague, The Netherlands, 2012.

58. Noorman, K.J.; de Roo, G. Energielandschappen: De 3de generatie. Over Regionale Kansen op het Raakvlak van Energie en Ruimte; Provincie Drenthe: Koekange, The Netherlands, 2011.

CHAPTER 4

Eco-Polycentric Urban Systems: An Ecological Region Perspective for Network Cities

André Botequilha-Leitão [1,2]

[1] University of Algarve, Faculty of Science and Technology, University of Algarve (UAlg), Campus of Gambelas, 8000-062 Faro, Portugal
[2] CVRM-Geo-Systems Center of IST, Technical University of Lisbon, Avenida Rovisco Pais, 11049-001 Lisbon, Portugal

ABSTRACT

The research presented in this paper is a work in progress. It provides linkages between the author's earlier research under the sustainable land planning framework (SLP) and emergent ideas and planning and design strategies, centered on the (landscape) ecological dimension of cities' sustainability. It reviews several concepts, paradigms, and metaphors that have been emerging during the last decade, which can contribute to expand our vision on city planning and design. Among other issues, city form—monocentric, polycentric, and diffused—is discussed. The hypothesis set forth is that cities can improve the pathway to sustainability by adopting intermediate, network urban forms such as polycentric urban systems (PUS) under a broader vision (as compared to the current paradigm), to make way to urban ecological regions. It discusses how both the principles of SLP and those emergent ideas can contribute to integrate PUS with their functional hinterland, adopting an ecosystemic viewpoint of cities. It proposes to redirect the current dominant economic focus of PUS to include all of the other functions that are essential to urbanites, such as production (including the 3Rs), recreation, and ecology in a balanced way. Landscape ecology principles are combined with complexity science in order to deal with uncertainty to improve regional systems' resilience. Cooperation in its multiple forms is seen as a fundamental social, but also economic process contributing to the urban network functioning, including its evolving capabilities for self-organization and adaptation.

Keywords: sustainable city-region planning; polycentric urban systems; landscape ecological planning; holism and systems thinking; resilience; urban metabolism and self-reliance; cooperation

1. URBANIZATION AND SUSTAINABILITY

Exponential growth of the world population has occurred only for the last 100 years, where it more than quadrupled: 1.6 billion in 1900, 2 billion in 1930, 3 billion in 1960, 4 billion in 1975, 5 billion in 1987, 6 billion in 1999, and presently approaching 7 billion [2]. Noteworthy is that the world urban population grew much faster. Population migration to live in cities and metropolises is a global trend. Presently about one in two people live in urban areas, which is estimated to increase to two out of three in 2050 [3]. Some estimates point to an even faster growth, where the urban population will reach about 61% in 2030 [2]. For example, in Europe approximately 75% of the population lives in urban areas and estimates point to approximately 80% in 2020 [4], representing the urbanization level of most industrialized nations today [5]. In the USA circa 80% of the population lives in urban areas [6]. One of the most urbanized nations in the world is Australia with more than 92% of its population concentrated in six State capital cities and other urban areas [7].

New megacities (>10 million) are growing in the developing world. The population in India (1.2 billion) has more than doubled during the last 50 years, but the urban population has grown nearly five times. These authors estimate that by 2021 the number of mega cities in India will increase from the current three (Mumbai, Delhi and Kolkatta) to six (including Bangalore, Chennai and Hyderabad), whereby India will have the largest concentration of mega cities in the world [8]. In China, since the "reform and openness policy" in 1978, urbanization has seen a tremendous boost, most prominent in the Pearl River Delta region during the past two decades, where urban areas have grown as much as 300% between 1988 and 1996 [9]. Economic growth and demographic changes will accompany growth in urban populations, especially in populous China and India, producing ever-greater demands on services that nearby and distant ecosystems provide [5]. Considering mid-sized cities (between one and five million inhabitants) urbanization rates have been steadily increasing globally, which will have profound impacts on natural and agricultural ecosystems, e.g., as reported by [10] to occur in China. "The merits of compact development were extensively debated in the 1970s. Critics questioned the claimed environmental, transport and costs benefits, and argued that was contrary to market forces towards sprawl, the decentralization of work and residents' desires. Debates focused largely on developed-country contexts and centrist approaches, but attention shifted to the merits of centrist versus decentrist compact development in the 1990s" [11].

Opposed to the concentration of urban population in large monocentric, high-density, and frequently compact cities is another important form of urban development—urban sprawl or the so-called "diffused city", which has increased during the last decades worldwide. It is broadly characterized by a dispersed spatial pattern of a mix of urban land uses, where four characteristics dominate: low-density, scattered development (i.e. decentralized sprawl), leapfrog development, and commercial strip development, and is associated with unplanned incremental urban development [4,12,13]. Typical in the USA in the early part of the 20th century, it was promoted by the utopian city vision of Frank Lloyd Wright' Broadacre City of 1935 [14]. Later this phenomenon proliferated to other parts of the world. In Europe, where cities were traditionally much more compact, urban sprawl is now a common phenomenon and regarded as one of Europe's major challenges [4]. And it is the most significant and urgent issue in American land use [15].

An alternative, intermediate form of urban development is through a polycentric or multiple-nuclei structure, which some define as being compact [13]. Polycentric development is a form of decentralized concentration of numerous small- and medium-size urban centers, frequently (but not restricted to being) organized around a compact city center, forming large urban agglomerations. This concept was introduced in urban geography by Harris and Ullman in 1945, representing an evolution from multi-center city to multi-center city region or polycentric city region. The process of sub-urbanization associated to a large city originated numerous settlements located in its surroundings. From this original concept a more complex urban pattern evolved, especially in Europe—polycentric urban regions, which are made up of numerous polycentric city regions [16]. Polycentricity can emerge from two distinct set of relationships: (1) intra-urban patterns of population and economic activity clusters, e.g., Los Angeles, London or Paris; (2) interurban patterns such as the Randstad-Green Heart complex in the Netherlands, the area of Padua-Treviso-Venice in Northern Italy, the Southern California urban region, and the Kansai area in Japan [17]. Other distinctions of polycentric forms are made according to its evolution process: some emerged as a result of households fleeing from the city center to the suburbs, followed by the relocation of firms, and services—the centrifugal mode; others via a coalescence of existent cities and towns of similar dimension into contiguous functional urban regions. Examples of the latter are the Randstad, the Rhine-Ruhr metropolitan region, and the Flemish Diamond [18]. This urban form "seems to have become one of the defining characteristics of the urban landscape in advanced economies" [17]. Since the last decade or so the polycentric approach has been widely implemented in the European Union as a cornerstone of its spatial development policy [19]. There is a sufficient agreement "about the desirability of a polycentric urban structure organised on small and medium-sized, compact centres, well connected through an efficient network of public transport" [12].

Landscapes are being subjected globally to dramatically significant changes due to the continuous urbanization process and a strong use (and misuse) of earth resources [20]. Urbanization is the most dramatic form of irreversible land transformation, affecting both landscapes and the people who live in and around cities [21]. Although urban population growth over the past century has occurred on a very small portion of the global terrestrial surface (<3%), the impact of cities has been global, with 78% of carbon emissions, 60% of residential water use, and 76% of wood used for industrial purposes attributed to cities, affecting energy flows, biogeochemical cycles, climatic conditions, biodiversity and ecosystem functioning and services far beyond its limits [2,5,21]. As Eugene P. Odum describes it: "Great cities are planned and grow without any regard for the fact that they are parasites on the countryside which somehow supply food, water, air, and degrade huge quantities of wastes" [22].

Novel approaches are needed to address the complex issues arising from increasing world population, depletion of resources and decreasing quality of human habitat. A more holistic way of thinking must be adopted to reduce global environmental stresses [23]. The sustainability paradigm has emerged from these global issues. Sustainability is a powerful but hard-to-define concept that confronts many disciplines, including planning. Sustainable planning is inherently multi-dimensional, aiming to assure the viability of ecological, social and economic systems presently and into the future [24]. Sustainability is the capacity of the earth to maintain and support life and to persist as a system [25]. This concept adopts a systems perspective being relevant to systems ranging from the global to the local scale. It strives for natural resource management consistent with the preservation of its reproductive capacity [22,26]. Recently sustainability science is emerging, focusing explicitly on nature-society interaction dynamics, and promoting inter- and transdisciplinarity perspectives, where landscape ecology should and would make significant contributions [27]. Many scientists believe that promoting sustainability is the over-arching goal of landscape (and regional) planning [28]. Cities must play a more central role when looking at global sustainability for several reasons [29], including the fact that they have increasingly sizeable ecological footprints [5,22,30,31], notwithstanding that "(...) cities epitomize the creativity, imagination, and mighty power of humanity. Cities are the centers of socio-cultural transformations, engines of economic growth, and cradles of innovation and knowledge production" [31], and that they represent arguably the most important habitats for humans [2]. "A sustainable city must achieve a balance among environmental protection, economic development, and social wellbeing. Urban sustainability requires minimizing the consumption of space and resources, optimizing urban form to facilitate urban flows, protecting both ecosystem and human health, ensuring equal access to resources and services, and maintaining cultural and social diversity and integrity" [31]. It is not surprising that one of the key research priorities in landscape ecology is the integration of ecological research into urban policy, planning, design, and management strategies [32].

This paper is centered on the (landscape) ecological dimension of cities' sustainability, with a particular focu s on horizontal or chorological processes from a regional perspective [33,34]. The hypothesis set forth in this manuscript is that cities can improve their sustainability by adopting intermediate, network urban forms such as polycentric urban systems under a broader vision (as compared to the current paradigm), to make way to urban ecological regions. This regional vision considers three main components: a network of cities, towns, and rural villages linked by corridors—ecological, e.g., hydrological networks, cultural, i.e. transportation and information infrastructures, and multifunctional (ecological + cultural); a multifunctional hinterland of rural and natural resources aiming at increasing regional self-reliance, structured by a network of ecological systems that provides for key-ecological services (the region´s "ecological backbone"); and the interrelationships between cities and their functional hinterland. Landscape ecology principles such as holism and systems theory, and its basic tenet—the relationships between ecological and cultural patterns, processes and change, are combined with complexity science in order to cope with uncertainty to improve regional systems' resilience. Cooperation in its multiple forms is seen as a fundamental social, but also economic process to the urban network functioning, including its evolving capabilities for self-organization and adaptation.

2. EMERGENT METAPHORS, CONCEPTS, AND PARADIGMS FOR ECOLOGICAL CITY PLANNING AND DESIGN

In the last decade several concepts, metaphors, and paradigms have been emerging, which can contribute to expand our vision on city planning and design. Some of those described below have been approached in earlier publications of the author [24,33,34,35,38,39,40], further developed and or summarized for the purpose of this section: holism and systems thinking; autonomy or self-reliance; urban metabolism; ecological footprint; uncertainty; adaptation; redundancy; the "form and function" principle; the "interdependence" principle, landscape context and chorological relationships; sustainability; sustainable landscape planning (SLP); landscape as an appropriate planning unit; strategic urban and landscape planning; connectivity; cities' networks and polycentric urban systems, and their hinterland; ecological infra-structure; dual perspective for landscape management; learning-by-doing and landscape monitoring; co-operation; and disciplinary convergence, and inter- and transdisciplinarity. Others had been proposed by several authors in the context of biological and ecological theory [23,41,42,43], complexity theory and theory of change [44,45], landscape ecology [46,47,48,49], urban ecology [26,30,32,46,50,51,52,53,54,55], landscape ecological planning ([25,56,57], green urbanism [29], regional, urban and open space planning [12,13,17,19,30,58,59,60,61,62,63,64,65,66,67,68,69,70,71,72,73], landscape urbanism [74,75,76,77], planning and design of green infrastructures [78,79], ecological urbanism [77,80], landscape ecological urbanism [80], and sustainability science [27,81]. Among these is auto- or self-organization and emergent properties; panarchies, resilience, regime shifts and critical transitions; variability; social ecological systems (SESs); ecosystem services, and landscape as a service matrix; sustainable regionalism; safe-to-fail; and translational research.

2.1. Holism

Holism states that the whole is more than the sum of their parts. It provides a new way to analyze landscapes, and argues that landscape elements receive their meaning or significance by their context, or their position within the whole [47]. An ecosystem's external "linkages" with the landscape are as important to proper functioning as the internal ecosystem environment [23]. Some even argue that context is more important that content [42]. This recognition of the importance of context emerges from systems thinking [42]. Besides landscape elements per se it is important to account for the (spatial) relationships between the elements that make up a landscape. All landscape elements, regardless of their specific land cover type, influence landscape functions through their spatial characteristics. This is a fundamental inter-relationship applicable to any landscape type, urban, rural, or natural. Thus, looking at landscapes holistically provides a common way of thinking about functions and processes, and how structure affects, and is affected by them [24].

The different systems that comprise our global habitat are highly interdependent. Indeed little is completely isolated from its surroundings, including people and cities. Urban landscapes are formed by a series of landscape elements such as houses and buildings, roads and highways, gardens and parks, etc. These elements are not isolated. They establish a number of relationships between each other. For example, housing is more expensive near urban parks because these generally provide for several urban functions, services, or amenities that are looked for by urbanites, e.g., urban climate is more amenable nearby parks which has a significant influence in bio-comfort, they provide for recreation opportunities, and a close contact between people and nature. Additionally cities are not isolated, and establish relationships with the surrounding rural landscapes and other cities.

2.2. Systems Thinking

According to Capra "(...) to understand things systematically literally means to put them into a context, to establish the nature of their relationships (...); the root meaning of the word "system" derives from the Greek "synhistanai"—to place together" [42]. "Systems thinking is a method of scientific enquiry that allows one to understand and investigate complex realities" such as landscapes [82], and can be characterized as an attempt to find common principles that apply at different levels of scale and across different types of phenomena [83]. The systems approach is hierarchical and views landscapes nested within larger systems (supersystems) and themselves composed of lower order systems (subsystems). A useful analogy can be made with the human body. Consider human cells as building blocks, which are organized as tissues, organs, organs systems (circulatory, respiratory, etc.) and ultimately as an organism. The human body is comprised of a group of systems. Similarly, individuals are part of communities that together form towns, states, and so on. Landscapes can be understood as groups of ecosystems, and regions as groups of landscapes [24]. The concept of networks introduced by early ecology enriched the systemic worldview where ecosystems are understood as networks of individual organisms [42]. Landscapes can also be viewed as networks of interacting ecosystems [46], cities as networks of urban elements (neighborhoods, buildings, infrastructures, etc.) [52], polycentric urban structures as networks of cities [19,60], and so forth. In the above context it is most important to acknowledge that when the notion of hierarchy (between levels) was introduced into ecological systems theory, it was not originally intended to portray a top-down, rigid structure involving a vertical authority and control, as tends to dominate in its everyday definition; the dynamic, adaptive nature of nested structures tended to be lost [45]. The latter introduced a new term to emphasize the latter interpretation: "Panarchy captures the adaptive and punctuated evolutionary nature of adaptive cycles that are nested one within the other across space and time scales" [45]. Here, the lesson to take home is that although living systems do present an organizational structure in hierarchies, it does not implies a top-down, vertical, and rigid but rather an adaptive, dynamic, network structure, whose elements work in complement with one another (see next section on autopoeisis).

In general, different systems levels have different levels of complexity, and each exhibits systemic properties that do not exist at lower levels—the so-called emergent properties, since they emerge at that particular level. Most important in contextual thinking is that the properties of the parts can only be understood within the context of the larger whole. This reverses the Cartesian paradigm where the dominant belief is that in systems the behavior of the whole can be understood entirely from the properties of its parts, leading to the Descartes's analytic method, an essential characteristic of modern scientific thought, in contrast with the ideas synthetized hereby on holism and system thinking. A crucial point in systems thinking is the ability to shift from one system level to another back and forth [42]. In some instances, is useful to perceive the larger context of a specific locale, or of a specific issue of concern (the forest; the "big picture") in order to recognize, understand and integrate the relationships between that place with its surroundings—the flows of energy and matter that crosses through and influence decisively its functioning. Complementarily we focus on the place in itself, in its parts (e.g., the trees of a forest) and most important in the intra-relationships between the parts (interaction between different trees); often we need to go even deeper and approach the specifics of each component (the functioning of a tree, its root system, the canopy, etc.), which in turn are systems by themselves.

Systems thinking is an emerging field. It was in the 50s and 60s that Ludwig von Bertalanffy, a biologist from Vienna, established his general systems theory or GST [84]. According to Steiner "(…) in GST control is maintained through the feedback received by what is dubbed the "control mechanism." The control works like a homeostat (…), and the result is a regulatory action (…) that keeps the system in a dynamic equilibrium" [85]. Since then it has developed over the last 40 years in many different disciplines and through a range of applications [83]. According to these authors "(…) the report The Law of Sustainable Development, produced by the European Commission, states: "Today, no serious study and application of the principles of sustainable development is possible without the help of systems science". (…) Concepts emerging from systems thinking have had a profound influence (…) in helping to understand "the complexity of ecological and organizational systems" [83]. Based on a comparative review of existing participatory and ecological planning methodologies the latter realized that "(…) the ecological planning methodologies have been developed in a whole within the last three decades reflect an increasing interest in the insights of systems thinking methodologies. Several of the key thinkers in these areas cite systems thinking and living systems biology as an inspiration in the development of the methodologies" [83].

System thinking appeals also to city planning theorists [85], and cities can be seen as systems. Inspired by Urban et al., Grove et al. adopt a hierarchical approach where households are a part of larger systems—neighborhoods, which in turn are part of a larger system—the city [86]. Cities can also be a part of larger systems, e.g., metropolitan areas [33] or other type of urban agglomerations worldwide, e.g., polycentric urban regions and network cities [19,60]. Closely linked with holism and systems thinking are the concepts of self-organization and autopoeisis presented below.

2.3. Self-Organization and Autopoiesis

Self-organization is a most important and distinguishing characteristic of living systems that explicitly or implicitly incorporates in itself several important concepts to understand complex systems, such as socio-ecological systems (SESs) that constitute cities and metropolitan systems. "Self-organization is a process in which pattern at the global level of a system emerges solely from numerous interactions among the lower-level components of the system. Moreover, the rules specifying interactions among the system's components are executed using only local information, without reference to the global pattern. In short, the pattern is an emergent property of the system, rather than a property imposed on the system by an external ordering influence. (…) Critical to understanding our definition of self-organization is the meaning of the term pattern. As used here, pattern is a particular, organized arrangement of objects in space or time. (…) Emergent properties (…) are features of a system that arise unexpectedly from interactions among the system's components. An emergent property cannot be understood simply by examining in isolation the properties of the system's components, but requires a consideration of the interactions among the system's components. (…) Systems are complex not because they involve many behavioral rules and large numbers of different components but because of the nature of the system's global response. Complexity and complex systems, on the other hand, generally refer to a system of interacting units that displays global properties not present at the lower level. (…) Complexity in a system does not require complicated components or numerous complicated rules of interaction" [87].

Some diverse phenomena have been described as self-organizing in biology, such as homeostasis (property of a system that regulates its internal environment and tends to maintain a stable, constant condition of properties like temperature or pH, in a dynamic equilibrium), and flocking behavior (such as the formation of flocks by birds, schools of fish, etc.) [45,87]. "Prigogine and Stengers (1987) showed that the evolution of settlement patterns and urban networks behave like complex systems out of equilibrium and that self-reorganisation of the spatial structure to adapt to the changing functional needs is characteristic." [47]

Poiesis is a Greek term that means making or production. Autopoiesis means self-making or self-production. Under this concept living beings are seen as systems that produce themselves in a ceaseless way through a network of interactions or production processes (the system's metabolism). The function of each component is to participate in the production or transformation of other components in the network through the relationships that specify the system. An autopoietic system is at the same time the producer and the product, in a circular organization (e.g., the nervous system) [26,41,42]. Notably the organization of a living system is always a network pattern [42].

Autopoeisis is a general organization pattern common to all living systems, whichever the nature of its components, i.e. the organization is independent of the properties of its components. The system´s structure is the physical embodiment of its organization, comprising both the system´s components and the functional relationships between components. Capra uses a bicycle to illustrate this concept: the systems' components are the frame, pedals, handlebars, wheels, chair, etc., which have a set of functional relationships between them; the complete configuration of the functional relationships constitutes the bicycle's organization pattern (all of these relationships must be present to give the system the essential characteristics of a bicycle). Additionally to organization (pattern) and structure, this author argues for a third criterion when describing the nature of life. Process, as the link between organization and structure, regards to the activity involved in the continued embodiment of the system's organization pattern. Using a designer's metaphor, organization is the design sketches that are used to build the bicycle; the structure is a specific physical bicycle; and process, the mind of the designer [42].

Recently it was proposed to substantially expand long term ecological research (LTER) by including the human dimension focused on coupled socioecological systems (SESs). Long-term socioecological research (LTSER) regards society-nature interaction as a dynamic process in which two autopoietic systems, society, and nature interact, an approach particularly relevant to understand the relationships typical of complex urban environments [55].

Note that, from a self-sufficiency perspective, there is no such thing as sustainable cities [88]. Cities by themselves are not autopoietic since they are highly interdependent on the surrounding landscapes [38] continuously importing energy, food, materials, etc. and exporting the products of its metabolism, e.g., waste [22] (see section below on urban metabolism).

2.4. Resilience

The capacity of a system to maintain its self-organization is closely related to the concept of adaptation (see above), and of resilience. Resilience comes from the Latin resilire, which means to rebound or recoil. This concept was first introduced to ecology and the environment in 1973 by Crawford (Buzz) Holling, who promoted, among others, the use of systems theory [43]. Resilience is the ability to absorb disturbances and reorganize while undergoing change, while retaining the same function, structure, identity, and feedbacks, the capacity for self-organization, and the capacity to adapt to stress and change [89]. Adaptive capacity resides in aspects of memory, creativity, innovation, flexibility, and diversity of ecological components and human capabilities [90]. It is important to distinguish two paradigms where resilience emerges that may be labeled equilibrium and non-equilibrium [51]. The first is focused on stable equilibrium conditions and is presently applied only to very particular situations; the second is more inclusive and deemed useful for urban planning and design, focusing on systems' dynamic and evolutionary capacity to adapt and adjust to internal or external change. Hereafter this second meaning is adopted.

The concept and theory of resilience have a growing appeal in the disciplines of ecology and planning [80], and one can identify an increasing dialog between these two disciplines in addressing urban environments. In this context an urban planner and an ecologist proposed a new metaphor, "cities of resilience", that both disciplines can share [91]. Ahern argues for an adaptive approach to planning and design, including monitoring and "learning-by-doing" [56,57,78,79], much attuned to the proposals of "designed experiments" [54]. Resilience capacity together with innovation can play an important role via "responsible experimentation, developing a culture of monitoring, and learning from modest failures" [79]. According to this author "resilience capacity can be strengthened by biodiversity, modularity, tight feedbacks, social capital, acknowledging slow variables and thresholds, and innovation "[79]. Resilience is at the core questions to be approached by the emerging science of sustainability [81], including also self-organizing complexity, inertia, thresholds, complex responses to multiple interacting stresses, adaptive management, and social learning [27]. There is common agreement in the literature that systems, organizations and people who are able and willing to adapt tend to be more resilient [43].

2.5. Redundancy

""Redundancy" is defined by the Oxford English Dictionary (OED) as "the state or quality of being redundant; superfluity, superabundance;" (...) "redundant" is further defined as "excessive, abounding too much." [92]. The word, paradoxically, has substantially different meanings in the fields we survey, yet most, like the definition from the OED, carry a negative connotation. In their work these authors refer to redundancy of multiple units (building blocks) within some larger system and provide a thorough discussion of different kinds of redundancy, ranging from genetic to engineering systems. This review includes ecological systems where "redundancy is typically of the "multiple non identical copies" sort within ecosystems, or across ecosystems" and is associated with biodiversity, ecosystem function, and resilience [92].

Functional ecological redundancy is basically the degree to which organisms have evolved to do similar things [93]. Several species fill similar ecological roles increasing the number of potential community organizations that can uphold similar ecosystem functions. By maintaining the distribution of redundant species across multiple time and space scales it is possible to maintain key-functions of the ecosystem in the face of change, which makes the system resilient [53,88]. On a different note Rosenfeld argues that in terms of practical conservation issues, the concept of functional redundancy is a double-edged sword. For example, it is important to prioritize species protection, but at the same time it postulates that certain species perform similar roles in ecosystems and thus redundant species can be expendable [92]. He recognizes, however, that this interpretation was not intended by its original proponent, where redundant species were seen as necessary to ensure ecosystem resilience in face of perturbation [94].

In the context of landscape ecological urban planning and design, Ahern argues for the advantages of redundancy (and modularization), and distributed or decentralized systems as opposed to concentrated. Here redundant elements or components provide for the same or similar urban functions, which help spreading risks, and thus constitute "strategies to avoid putting all your eggs in one basket," and for preparing and pre-planning for when (not if) a system fails [79]. It represents a "humble" design tactic where one acknowledges that it is not possible (and desirable) to exert total control over socio-ecological processes, and just try to mediate indeterminacy: "(...) since we can never be completely certain of how water flows and other ecologies work, one needs to build in redundant systems to make sure it works (...)"[95].

2.6. Urban Metabolism

Urban metabolism is a metaphor that looks at the city as a system, which requires inputs and outputs; if we look at it as an organism, it requires food and other resources, e.g., water, energy, materials, etc. and releases the byproducts of its metabolism to the environment, i.e. waste. This metaphor was developed earlier in the 20s and 30s by the human ecological approach of the so-called "Chicago School" where the city was conceived as a closed and functional system that could be treated as an organism or "superorganism" [96]. Later, in the 60s, a few academics also adopted this perspective but rarely if ever used it in policy development in city planning: "by looking at the city as a whole and by analyzing the pathways along which energy and materials including pollutants move, it is possible to begin to conceive of management systems and technologies which allow for the reintegration of natural processes, increasing the efficiency of resource use, the recycling of wastes as valuable materials and the conservation (and even production) of energy" [63]. Since this metaphor was essentially biological in nature, the latter extended the original idea to "include the dynamics of settlements (transportation, economic and cultural priorities) and livability in these settlements (health, employment, income, leisure, etc.), which he called the "Extended Metabolism Model of the City".

This emergent notion has been very useful in quantifying the horizontal (chorological) relationships and trends in consumption and waste generation of expanding cities. Over two decades several studies have shown large increases in the output of materials of cities, e.g., food and building materials, and outputs such as food wastes, paper and plastics. As an example, in Beijing "total carbon emitted from solid-waste treatment increased by a factor of 2.8 from 1990 to 2003." [5]. Additionally the metabolism approach provides a way to integrate biophysical and socioeconomic processes [55]. Analyses of urban metabolism include the analysis of the pathways along which material and energy [22,63], and more recently information flow [55]. Urban metabolism is consistent with the holistic and systemic approaches to cities, and new city planning approaches that consider the relationships established between the built environment and its wider landscape (spatial) ecological context, as in the ecological footprint approach [33].

2.7. Cooperation and Competition

Capra argues for a change in the XXI century that brings new thinking and values, shifting from self-assertion to integration [42]. The author points out that neither tendency is good or bad and both are essential aspects of all living systems. However the Western industrial culture overemphasized the former and neglected the latter. For example, the competition paradigm, a self-assertive value, needs to be replaced, or better, complemented by cooperation in order to enable sustainable human ecological systems. In order to provide support for the above stated argument we can look at living systems and its organization once more to provide insight and analogies that arguably could be useful for the core theme of this paper—sustainable city planning and design.

An important concept in the context of autopoeisis is structural coupling [98], which is closely related with the notion of interdependence. It occurs whenever there is a history of recurrent interactions leading to the structural congruence or compatibility between two (or more) systems (e.g., between two organisms) or between a system and its containing environment. Note that in this context the structure of the organism will not change as specified or instructed by the environment's structure, and vice versa—these interactions only "triggers" structural changes in one another. As long as a set of nondestructive, compatible or congruent interactions exist between a system (e.g., a city) and its environment (e.g., the surrounding landscapes, and or the hinterland) these two act as mutual sources of perturbation, triggering changes of state [98]. These authors provide an example in the context of cities: "Thus for example, in the history of structural coupling between the lineages of automobiles and cities there are dramatic changes on both sides, which have taken place in each one as an expression of its own structural dynamics under selective interactions with the other" [99]. In sum, structural coupling is always mutual; both organism and environment undergo transformations, and for example changes in cities trigger influences in its "region of influence" (see below) in multiple ways, and vice versa. Expanding on the example above mentioned on structural coupling and the influence of cars in cities, the urban form of the "diffuse city" (or sprawl) was facilitated by an increased urban mobility provided by individual transportation. A similar effect can be seen in the star-shaped urban form of post-industrial cities that expanded along the main axis of transportation infrastructure—railways, roads and highways, or both. Reciprocally cities congestion, partially due also to urban form, induced also the appearance of smaller cars—city cars. Both phenomena are intrinsically connected, reflecting a strong interdependence. The corollary is that sectorial approaches (in this case the city, for one side, and the car industry for the other) are bound to influence each other, significantly. Thus the need to look at the several dimensions of cities as a whole, where systems interact with other systems, at different levels, rather than from a reductionist perspective, focusing on one single dimension or on particular places as isolated features.

Maturana and Varela explained above how organisms interact to each other and with the surrounding environment [41]. The notion of structured coupling is useful to provide for insights on how organisms adapt to the environment. Below the authors continue this explanation by shedding light to "competition" and "natural selection" as the mechanisms that are traditionally related to "survival" of some species over others, and thus to species evolution. "The maintenance of the organisms as dynamic systems in their environment is centered on a compatibility of the organism with their environment which we call adaptation. The adaptation of a unity to an environment (…) is a necessary consequence of that system's structural coupling with that environment. (…) Conservation of autopoeisis and conservation of adaptation are necessary conditions of the existence of living beings". Important to the concept of cooperation (versus competition) is the above mentioned authors' observations on how Darwin supposedly proposed the process of "natural selection": "We often hear that what Darwin proposed has to do with the law of the jungle where each one looks out for himself, at the expense of others in unmitigated competition. (…) This view of animal life as selfish is doubly wrong: (a) instances of behavior which can be described as altruistic are almost universal in natural history; (b) living organisms existence is not geared to competition but to conservation of adaptation, in an individual encounter with the environment that result in the survival of the fittest" [100]. At this point it is important to clarify the distinctions between the three basic types of interactions between species: competition "leads to negative outcomes for both groups involved", whereas symbiosis "benefits both participants", and predation, or parasitism "benefits one and is detrimental to the other" [101]. Complementarily it is important to note that "(…) in nature there is no competition. What exists is competence". As noted by Maturana, when two animals meet before the same piece of food and only one eats, this happens because in that specific moment one of them was the most competent to do so. But this does not mean that the animal that was unable to eat is doomed to be, from that moment on, forever forbidden to eat until death arrives. This does not happen in nature. However, when circumstances involve competition in human culture, the individual who succeeds to eat does not satisfy himself with this fact: he or she needs to make sure that the one who was not able to eat must cease forever to be a threat. In other words, competitive men usually do not feel sure of their competence, so they have the need to get rid of whoever could jeopardize them. In other words, when men cannot trust in themselves as living beings, their peers must be eliminated as soon as possible. But even so—let us insist on this point—this cannot be ascribed to the cultural dimension in itself: it plays such a role in a culture like ours, which does not know how to deal with aleatory and ceaseless change. And these conditions, as we know, constitute the very essence of life. In other words, we do not know how to deal with autopoiesis—that is why we feel ourselves in need to aggress it and to deny its reality" [102]. As part of the paradigm shift in sciences presented in earlier sections the "neo-Darwinian conception of evolution" is challenged by the notion of co-evolution, "(…) that emphasizes cooperation as the creative play of an entire evolving universe" [103].

Autopoesis, self-organization, adaptation, complex systems and the above discussed concepts have been increasingly integrated in the last decade in social sciences research. "In the literature on complexity theory applied to social systems, 'self-organization' has a more specific meaning, for example, 'a process in which the components of a system in effect spontaneously communicate with each other and abruptly cooperate in co-ordinated and concerted common behaviour" [104].

As an example in spatial planning, cooperation between urban areas is at the very core of the polycentric regions paradigm adopted in the European Union. Here cooperation is viewed as a competiveness factor for the intervening cities: "Promoting complementarity between cities and regions means simultaneously building on the advantages and overcoming the disadvantages of economic competition between them" [19]. In the spatial planning policy framework for the European Union the polycentric paradigm plays a pivotal role. Complementarity is approached from a broad perspective and should focus not merely on economic issues but together with other urban functions such as environmental quality and social well-being. As an example, when one considers the emergence of polycentric communities foreseen to result from the implementation of this spatial planning approach, if we are to assure those to be socially viable, then co-operation should be fostered and built on common interests of all participants, as to re-integrate the n cities of the urban ensemble into one single community [65].

Another dimension of cooperation is the most needed collaboration between ecologists and social scientists, and planners and designers in inter- and transdisciplinarity studies focusing on urban environments, which are found critical to the emerging sustainability science [32]. Arguing for a stronger emphasis in the human dimension of sustainability and for both inter- and transdisciplinarity studies in planning Botequilha-Leitão emphasized the need for a symbiosis, within the SLP framework, between natural sciences (e.g., landscape ecology), social sciences (e.g., collaborative methods), and humanities (e.g., landscape history) as they all hold much value to an integrated, transdisciplinary planning approach [38]. It also argued for bridging the gap across a (too often) fragmented and divorced science (and knowledge as a whole). This holds true also between science and planning. In a world still dominated by reductionist thinking different areas of knowledge are reluctant to knowledge sharing and cooperation. It does not facilitate a proper, efficient, and true multidisciplinary integration much needed when planning for sustainability. In similar terms Musacchio argues to link sustainable design to sustainability science and calls for an expanded definition of translational research "[The process medical researchers use to bring scientific discoveries from research into clinical practice] (…) directed toward environmental professionals: a collaborative learning process between scientists, designers, planners, and engineers who seek to solve complex environmental problems by connecting scientific theory, concepts, and principles to the design and planning of the built environment. This definition of translational research assumes that such approaches and methods are transdisciplinary—not only are interactions among scientists, designers, and planners important, but public participation is a vital part of the process that should include practitioners, elected officials, local residents, and others" [105].

In the above context it is most important to develop collaborative ecological planning. As argued before public participation in the planning process is essential to successful planning [38,39]. Failure of former planning approaches led to the increasing recognition that collaborative methods are crucial in order to promote more and better citizen participation. Research has shown that people are more likely to accept an issue resolved when they have had a voice in the decision-making process [106]. Landscape planning and design professions have acknowledged this fact and incorporated participation in most methodologies. Meaningful and informed stakeholder and public participation is viewed as a most important dimension in a sustainable land planning process. Bottom-up approaches are needed and citizen participation is a key issue for successful planning, design and implementation of sustainable landscapes. Collaborative methods, e.g., collaborative design, are most useful to understand the cultural dimension and its interface with natural processes [38,107]. Public participation increases acceptance and the implementation success of plans, by increasing plan's legitimacy. It empowers citizens, decreases the participation deficit, and thus contributes for a better democracy, and promotes social connectivity. It also increases citizens' self-esteem and confidence. Finally it helps in accounting for uncertainty in planning by sharing responsibility and involving citizens that will be affected by decisions in the decision-making process, which contributes also for sharing power and thus increases decentralization and shortens the distance between decision fora and the receivers of policies [38].

I finish this section by stating "Axelrod's (1984) principles of cooperation (...)—co-operation can get started by even a small cluster of players who are prepared to reciprocate, can thrive even in a world where no one else will cooperate and can protect itself once established—so long as the co-operation is based on reciprocity and the shadow of the future is important enough to make this reciprocity stable" [108].

3. ENVISIONING THE CITIES OF THE FUTURE

3.1. A Paradigm Shift

More than half of the world's population is urban and this phenomenon will continue to grow, with an emphasis on coastal areas where natural resources, e.g., biodiversity, are particularly concentrated; it is in cities that most of the environmental problems concentrate; moreover cities constitute highly vulnerable situations to the effects of global climate change [5]. Cities are highly vulnerable also due to the almost total reliance on external inputs: some commodities and manufactured goods travel thousands of kilometers between the point of production and the point of consumption [22], which is arguably made possible by an oil-based economy. Industrial societies in general and cities in particular, are the product of petroleum and may implode without it [109].

Not surprisingly "(...) most, if not all, our cities are unsustainable" [31]. However urban spatial planning is primary concerned with the degree of segregation or aggregation of different economic and social functions, efficiency of transportation and delivery of utilities, and efficient filling of undeveloped space [110]. The environmental dimension is usually a secondary consideration (if considered at all). Urban regions' planning targets are traditionally focused on the several dimensions of socio-economical systems such as economics, transportation, housing, industry and so forth and not so much on climate, water, biodiversity, and other ecological systems dimensions [46]. Therefore we need a novel approach in city planning and design to lead us into the path of sustainability.

I argue that one of the key-challenges to Man today is to envision human habitat from a broader perspective (both thematically and spatially) than just the built-up space, i.e. urban areas, particularly when considering large urban agglomerations. Planners should mesh both socio-economic and ecological dimensions, and acknowledge the horizontal relationships of both processes with its context, into a unified approach. When planning and designing the cities of the future one should include all of the other landscape functions and processes that sustain them. These include those landscapes that provide for the necessary inputs needed for urbanites not only to survive but to live fully, namely the landscapes of production (and recycling) and recreation, and those that provide for the essential services that forms the ecological backbone that sustain all of the other functions [38] together in regional, cohesive, multifunctional, and resilient landscapes [22,61,73]. It is a shift from the parts to the whole, i.e. from the city, the economic-financial dimension, and mostly sectorial-based policies, to considering the city and its "region of influence" (see Section 3.3), and the three dimensions of sustainability—economic, social and ecological—supported by truly integrated policies, and governance institutions.

In a recent past the dominant conservation paradigm was focused on ex situ solutions such as zoos to preserve endangered species, and segregation-based, museum-like approaches for preserving the last natural ecosystems from Man's negative influence. Today a paradigm shift in conservation biology is undergoing from single-species management to ecosystem management and from isolated reserves to managing the entire landscape [15,111]. The implementation of wide range ecological networks can be seen in Europe, i.e. the ecological network "NATURA 2000" at continental level [19,29,38], at the national level in some European countries such as in the Netherlands [29], among others, at the regional level, e.g., the Regional Ecological network for the Algarve, Portugal [34], or at the metropolitan level, e.g., in the Metropolitan Area of Lisbon [38]. As conservation is broadening its objectives to encompass the entire landscape as a whole so should humans consider the entire globe as their habitat. The former, narrow perspective of human habitat is rooted in the 17th Century Western culture that viewed Nature, not as its home (habitat) as the ancient Greek did, but solely as a reservoir of natural resources for Man's own benefit, as implicit in Bacon (1624) and Descartes (1636) writings [112]. A legacy of the Enlightenment, this "Cartesian dualism" that maintains the psychoseparation of humans from their natural roots where human enterprise is somehow seen as separate from and above the world. In this period of "transition to sustainability" [81] it must be counteracted and overcome.

To cope with such complex, multidimensional issues that cities are facing entering the XXI century, the cities of the future need novel planning perspectives informed by holistic and systemic thinking. "The major problems of our time cannot be understood in isolation, since they are systemic, meaning they are interconnected and interdependent and must be seen as different aspect of one single crisis—a crisis of perception. This results from an outdated worldview—a perception of reality inadequate to deal with these problems" [42]. The new logic introduced by the information society with its high rates of change and the emergence of new, complex environmental issues calls also for a long-term planning vision that ought to frame everyday decisions. These broad visions would contribute to counteract those small, piece-meals, non-concerted actions decisions that have a particularly high impact in urban and suburban landscapes [111]. Part of the problem is that science has become so reductionist that society is victimized by a "tyranny of small technologies" (deriving from small decisions). "Piece-meal" or "quick-fix" approaches often work well in the short term of economic and political worlds, and when done independently as they often are, the central problem is not properly addressed [47]. According to T. Kuhn we have been experiencing in the turn of the XX century a paradigm shift, both scientific and social [49,113]. "Such a scientific revolution has occurred in the last 20±30 years with the emergence of the new field of what could be called `complexity science'. It has been enabled by the major paradigm shift from parts to wholes, leading from entirely reductionistic and mechanistic toward more holistic and organismic approaches, (…) and systemic thinking" [49].

In the former section I have explored several useful metaphors, concepts, and paradigms for sustainable city planning and design. These are emerging as a response to the challenges for urban planning of the new century, in the context of the abovementioned paradigm shift. Below I will elaborate on the role of the science of landscape ecology for a regional approach to city planning, focusing on some key-concepts and how those new emergent ideas described above can together contribute for better planning the cities of the future.

3.2 The Role of Landscape Ecology for Regional Planning of Cities. The SLP Framework

Landscape ecology, the scientific pillar of the sustainable land planning framework (SLP) [24,38,39], is increasingly relevant to sustainability in general and urban development in particular [31,46]. In order to promote more sustainable approaches to planning in the last decades of the 20th century we observed a transformation of the landscape planning paradigm to incorporate an explicit ecological approach, namely in Europe and in the USA [38,39]. More recently landscape planning begun to adopt landscape ecological principles and tools [24,25,38,39,40,56,114,115] and extending its principles more or less explicitlyto urban planning and design [24,31,32,33,34,35,36,37,38, 39,40,46,51,53,54,57,74,75,76,77,78,80,85] inter alia.

In earlier works I proposed and explored a framework for sustainable land planning (SLP) [24,38,39]. The SLP framework is a strategic landscape planning approach, based on the science of landscape ecology and several associated key-principles derived from holism, systems theory, and complexity theory described in earlier sections. SLP is proposed as both an art and a science by promoting the integration of the ecological, social, and cultural dimensions into design, planning and management of sustainable landscapes; it thrives for ecological and social equity and for the involvement of citizens and stakeholders at large across the entire planning (circular, iterative, continuous, learning) process (see PROBIO case-study below); and it adopts an adaptive planning approach under a "learning-by-doing" attitude which includes a continuous cycle of implementation-monitoring-evaluation [44] "(...) by treating the adopted planning solution as a working hypothesis rather than a 100% full proof solution (as done traditionally), that should be tested and closely monitored for its consequences" [38]. "Sustainability is a goal that no one as yet knows how to achieve. The act of sustainable planning and design is a heuristic process; that is, one in which we learn by doing, observing, and recording the changing conditions and consequences of our actions" [117]. Together with scenario techniques, and public participation this adaptive approach also contributes to deal with uncertainty both in science and planning [48,57], and thus to increase system's resilience [44,83]. It integrates ecological with social and cultural processes, e.g., by promoting a participated process and by incorporating landscape history, which both contribute to formulate landscape visions. Finally SLP encourages intuitive and creative thinking by incorporating the development of shared planning visions, and the design of spatial planning concepts. The latter are useful to explore possible future directions, and to support the discussion of a broader range of ideas, the development of scenarios and the engagement of both decision-makers and the public at large into the planning process. SLP is arguably appropriate to all planning realms, e.g., waterresources [24,40], conservation planning [24,38,39], and urban planning [24,33,38,39].

The research project "Decision Support System for Planning and Management of Biodiversity in Protected Areas (acronym PROBIO, Ref. no. POCTI/MGS/36580/99)" (1999-2003) is an example of the application of the SLP framework [38]. The following description of PROBIO will focus on the integration of the social component via the process of collaborative planning and design implemented.

The major goal of PROBIO was to develop a Decision Support System (DSS) for Planning and Management of Biodiversity in Protected Areas. The DSS integrated landscape metrics, scenarios, and a Multi-Agent System (MAS) in a GIS environment, supported by collaborative planning and design. The study area was the Natural Park of Sintra-Cascais (PNSC), located in the Lisbon Metropolitan Area, in Portugal. The PROBIO project used alternative future planning scenarios in order to anticipate and prevent or minimize environmental impacts on protected areas, and to attract stakeholders' participation, and the public in general to the planning process. Biodiversity indicators based on landscape metrics were useful to evaluate the different planning scenarios or management alternatives. The MAS was developed to model the socio-economic component and to help multi-purpose negotiations between the several (social and economic) agents in the Park, forming the central core for scenario generation. The assumption was that agents modeling could help the PNSC staff enhance its relationships with the social-economic agents involved in the park area. A series of workshops were promoted as to involve public and stakeholders participation. Before the workshops individual meetings were conducted with more than 40 stakeholders (individuals and groups) who supplied important information about the activity and vision of each participant. This information allowed identifying the main strategic vectors for the development of the region, which were translated into the scenario themes. The first workshop aimed at a pre-diagnosis using a SWOT procedure. At the end a questionnaire was distributed that allowed us to evaluate stakeholders' opinion about the workshop. They thought it to have been relevant, stimulating, informative, and efficient. They mentioned they learned about the issues that were debated and about the other stakeholder' opinions. The goal of the second workshop was to translate the planning issues identified by the stakeholders into a more spatial and detailed form, i.e. to reference geographically (present and potential) land use conflicts. This method is referred to as collaborative design. We produced a set of maps in transparencies that allowed to overlap them as in the McHarg method, e.g., natural resources, cultural heritage, zoning plans, etc. In the last workshop we presented the results to the stakeholders, in a nontechnical language. Previously we sent to the stakeholders a list of criteria (social, economic and ecological) asking them to rank these according to a dual goal: allow urban development, and protect wildlife. We discussed these criteria with them and produced a new ranking, as a product of a group decision. For this session we divided the stakeholders in groups as done in the other workshops. Then we aggregated the individual rankings (surveys were anonymous) and debated a final ranking with them. These criteria served as input for the MAS, to define rules for the simulation of urban growth in the PNSC. According to both the PNSC director at the time (O. Knoblich, pers. com), and the one that followed a few years later (C. Albuquerque, pers. com.), the experience was highly useful to the ongoing process of the PNSC new zoning plan, and to establish a strong linkage with the stakeholders and the public in general in this protected area. The integration of social sciences with landscape ecology, history, and planning, and computer modeling was crucial to approach the management of protected areas more sustainably by integrating the social-economic with the ecological dimension.

3.3. The Ecological Urban Region

3.3.1. A Chorological Perspective for City Planning
In the last decade studies on urban landscapes adopted a broader perspective by looking at cities as ecosystems, including the relationships established with the surrounding landscapes [5,22,31,33,34,46,70,73]. I argue this a fundamental issue to be taken in consideration when aiming at increasing resilience and thus sustainability in cities, metropolitan areas, polycentric urban regions, and urban agglomerations at large. A key-principle of the SLP framework is the "interdependence principle" [118], which stems from holistic and systemic

thinking. It implies the recognition of important interdependencies between ecosystems and human culture. Additionally, the spatial dimension of sustainability is strongly related to the interdependence of land uses, and spatial processes. Implicit in this principle is another system's key-concept embedded in SLP—"context". From a spatial perspective both concepts are strongly related with an important dimension of landscape ecology—the horizontal or chorological approach.

"Traditionally, resource planners and managers do not consider horizontal relationships, for example by ignoring the (ecological and environmental) context in which exploited resources are often, if not always, embedded (De Leo and Levin 1997, p. 9)". (…) Prior to the advent of landscape ecology, even ecologists would not consider context to be a major factor for studying ecological systems. An ecosystem, or a site based approach, was followed. However ecosystems, and sites, are not isolated. Horizontal (chorological) natural processes or ecological flows are a fundamental component of ecological systems. It is therefore crucial to approach a site from a chorological perspective, considering horizontal processes. Such processes are flows and movements that cross local ecosystems or land uses" (Forman 1999). (…) We have to enlarge our lens of planning to encompass these processes occurring in the (ecological) context where the activity we are planning for takes place. (…) From a systems, hierarchical perspective, the site in itself is a system integrated within a higher system. It is therefore important to explicitly consider the horizontal relationships between a site and the system(s) within which it is integrated" [38].

3.3.2. A Spatial Conflict

Because cities are not planned and managed considering the chorological dimension spatial conflicts arise that need urgently to be addressed. Most cities are located in strategic areas from the natural resources perspectives, e.g., water, seashores, fertile soils, minerals, or combinations thereof. Across its history human populations tend to aggregate in places where there is a high concentration of resources—ancient civilizations appeared in the fertile valleys of most important rivers, e.g., Nile, Tigris, Euphrates and Indo. Presently if one looks at Europe, the United States, and the world at large we can see large cities and or continuous urban agglomerations located nearby the coasts or in rivers' valleys, deltas, and estuaries e.g., Hamburg, Amsterdam (and practically the entire Netherlands), Paris, London, Barcelona, Seville, Lisbon, New York City, Boston, Chicago, San Francisco, etc. Regardless, cities are continuously growing. They expand frequently into the productive landscapes that surround and frequently sustain or can sustain them in the future. This causes a spatial conflict, where urbanization usually prevails. For example, the city of Lisbon alone depends for its biological metabolism on an area almost three times the Lisbon Metropolitan Area (LMA) per se, i.e. 8000 sq.km [62]. Its main water supply comes from a dam (Castelo de Bode) located circa 120 kilometers northeast of Lisbon. This is similar to Boston and its Quabbin Reservoir, New York City, San Francisco, and many other cities. A noteworthy aspect is that the LMA has a wide reserve of underground water underexplored, namely in the Setúbal Peninsula located nearby the city of Lisbon. This is the expansion area for the southern part of the LMA, causing among other urbanization effects soil sealing which reduces infiltration into the aquifer systems. In fact the most important aquifers in Portugal are located along the coastline, where the two Portuguese metropolitan areas (Lisbon and Oporto) are located as are large urban agglomerations, e.g., in the region of the Algarve located south of Portugal.

This phenomenon is augmented by growing annual rates of land consumption per capita. For example, in American cities and metropolitan areas, the amount of land consumed by urbanization far exceeds the rate of population growth [120]. To counteract this trend in the USA movements such as "New Urbanism" and "Green Urbanism" promote, among other concepts, more compact urban development and walkable communities, similar to the design of a large number of European cities [29,77]. The fact is that cities tend to expand by frequently destroying their local resource base, increasing dependency on remote areas.

3.3.3. Post-Oil Cities

I argue that planners should be able to envision what can be called "post-oil cities", anticipate the effects of such transformations, and act on it in order to provide solutions that can deal with a new, oil-scarce world that sooner or later will be a reality. Oil was for a long time, and still is an (presently relatively) abundant and cheap, and most of all highly "portable" source of energy (i.e. very easy to store and transport, when compared to other alternative energy sources such as electricity and natural gas, among others). It supported a diffuse pattern of urbanization, such as urban sprawl and associated urban mobility levels, mostly with individual transportation as cars, that we can observe today in most cities around the world—the "automobile cities" [70]. It also supports the transportation of many goods that, in the present era of globalization, travel thousands of kilometers from their origin to cities' destiny making ""the operationally inseparable" (primary production is spatially removed from consumption and consumption from most subsequent decomposition)" [26]. However, oil is an increasingly scarce resource. "The world supply of oil is projected to last approximately 50 years at current production rates." [121] "(…) These estimates, however, are based on current consumption rates and current population numbers. If all people in the world enjoyed a standard of living and energy consumption rate similar to that of the average American, and the world population continued to grow at a rate of 1.5%, the world's fossil fuel reserves would last about 15 years" [122]. One cannot avoid wondering how long the world will have abundant and cheap oil. Furthermore oil has no known substitute which bring together those three characteristics that made it such a popular source of energy: abundant, cheap, and most of all highly "portable". "For transport, it is very difficult to find viable alternatives to oil in sufficient quantities to meet current and future demands" [70]. When considering gasoline consumption from the transportation viewpoint, based on a considerable sample of cities worldwide, studies suggest that urban structure within a city is a fundamental factor, with a clear link to urban density [58]. As an example, when looking at Toronto, Canada and the five U.S. cities with lowest gas consumption among their sample, all have a strong inner city area. However Toronto outer area is more compact in population and jobs by, on average, nearly three times. The existence of strong subcenters developed in the suburbs around transit stations seems to play an important role. As a result Toronto presented an annual gasoline use per capita of 265, where the average for the five U.S. cities is circa 400. These authors suggest that "(…) subcenters could be the means for more intensive outer area land use", which would decrease gasoline consumption, probably due to the combination of higher concentration around transit stations and the more intensive use of transit as compared to automobile. More recent data on 84 cities across all continents points out to a clear increase in car use (and transport energy) as a city sprawls [70].

3.3.4. Polycentric Urban Structures and Network Cities

The path to cities' regional sustainability is complex and multi-dimensional. A pivotal issue is the ongoing discussion on the importance of urban form to city sustainability. Noteworthy is the debate between (a) compact, monocentric, and high-density, (b) diffused, low density, and (c) intermediate forms of urban development [12,13,29,68,70,71]. I argue that polycentric urban structure holds promise as an intermediate, alternative to the present urban form of compact monocentric cities or to diffused city patterns. The "compact city" metaphor "has been put in question by some scholars as too broad, generic and ideological [123]. A relevant issue raised is at which urban scale it should apply, arguing that "beyond certain levels of density and size, it could produce 'town cramming' and scale diseconomies which are among the main causes of present suburbanization tendencies" [124]. In a different tone studies relating social and health problems and concentration of people have never pointed to a clearly negative connection; "the top four "alpha" cities of the global economy— London, Paris, Tokyo and New York—still appeal to their residents and visitors despite their being large and dense". However the optimal population size of cities in order to attain various social goals has not yet been determined [71]. A study on Baltimore in the 1980s points out that it is not density per se but population size that influences urban dwellers, e.g., emotional stress and other negative psychological conditions. It concludes urban compactness is neither a necessary or sufficient condition for city sustainability, claiming that there is an overemphasis on urban form strategies, which should be refocused into a new dynamic conception of urban planning, where process must have the final word [69].

On the other hand, sprawl has tremendous effects on the surrounding natural resources and the environment, and on rural landscapes and the people who live in it [2,4,5,10,58,69,70,71]. The disadvantages and costs associated have been incremental—increased travel time, transports costs, pollution, degradation of the rural landscapes, and so on [125]. The "dispersed city" requires costly infrastructure, intensifying financial burdens to communities [73], it is blamed for creating suburban gridlock and amplifying social polarization [126], intensifying political fragmentation, increasing homogeneity (including physical character) by the proliferation of "mass culture" (via the traditional industrial service chains, e.g., Wall Mart, McDonald's in the US, with parallels in Europe) thus eroding regional and local "sense of place" [73]. On the other hand suburban areas (I presume for some high-income urban dwellers) often offer better life quality than the inner city and relatively lower costs to fulfill the dream of a "home of one's own" can often only be realized there [19]. Some even argue that large scale planning aiming at controlling urban sprawl is above all socially undesirable. The argument is that this type of unplanned development offers an increasing opportunity to people to express their own free will and design their own spaces [12]. The discussion is also about the merits of low- versus high-density living and generally "living green", deemed possible only in low-density rural or semi-rural context; it revolves around two opposite views: the "rural commons", a view stemming from strong-urban sentiments that are opposed to "density". The "urban commons" view, which is pro-urban and values the city, promotes an overall "greener" functioning, as in Zurich, Stockholm, Helsinki and Freiburg [70]. In the city of Milan responses to urban sprawl can be illustrated by recent proposals that imagine alternatives within the compact inner city that offer the living conditions, comfort, and equivalent costs to those offered by the suburbs; it envisions limited, governed and selective densification in nodes where public transport is a deterrent for car use and consequently does not imply an increase in private traffic (much in tune with the polycentric mode); finally, under development since 2009 within the heart of the city, the project "vertical forest" aims to provide the equivalent of four hectares of forest in a limited urban space, conveyed by two towers, together with 43 floors, with 2100 plants both in the interior and in tree-shaded terraces [127].

This debate has been enriched by looking at alternative, intermediate forms between these two opposite spatial strategies, such as polycentric urban structures (PUS) [12,13,68], also called network cities [60]. Network cities combine various nodes to form a unique yet flexible exchange (economic, creative) environment. Creative network cities promote "a creatively diversified environment for all citizens through the amalgamations of urban functions for living, working, learning and playing", and "the formation of cultural and knowledge "corridors" to stimulate interaction among creative minds". Note that in the late 80s, seven of the ten most creative European regions were corridor or network cities [60].

A concept of PUS was introduced in European spatial planning in the last two decades or so by the European Spatial Development Perspective (ESDP) as one of the three policy guidelines aiming at a more balanced (sustainable) development: development of a balanced and polycentric urban system and a new urban-rural relationship, overcoming the outdated dualism between city and countryside [19]. However the emphasis is on economic development and cohesion to promote global economic integration in the EU, although it strives also to consider (within PUS) the corresponding rural areas and their small cities and towns. Indeed it aims for an integrated treatment of the city and countryside as a functional, spatial entity with diverse relationships and interdependencies and acknowledges that small and medium sized towns and their inter-dependencies form important hubs and links, especially for rural regions. As presented the focus of ESDP is also on promoting the concept of the "compact city" (the city of short distances). Sufficient agreement exists about the desirability of PUS composed of small and medium-sized, compact urban centers, combined with strong connectivity provided by an efficient network of public transport [12].

The planning principle or spatial concept behind polycentric urban structures is concentrated deconcentration [60]. This concept was originally introduced in the 1966 in the Netherlands, resembling the expanded-towns policy in Britain. The city region concept became the complement of concentrated deconcentration. Concentrated deconcentration was included in the Second National Physical Planning Report (1966) "as a chief innovation", being the Dutch government' planning solution to put an end to uncontrolled suburban growth in the Randstad-Green Heart complex (a classical network city), a large territory located between Amsterdam-The Hague-Rotterdam-Utrecht. The intent was not to reject altogether suburban-type development, but to concentrate new development in and around existing towns and cities to relieve pressure on central cities [59].

There are strong advantages of a planned form of urban development (growth management and wiser planning and design) as compared to market-guided suburbanization, e.g., savings of 20–45% of land resources, 15–25% of costs in providing local roads, and 7–15% for water and drains [128]. Not surprisingly, according to other studies cited by the latter, the lower the density of development and the greater the distance to the metropolis center the higher the public costs of road construction, public services, and school management. Studies on urban development patterns in the Barcelona Metropolitan Region (BMR) refute the polarization between absolute versions of the dispersed and the compact urban forms raising arguments favoring alternative, intermediate urban environments as for example those described for the BMR where dispersion was attenuated by polycentric urban structures [68]. However this is not to say that the BMR has solved entirely this issue as some of its landscapes affected by sprawl are described as "territories without speech" and "landscapes without imaginary" [129]. In the words of Josep Acebillo (Chief Architect for the Mayor of Barcelona, in 2000) "We're wasting land!", particularly in the urban region [46]. A slightly different perspective are provided by the results of a research project (1997–2001) entitled 'Housing as a basis for sustainable consumption' [67]. This study was based on two large surveys in the Norwegian towns of Greater Oslo and Forde, using ecological footprinting as an analytical tool: "These ecological footprint analyses suggest that sustainable urban development points towards decentralized concentration, i.e., relatively small cities with a high density and short distances between the houses and public/private services" [130]. Finally the polycentric approach increases the interface between built-up structures and the natural-rural environment as compared to compact cities, and in a more balanced way than in the "diffuse city" —it's a middle term, providing also for an increased sense of place.

3.3.5. Towards an Extended Perspective for Cities' Regional Planning

I argue that from a sustainable city planning and management point of view it is important to consider a wider perspective that includes explicitly both cities and their hinterland or region of influence, both spatially and functionally. Again it is a shift from parts—the built-part of cities, and a notion of cities as confined by the limits of the built-part, isolated from its context, to the whole— "city+hinterland". Therefore it is crucial to acknowledge the most important role of the hinterland or "region of influence" of cities and its reciprocal relationships. In the last decade or so a 19th century concept—the "city region" has been incrementally "reclaimed" by advocating for a broader concept of the city, to include its "influence" zone. As discussed previously cities depend critically on inputs from and outputs to a space outside the city, establishing the so-called chorological relationships with its "context" —the hinterland or "region of influence". The term "hinterland" literally means back country (hinter = behind, land = land). The word includes any area under the influence of a particular human settlement, e.g., by providing necessary energy and materials, and to absorb the waste generated by that settlement. Thus, an aggregated human settlement and its hinterland are often bound by their production-consumption relationship [131]. "In agricultural societies, limited by technology and high transportation costs, locales or regions, i.e., "hinterlands," had to provide all or most functions necessary for the everyday life of the local population. Under industrial conditions, the spatial division of labor increased. This spatial expansion improved people's ability to meet their needs and fulfill their social functions because the supply of goods and services was no longer constrained by local resource availability or costly transport. The connections between people's way-of-life and their cultural landscapes weakened, and it became increasingly difficult to link local and regional ecologies with the behavior and consumption patterns of their human inhabitants." [132] Most important is that the relationship between settlement and hinterland is one of dominance of the former over the latter. This view is however changing, especially through the discussion of sustainable urban development [131]. For example, the European Union spatial planning policies embedded in the ESDP promotes an "integrated treatment of the city and countryside as a functional, spatial entity with diverse relationships and interdependencies. A sharp distinction between city and countryside within a region ignores in most cases the fact that only regions can form labour, information and communication markets. The region is, therefore, the appropriate level for action and implementation" [19].

In order to plan and manage the "extended city" it is important to be able to define its boundaries. In theory this hinterland or "region of influence" is constituted by the surrounding rural landscapes located at a relatively closer distance and or with places located very far, even in other continents due to global markets. From this perspective the region of influence is a functional hinterland [19] that together with cities and towns involved form a functional urban region [46]. The latter adopted a set of criteria to define the boundaries of urban regions, ranging from landscape physical features (e.g., mountain ranges), outline of major drainage basins around major water supplies, major biodiversity areas, one day recreation and tourism sites, to major political/administrative borders combined with the relative size of the core city and the other(s) cities in its surroundings. When none of those criteria seemed appropriate, a radius of circa 100 km was adopted, partly reflecting a typical maximum distance on paved highway that people would travel in one day shopping travel. "Several attributes initially thought to be important turned out not to be so, because they usually did not extend very far beyond the metropolitan area (ends of commuter rail lines, communities with substantial commuter populations, airports, sewage-treatment facilities, solid-waste disposal sites, reduced air-quality sareas)" [46].

Advancing the principles of sustainable regionalism, Ndubisi proposed to restore the concept of ecological region to manage metropolitan growth, enlarging the urban planning "lens" to a regional perspective; the boundaries, the latter argues, should be defined by watersheds or interacting mosaics of watersheds, modified by political-regulatory boundaries [73].

When considering the sustainability of urban agglomerations, such as metropolitan areas and polycentric cities and urban regions the relationships between a city and its context could be seen at two levels: (1) intra-urban, considering the hinterland located within the space formed by the urban network, and (2) with the surrounding landscapes located outside the urban agglomeration boundaries (that again can be close or far). More or less isolated, compact cities fall into the latter. As a whole, this functional "region of influence" will necessarily reflect cities ecological footprint, which can be spatially defined in combination with the carrying capacity of the "region of influence" that enables the production of the several inputs cities need and the absorption of the products of its metabolism.

An important point to make at this stage is that, according to the conducted literature review, the planning and management of polycentric cities or urban regions in the EU are focusing mostly on socio-economic development. Despite the broader goals stated in the ESDP policy document, at least for the European context, and despite the debate and research on the role of multifunctionality, the implementation of this planning concept has been relatively shy regarding the explicit involvement and incorporation of landscape resources and ecological services, reflecting the dominant attitude of the city over its hinterland, which derives essentially from the narrow perspective of Man over Nature (see Section 3.1). According to a new vision for European landscapes explored in the Report "Blueprint for Euroscape 2020—Reframing the future of the European landscape" [133] "(…) the focus on land use only is lacking a spatially coherent vision and regional focus. The concept of European Polycentric Regions is an answer to that as a meta-scale regional planning instrument for integrating multi-functional land use into a spatial framework based on landscape functions. Polycentric Regions can be characterized by: designation of region-specific resilience centers that provide essential compensation and buffer functions for adjacent high agglomeration and that can support structurally weak zones; spatial distribution of landscape services that reflects the bio-physical structure as well as socio-economic necessities at various levels of scale (…); governance structures that build upon bottom-up civil society initiatives (…); awareness of the importance of linking regional identity with global sustainable development objectives (…)". I believe this vision could advance the concept of PUS towards a more balanced approach.

In the above context it is important to acknowledge the important contribution of several spatial concepts (or metaphors [54]) in territorial planning throughout the last two centuries. A spatial concept expresses through words and images an understanding of a planning/design issue and the actions considered necessary to address it [39]. Some of these spatial concepts were already referred to before in this article—the "Randstadt" and its counterpart the "Green Heart". These translate the organic urban architecture perspective of the 1950s, interpreting cities as living organisms [65]. Together these spatial concepts have also close relationships with those of the "Garden City" and the "Green Belt" introduced at the end of the 19th century by city planners. These have proven to be very powerful and effective and some endure still until today. For example the Green Belt it is a major instrument of land use and urban planning in the U.K. Not surprisingly the ideas behind these concepts have been revisited and re-used under the new perspectives of urban ecology, where cities are approached as living organisms, with strong relationships with the surrounding landscape [22]. Here urban metabolism plays a central role to understand those co-dependencies and is modeled to understand the fluxes of energy, materials and organisms that flow between built-areas and the hinterland that permeates between those.

Inspired on landscape ecology I proposed a spatial concept aiming to contribute to address uncertainty and the need of flexibility in planning and thus to increase landscapes adaptive capacity and resilience [24]. It is based on the idea that cultural landscapes needs a dual approach—a more deterministic, for areas where critical ecological resources concentrate, and a more flexible approach to the remaining areas. The former is supported by an "ecological backbone", an ecological infrastructure that supports the overall functioning of the landscape [38]. "This idea asserts that for sustainable human development, planning must recognize those ecological structures that are most fundamental to assure overall ecological sustainability, including abiotic, biotic, and cultural functions and processes, and to provide the capacity for the landscape to compensate for impacts caused by human uses and activities" [134]. A similar approach was proposed under the "casco" or framework concept [135] representing a systematic decoupling of functions, where low-dynamic functions or slow-change variables [44,79], i.e. long-term ecological processes, such as groundwater recharge or soil formation are combined into a coherent spatial framework, and the high-dynamic functions (i.e., production agriculture, extraction industries, urban development) are located in other spaces providing them with the essential spatial flexibility and freedom they operate under [135,136]. From an operational perspective the "ecological backbone" concept is complemented with the approach of differential prioritization [24] proposed by Haaren: stricter, mandatory goals for the areas where to implement the ecological backbone, and flexible rules for the remaining areas [64].

The notion of landscape as infrastructure or as a service matrix put forward by landscape urbanism is most appropriate to envision the multifunctional role of hinterlands. Landscape urbanism evolved from design theory, combining high-style design with ecology [80]. It looks at ecology as a meta-science that allows the integration of culture and art, and where the landscape is seen as a "hybridization of natural and cultural systems" [137]. Here the landscape is seen as the "background" of urban agglomerations, i.e. the matrix where the city is embedded. The hinterland is conceived as the infrastructure for the development of the human habitat under a broader concept allowing for the integration of infrastructures (water, energy, transportation, etc.) and public spaces [76]. Recently a related concept evolved that promoted ecological urbanism [77]. Although drawing heavily on the former, it pays little attention on advances of urban ecology [80]. The latter argues for a synthesis incorporating those advances to form a new, integrative approach under the term "landscape ecological urbanism". The concept of green infrastructure is closely related to the latters. Green infrastructure is an emerging planning and design concept that applies landscape ecology principles to urban environments [78]. Its main structure is supported by hybrid hydrological/drainage network, complementing and linking relict green areas with built infrastructure that provides ecological functions. This approach uses a suite of strategies intended to build urban resilience capacity: multifunctionality, redundancy and modularization, (bio and social) diversity, multi-scale networks and connectivity, and adaptive planning and design [79].

Not surprisingly both landscape (ecological) urbanism and green infrastructure concepts share a common interest on human habitat and man-nature integration. To do so they more or less explicitly draw from landscape ecology an interest not solely on landscape structures per se, but also and foremost on landscape processes, functions and services, its reciprocal relationship, and its dynamic nature. The latter spatial concepts briefly explored above build on closely related concepts proposed in a more recent or distant past. They do so to envision alternative solutions to accommodate the need for a more efficient planning and design for cities and metropolis where Man can find a truly satisfying habitat to live. These concepts hold per se a large potential to attain this purpose. I believe they can have an enhanced contribution when associated with which other and with others such as the polycentric urban form [34].

4. STRATEGIES FOR SELF-RELIANT CITIES

Strategies for self-reliant cities emerging from a new thinking context include local production for local consumption [22,33,29,65,70,73,77] local markets and ecological commerce [29,139], multifunctional, redundancy and modularization for the hinterland's ecological infrastructure [78,79] and maximizing circular organization (or closed loops) of inputs and outputs [22,29,41,70,71]. The keywords are "reduce, re-use, and recycle" (the 3Rs). The use of local materials and techniques boosts the regional economy [138]. Local markets bring together production and consumers, and the community as a whole, e.g., as proposed for London's 160 sq km of farmland by the Sustainable London Trust [29]. In this context a new understanding of economy is urgently needed, as for example the so-called "ecological commerce": "Economic development, the foundation for human settlements, seldom acknowledges ecological limits in either capitalist or socialist systems. However, the ecological footprint demonstrates the need for economic restructuring aligned with the natural world. Sustainable urban development therefore needs an ecology of commerce. Such an economic system would move beyond resource conservation to promote adaptive reuse of existing natural resource and built resources, emphasize renewable resources, and restore environmentally degraded areas such as brownfields. As an example in the USA, Chattanooga, Tennessee is committed to eco-commerce. It has created lucrative new industries such as electric vehicle production, ecotourism" among others [139]. Gauzin-Muller provide for extensive examples in Europe, namely twenty-three on the "environmental approach" to architecture of housing, public buildings, and commercial and service buildings, and six on urbanism and sustainable development [138]. Beatley provides also for numerous examples in Europe, from ecocycle balancing in Stockholm (Sweden), to Ecover—a sustainable factory in Oostmalle (Belgium), an ecological approach to commerce and economic development in Graz (Germany), or industrial symbiosis in the eco-industrial park (EIP) of Kalundborg (Denmark). Ecocyle balancing in Stockholm is promoted via sewage treatment plants that produce energy (biogas) and fertilizer (to be re-introduced in the farms nearby) [29]. The municipality of Graz contracted with farmers to accept and compost (source-separated) organic and lawn wastes collected in the city farms (located within a 60-km radius of Graz) and then apply to their fields. Farmers are paid providing an additional source of farm income, as well as a way to substantially reduce the city's composting costs. Finally, at the finer scale of houses and buildings bioclimatic design, based on site conditions and buildings shape and orientation, promote the rational use of energy [138].

5. CASE-STUDY—KALUNDBORG

Kalundborg is a city of 50,000 inhabitants located on the seashore of the island of Zealand, circa 100 kilometers East of Copenhagen, Denmark (Figure 1). Here we can find the first EIP formally identified as such, later followed by others, e.g., in Styria, the Austrian province where the city of Graz is located (see above), and in the Ruhr region (Germany) [140]. Despite its small population Kalundborg is the largest industrial center on the island with an industrial turnover similar to that of a middle-sized European city, and is still growing. The area includes e.g., two of the world's leading producers of enzyme and insulin (Novo Nordisk), the largest water treatment plant of Northern Europe and the second largest oil refinery of the Baltic Region (Statoil). On the other hand, due to the heavy-industry located here, there are, among others, pollution problems to be solved, e.g., the production of green-house gas (GHG) emissions: Kalundborg is responsible alone for circa 9% of the total Danish CO_2 emission. However, and according to the municipality, Kalundborg is striving to become a green industrial municipality by 2020; its policy is to make compatible its continued growth with the protection of the environment [141].

"An EIP is a community of firms in a region that exchange and make use of each other's byproducts, in the process improving their environmental and economic performance. The argument is that by working together, this symbiotic community of businesses achieves a collective benefit that is greater than the sum of the individual benefits each company would realize if it optimized its individual performance only" [140]. EIP are based on the concept of "Industrial Symbiosis". IS is a central concept in the industrial ecology literature, which describes geographically proximate inter-firm relationships involving the exchange of residual materials, water, and energy. Here one industry's residue is another industry's resource through a structured exchange of resources: water, energy and other industrial residues are exchanged across company boundaries [141].

Figure 1. Industrial symbiosis. The Eco-Industrial Park (EIP) at the coastal city of Kalendborg, Denmark. The photo shows the location of the major industries that incorporate the EIP.

All started in 1961 when Statoil, an oil refinery newly installed in Kalendborg started to use surface water from Lake Tissø in order to save the existent limited supplies of underground water. Mind that water is a scarce resource in this part of Denmark. This project was developed together with the municipality. The reduction in the use of ground water has been estimated in circa 2 M. m³/year. Later a number of other collaborative projects were introduced and the number of partners gradually increased. By the end of the 1980s, the partners realized that they had effectively "self-organized" into what is probably the best-known example of a working industrial ecosystem, or an industrial symbiosis. Note that the IS is based upon commercial agreements between independent partners [142].

Currently, the EIP is made up of seven key industries and Kalundborg Municipality. Hereby we describe four of them. Asnæs electric power station, the largest in Denmark, is at the core (Figure 2). It provides residual heat to the municipality that feeds up the district heating system, replacing highly polluting oil burning heaters in individual homes, and to another major player—Statoil, presently Denmark's largest oil refinery. Asnae produces other valuable by-products including 170,000 tons/year of fly ash, which is used in cement manufacturing and road building, e.g., by local construction firms. Finally it supplies also a fish farm. The power plant uses salt water, from the fjord, for some of its cooling needs, helping to reduce withdrawals of fresh water from Lake Tissø. The resulting by-product is hot salt water, a small portion of which is supplied to the fish farm's 57 ponds [143,144]. Gyproc, Scandinavia's largest plasterboard manufacturer, uses the power plant's fly ash to obtain gypsum, a by-product of the chemical desulphurization of flue gases. Gyproc purchases about 80,000 tons/year, meeting almost two-thirds of its requirement. By purchasing synthetic gypsum from Asnæs, Gyproc has been able to replace the natural gypsum that it used to buy from Spain. Statoil surplus gas, which used to be flared off, begun to be treated in 1993 by removing sulfur, which is sold as a raw material for the manufacture of sulfuric acid. The clean gas is supplied both to Asnæs and Gyproc as a low-cost energy source. Gyproc switch from oil to gas recorded a 90–95% saving in oil consumption. Finally it supplies its purified wastewater as well as its used as cooling water to Asnæs, thereby allowing this water to be "used twice" and saving additionally 1 M. m³/year of water. A large pharmaceutical company, Novo Nordisk has its largest production site in Kalundborg. The factory site is shared with Novozymes, Local farmers make use of Novo Nordisk's by-products (sludge) as fertilizers. Industrial enzymes and insulin are created through a process of fermentation, the residue from which is rich in nutrients. After lime and heat treatment, it makes an excellent fertilizer. Some 1.5 M. m³/year are delivered to local farmers, free of charge [144].

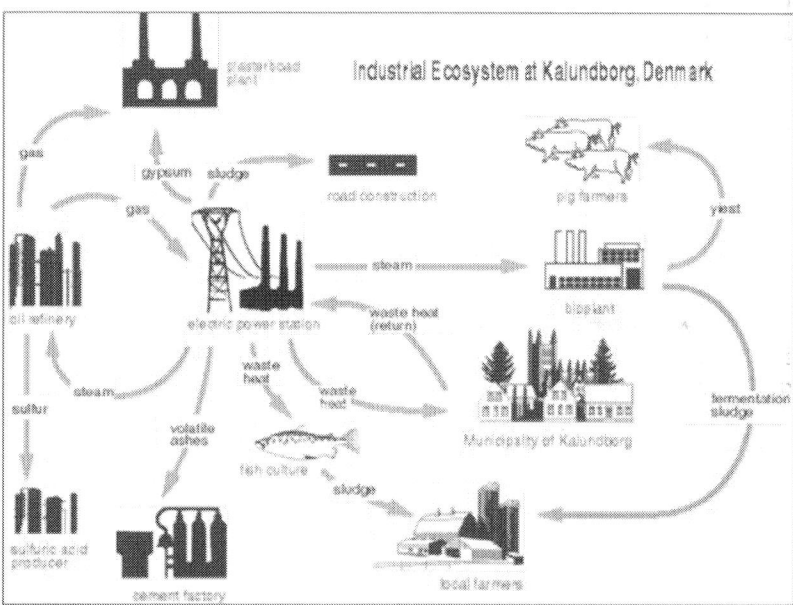

Figure 2. The several components of the industrial (eco) system at Kalundborg, Denmark and its interrelationships, including the flows of energy and materials between the several players [145].

The positive environmental impact appears to be substantial: on an annual basis, CO_2 emissions are reduced by 240,000 tons, 3 M. m³ of water is recycled, etc. In addition numerous by-products should be added, which are sold to industries located outside the industrial cluster. In addition to these reductions, the use of the excess heat from Asnæs for household heating has eliminated the need for about 3,500 oil-burning domestic heating systems [141].

The original motivation behind this industrial cluster was to reduce costs by seeking income-producing applications for unwanted by-products. Gradually companies realized that they were generating environmental benefits as well [141]. It is a win-win situation. The Kalundborg industrial ecosystem could serve as a beacon to sustainable planning aiming at increasing self-reliance of cities [144]. Many policy analysts argue that public planners can copy and even improve on Kalundborg. However some argue that "The planning of a community of companies in a region that exchange and make use of each other's byproducts has been advocated in many academic, business and political circles. The real world examples that justify such an approach, however, were entirely the result of market forces" [140].

Contributing to the gradual and evolutionary process initiated in 1961 there seems to be several keys for success [29]:

1.   The energy crisis of the 70s and 80s prompted many of the energy efficiencies;
2.   The economic benefits and an enhanced environmental image: "(...) environmental altruism has little to do with the symbioses that have been developed" ([29], p. 244);
3.   The flexible and cooperative Danish regulatory systems (see a discussion on the difficulties to develop such an approach in otherwise less flexible regulatory systems, e.g., in the USA [140,146]);
4.   Physical proximity of the companies involved in the process in an area of circa 4 km^2, and the fact that Kalundborg is a small town, with a strong sense of community (see below);
5.   5. Cooperation and complementarity between the companies, the municipality, local environmental NGO's, and others actors. The governance system led by Asnæs—the environmental "club" started in 1989 with the abovementioned main players, that promoted discussion and brainstorming in order to expand this symbiotic system.

There are further lessons that we can learn, derived from some comments from those directly involved:

o   All contracts have been negotiated on a bilateral basis;
o   Each contract has resulted from the conclusion by both companies involved that the project would be economically attractive;
o   Opportunities not within a company's core business, no matter how environmentally attractive, have not been acted upon;
o   Each partner does its best to ensure that risks are minimized;
o   Each company evaluates their own deals independently; there is no system-wide evaluation of performance, and they all seem to feel this would be difficult to achieve.

Jørgen Christensen, Vice President of Novo Nordisk at Kalundborg, identifies several conditions that are desirable for a similar web of exchanges to develop:

o   - Industries must be different and yet must fit each other;
o   - Arrangements must be commercially sound and profitable;
o   - Development must be voluntary, in close collaboration with regulatory agencies;

o - A short physical distance between the partners is necessary for economy of transportation (with

o - heat and some materials);

o - At Kalundborg, the managers at different plants all know each other" [143].

According to P. Desrochers "numerous EIPs have been planned in North and South America, Southeast Asia, Europe, and southern Africa" ([146], p. 345). THE EIP concept is also extending to developing a food and agriculturally focused EIP (Figure 3) [147].

This real-world case-study illustrates the need to further incorporate in city planning, design and management much of the emergent concepts presented in earlier sections in this manuscript. Cities will not be completely self-reliant in a near future. As we could see it takes time to build such systemic relationships as those in Kalundborg EIP. Furthermore some argue that the known examples of success were not planned activities (see above); Kalundborg EIP emerged as a result of self-organizing capacities of the many players involved: industry, municipality, agro and fish-farmers, NGO's, etc. It is important to acknowledge that, purposefully or not, the Kalundborg industrial cluster was envisioned as a system, where components (players) are interacting through a bundle of horizontal (or chorological relationships) flows of energy and materials. Across time a circular organization with a network pattern emerged—the EIP, showing similar characteristics as those in living systems discussed earlier (Section 2.3). Here we can recognize some of the principles of systems and complexity theories and of the science of landscape ecology (see Section 3.3).

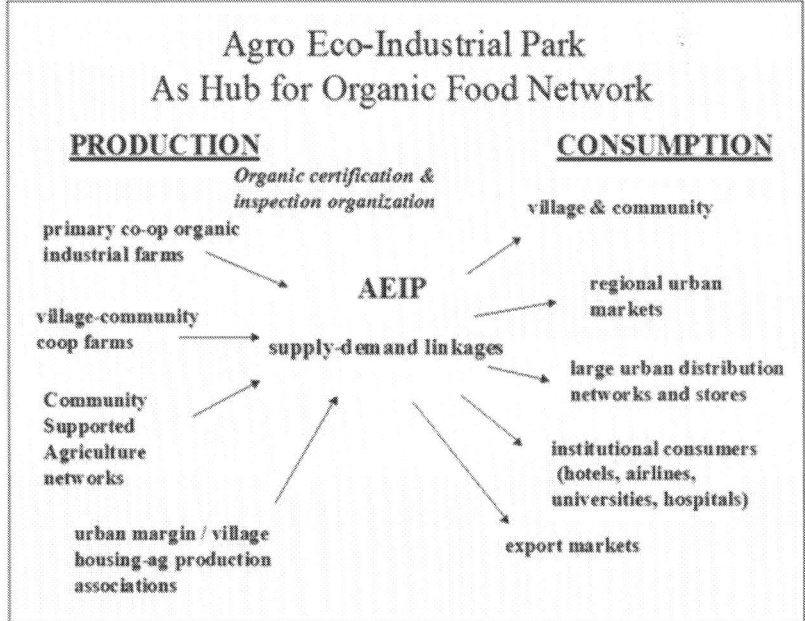

Figure 3. The concept of Agro Eco-Industrial Parks [147].

The metaphor of urban metabolism is also applicable to Kalundborg (see Section 2.6). Inputs and outputs are integrated to the benefit of all, reducing some of the exports coming from outside the city, sometimes very far such as the natural gypsum coming from Spain to Gyproc. Linkages were established between the city and its hinterland, e.g., via the input of fertilizers for local farming or fly ash to local construction firms.

Notably the IEP was achieved via cooperation between institutions (governance mechanisms) and complementarity across companies and other actors (actors with different functions which input-outputs feed into each other in quasi-closed loops). These are two most important characteristics of polycentric urban systems (see Section 2.7 and Section 3.3)

5. CONCLUDING REMARKS

Earth population is increasing substantially at large, and tends to concentrate in large cities and urban agglomerations. Urbanization growth patterns are unsustainable, leading to the degradation of water resources, productive land, biodiversity and quality of life, particularly for urbanites. Cities are increasingly vulnerable to global climate change and are highly dependent on landscape resources located outside cities, sometimes very distant. Forecasts on oil supply point to an increasing depletion of existent deposits. Modern mobility patterns are much oil-dependent. The oil crisis will arguably affect tremendously urban life as we know it. Many of the reviewed authors challenge us to think about the so-called "post-oil cities", and provide extensive examples globally on how to promote a more circular, closed-loop pattern in using energy and other natural resources in cities [22,29,70,71,77,138].

Following the tradition of many disciplines even today a reductionist approach to urban planning focuses on the (built-up) space within city boundaries. It does not look at cities as a whole, by ignoring or undervaluing the non-urban space in cities as an integral part and the linkages between ecological and cultural structures, processes and functions, and not acknowledging cities as social-ecological systems. Additionally it does not consider the horizontal or chorological relationships within cities taken as whole, and those with the surrounding landscapes and resources at large that support them. Most important for the argument of this essay, frequently the environment is a secondary consideration (if considered at all), dominated by economic development policies. Moreover it is mostly fragmented, sectorial-based, and often within the narrow time-frame of political cycles. Strategic planning in the 1980's onward argues for almost nonexistence of explicit urban image-planning, focusing on fragmented urban operations framed by a strategic planning framework that solely provides general guidelines for urban development. Some even argue against controlling urban sprawl through large scale planning in the grounds that is impossible, pointless, and most of all socially undesirable since "la ville a la carte" or the "ville aux choix" offers freedom of choice for people to design their own "life-spaces" [148]. Some adopt the "rethoric of uncertainty" in architecture and town planning as a reaction to legacies of determinism, legitimizing different forms of relativism: "if anything is uncertain then anything can be possible", warning that it seems "a device to keep out of the decision processes the weaker part of the society" [149].

Kuhn and Capra argue for a paradigm shift, both scientific and social, that is occurring in the turn of the last century and entrance to the XXI [150]. This paradigm shift is contributing to a new vision, where those emerging concepts and ideas presented in this article are important contributions emerging from varied sources of knowledge areas, from life and social sciences to planning and design. Most look at living systems functioning as a metaphor that enable building useful analogies to better plan and design human habitat—cities and functional hinterlands taken as a whole. Most important is the notion of context and interdependence, stemming from a landscape ecological chorological approach to planning, as the SLP framework [38]. System thinking is about context and relationships, and acknowledging networks as an organizing pattern for living systems. Networks of cities have been gathering consensus as alternative urban forms—the polycentric paradigm in Europe (but also in California, USA, China, Japan, etc.), forming complex socio-ecological systems (SESs). Self-organization of spatial structures is characteristic of urban patterns evolution. It is fundamental for its adaptation to disturbances and catastrophic events such as those triggered by climate change, e.g., floods and draughts, or more dramatically earthquakes, tsunamis, hurricanes and so forth [5]. SESs consider cities as the arena where two autopoietic systems (nature and society) interact in a dynamic process [55]. From a self-reliance viewpoint cities are not sustainable or autopoietic. One of the challenges posed to cities is to make cities more resilient [30,51,53,55,79,80,85], i.e. more adaptable to overcome, sometimes, critical situations. I concur with [79] inter alia that we should manage for resilience, and learn to live within SESs instead of trying to control them, recognizing that systems are safe to fail. In this context we need adaptive urban strategies, e.g., by engaging into a "learning-by-doing" process [38,44,56,57] via experimental design [54,79] or reflective learning through practice [80], by promoting diversity [79,80], multifunctionality [79] or by built-in redundancy into urban systems [79].

I argue that acknowledging explicitly the relationships between built-up and its functional hinterland contributes also to the overall system's resilience, where adaptive capacity is supported by local landscape resources within an urban ecological region (UER). The spatial conflict caused by existent urban forms, particularly by sprawl but also by the expansion of compact cities needs to be resolved by taking into account the land capabilities for production of food and fibers, waste re-absorption, recreation, and more intangible functions as aesthetics and contact with nature, and a diverse set of ecological services essential for human quality of life that are provided by the functional hinterland of a polycentric city-region. City planning, including PUS, needs to balance its present focus on mostly economic concerns to consider other important roles of the hinterland [34], e.g., as the landscape matrix of the UER, functioning as an multifunctional infrastructure to human (urban) habitat.

The discussion on urban form—compact, intermediate and diffused settlement patterns provided the stage for arguing for an urban network, polycentric solution, which is the cornerstone of the European spatial planning policies. It argues for the explicit consideration of a chorological perspective when planning cities and urban agglomerations at large from a regional perspective, in the context of extending the present concept of polycentric urban structures and network cities to include explicitly, and plan and manage for a multifunctional hinterland of rural and natural resources aiming at increasing regional's self-reliance, structured by a network of ecological systems that provides for key—ecological services (the region's "ecological backbone"), considering the interrelationships between cities and its functional hinterland. It does so by combining principles of landscape ecology with those deriving from complexity science, such as self-organization and autopoeisis, resilience and adaptation, and others close related such as self-reliance and cooperation. In order to plan for this broader concept of PUS, tools as urban metabolism and land suitability analyses, ecological footprinting, strategic environmental assessment and more recently sustainability assessment [71], combined with other methods [151] can contribute to support the spatial delimitation of the functional urban ecological region. In a time of uncertainty [48,57,90] a dual planning approach [64] could arguably be appropriated-the ecological backbone [24] or a green infrastructure [78,79] on smaller, strategically located portions of the landscape matrix, a space-efficient strategy based on the resource concentration theory [56], focusing on slow variables or more stable landscape structures that are crucial to resilience management [44,45,46], where more mandatory rules could or should be adopted. On the remaining parts of the landscape, including areas for urbanization, industry, agriculture, recreation and so forth, one could adopt a more flexible planning approach [24,135,136].

To implement eco-polycentric urban systems (Eco-PUS) I argue that most of all we need cooperation in city planning, design, and management. The above examples reveal innovative approaches to city planning. These are supported by close cooperation between the several agents involved—farmers, industry, and local governments. Cooperation is needed between (networks of) cities, as in PUS, where it constitutes a basic premise [19]. We need also cooperation across the scientific community (life and social sciences), and of the latter with planners, designers and engineers, under a translational research paradigm [30]. Transdisciplinary efforts are also needed, e.g., collaborative planning and design [83], and collaborative resilience management [90] to increase cities' ability to adapt, including governmental across institutions at its different levels, and the private sector. Inter-institutional cooperation and (planning) legislation is crucial to support integrated planning of human activities sectors. In the last decade the European Union (EU) has pushed this agenda based on its spatial planning policies: "Spatial development issues in the EU can, in future, only be resolved through co-operation between different governmental and administrative levels. (...) New forms of co-operation proposed in the ESDP should, in future, contribute towards a co-operative setting up of sectorial policies—which up to now have been implemented independently—when they affect the same territory. The Community also requires the active co-operation of cities and regions in particular to be able to realise the objectives of the EU in a citizen-friendly way. This is how the subsidiarity principle, rooted in the Treaty on EU, is realised." [152]. It is also needed increased cooperation within local communities. For example, cooperative housing in Scandinavia accounts for 20% of national housing stock, but only 1% in the USA and in the UK [153].

Housing projects in the Netherlands and in Germany are being developed aiming at social cohesion [29]. Local Agenda 21, a tool for community involvement and partnership (e.g., Middlesborough, UK; Den Haag, NL or Helsinki, Finland) represents a considerable effort to engage citizens in thinking about what sustainability might mean for their neighborhoods and communities [154].

Fortunately a convergence is undergoing, evolving into integrated approaches stemming from both directions, i.e. ecology, including landscape (and urban) ecology [27,32,51,54], and from (landscape) planning and design [24,33,34,35,36,37,38,39,40,56,76,77,78,79,80] among many other contributions across the last decade or so. Some argue that if we are to promote an integration of biophysical and cultural approaches, we should be focusing on commonalities instead of focusing on differences [31,38]. The landscape (ecological) spatial dimension provides a common platform with other landscape dimensions and disciplines [38], including integrating the relationships between different resources in landscape modeling [40]. Scientists need to develop more practice-oriented research and orient research to the development of operational tools. Consequently they need to understand planners' goals. Planners, designers and engineers need to work harder in promoting effectively the integration of several disciplines, and to incorporate ecological theory appropriately.

Interestingly, a review showed that it is not the form itself that is sustainable or not but the processes associated to a particular form (or structure) and the reciprocal relationships that are established in a dynamic interaction between form and these processes [69]. Therefore it is paramount to understand urban processes of different nature, and the drivers that are associated with those processes. For example, in the context of environmentally led urban design environmental metrics are seen as potentially beneficial to a "kind of "performalism"—an initially direct relationship between environmental performance and urban form—that can knit up culturally uninformed environmental design and environmentally uninformed urban design, particularly in cities that are growing too rapidly" [155]. Landscape ecology is well posed to address these structure-functions or processes relationships in the context of sustainable landscape planning [24,38], and sustainability science [27]. According to the latter, to develop a rigorous science of sustainability one needs to quantify whatever sustainability means, and landscape metrics can be used for this purpose. A major concern in spatial planning is the study of the spatial characteristics of urban processes, such as urban sprawl [4]. In the last ten years, spatial metrics derived from landscape ecology have been increasingly used to study the spatial characteristics of urban processes, namely the spatial characteristics of urban patches, including their size, shape, and spatial distribution. These useful tools can be very valuable for planners who need to better understand and more accurately characterize urban processes and their consequences [36]. Since the late 90s research have been produced focusing on applying landscape ecological principles and metrics to sustainable landscape planning, with a focus on urban processes and metropolitan areas [2,8,9,10,21,24,33,34,35,36,37,38,39,40,47,156,157]. In order to explore the connection between aspects related to spatial processes and their environmental and territorial consequences, analyses such as those explored in the above mentioned articles could be integrated in metropolitan observatories. This would enable to study the relation between the processes and spatial patterns identified by the metrics as well as the changes in these aspects measured by available indicators (e.g., energy consumption, automobile dependence, use of public transportation, alteration of environmental processes, etc.) [36], and thus potentially contribute to sustainability assessments (e.g., the SA process was implemented in 2005 in Sydney, Australia) [71]. These applications are deemed important to contribute to better understand the complexity of cities and to address some of the challenges posed to the planning and design of the cities of the future.

Polycentric urban systems are being implemented for some time across Europe. I argue one of the major challenges today is to move the present concept towards Eco-PUS. Taken as a whole it remains a hypothesis to test. Therefore it lacks empirical evidence and case studies. However, some of the ideas discussed in this manuscript that are encapsulated in the proposed concept of Eco-PUS are already being put in practice in Europe and elsewhere, e.g., Australia. As described above there is empirical evidence that these ideas can work, e.g., in Barcelona (Spain), Stockholm (Sweden), Kalundborg (Denmark), Portland (Oregon, USA), Curitiba (Brazil), Surabaya (Indonesia), and many others cities across the world [29,68,70,138], arguably helping cities move towards a more sustainable path. Moreover, general principles of landscape ecology have been proposed to be put in practice to plan metropolitan areas such as in the Barcelona Metropolitan Region by request of the BMR planning authorities [157]. Finally, and as stated from the beginning, the proposal described in the present article is still a work in progress. It will be further developed in new articles of the author to be published subsequently, namely by testing the potential of some of the ideas described to Mediterranean coastal regions in the Iberian Peninsula.

"For now, the most important lesson of complexity theory is that it counsels us against placing too much confidence in deterministic models of economic, social and political behaviour and against over-elaborate analysis of single agency interventions in policy making, strategic management and public governance within policy systems whose interactions are, at best, only partially understood." [72]. As the former Secretary-General of the United Nations, Kofi Annan, said: "The future of humanity lies in cities." [31] However, the core of all the troubles we face today is our very ignorance of knowing [41]. I strongly believe the path of sustainability is greatly dependent on the levels of empathy and cooperation we all—scientists, planners and designers, institutions, the society at large, etc. —can achieve to surmount the present "crisis of perception" [42] by bringing forth a new social and scientific paradigm and thus deal with the tremendous task that lies ahead of Mankind in general, and all of us engaged in bringing forth a new sustainable way of life, in or outside cities.

REFERENCES AND NOTES

1. Jacobs, J. The Life and Death of Great American Cities; Modern Library: New York, NY, USA, 1961.

2. Wu, J.; Jenerette, G.D.; Buyantuyev, A.; Redman, C.L. Quantifying spatiotemporal patterns of urbanization: The case of the two fastest growing metropolitan regions in the United States. Ecol. Complex.**2011**, 8, 1–8.

3. UNFPA, State of the World Population 2011 Report; United Nations Population Fund, Information and External Relations Division: New York, NY, USA, 2011.

4. EEA, EEA Report Nr.10/2006. In Urban Sprawl in Europe. The Ignored Challenge; European Environment Agency (EEA): Copenhagen, Denmark, 2006.

5. Grimm, N.B.; Faeth, S.H.; Golubiewski, N.E.; Redman, C.L.; Wu, J.; Bai, X.; Briggs, J.M. Global change and the ecology of cities. Science**2008**, 319, 756–760.

6. Alig, R.J.; Kline, J.D.; Lichtenstein, M. Urbanization on the US landscape: Looking ahead in the 21st century. Landsc. Urban Plan.**2004**, 69, 219–234.

7. Bohnet, I.C.; Pert, P.L. Patterns, drivers and impacts of urban growth—A study from Cairns, Queensland, Australia from 1952 to 2031. Landsc. Urban Plan.**2010**, 97, 239–248.

8. Taubenbock, H.; Wegmann, M.; Roth, A.; Mehl, H.; Dech, S. Urbanization in India—Spatiotemporal analysis using remote sensing data. Comput. Environ. Urban Syst.**2009**, 33, 179–188.

9. Yu, X.J.; Ng, C.N. Spatial and temporal dynamics of urban sprawl along two urban-rural transects: A case study of Guangzhou, China. Landsc. Urban Plan.**2007**, 79, 96–109.

10. Schneider, A.; Seto, K.; Webster, D.R. Urban growth in Chengdu, Western China: Application of remote sensing to assess planning and policy outcomes. Environ. Plan. B**2005**, 32, 323–345.

11. Caves, R.W., Ed.; Todes cited in Encyclopedia of the City; Routledge: Oxon, Canada, 2005; p. 94.

12. Cagmani, R.; Gibelli, M.C.; Rigamonti, P. Urban mobility and urban form: The social and environmental costs of different patterns of urban expansion. Ecol. Econ.**2002**, 40, 199–216.

13. Tsai, Y.-H. Quantifying urban form: Compactness versus 'sprawl'. Urban Stud.**2005**, 42, 141–161.

14. Lewis cited in Caves 2005 [11], p. 426–427.

15. Saunders, D.A.; Hobbs, R.; Margules, C.R. Biological consequences of ecosystem fragmentation: A review. Conserv. Biol.**1991**, 5, 18–32.

16.   Zonneveld cited in Caves 2005 [11]

17.   Kloosterman, R.C.; Musterd, S. The Polycentric urban region: Towards a research agenda. Urban Stud.**2001**, 38, 623–633.

18.   Dieleman cited in Caves 2005 [11], p. 321.

19.   European Commission, European Spatial Development Perspective (ESDP). Towards Balanced and Sustainable Development of the Territory of the European Union; European Commission (EC), Committee on Spatial Development: Luxembourg, Luxembourg, 1999.

20.   Vitousek, P.M.; Mooney, H.A.; Lubchenco, J.; Melilo, J.M. Human domination of earth's ecosystems. Science**1997**, 277, 494–499.

21.   Luck, M.; Wu, J. A gradient analysis of the landscape pattern of urbanization in the Phoenix metropolitan area of USA. Landsc. Ecol.**2002**, 17, 327–339.

22.   Rees, W.E. Understanding Urban Ecosystems: An Ecological Economics Perspective.In Understanding Urban Ecosystems. A New Frontier for Science and Education. In Proceedings of the 8th Cary Conference, Milbrook, NY, USA, 27–29 April 1999; Berkowitz, A.R., Nilon, C.H., Hollweg, K.S., Eds.; Institute of Ecosystems Studies: Milbrook, NY, USA, 2003; pp. 115–136.

23.   Odum, E.P. Input management of production systems. Science**1989**, 243, 177–182.

24.   Botequilha-Leitão, A.; Miller, J.N.; McGarigal, K.; Ahern, J. Measuring Landscapes. A Planner's Handbook; Island Press: Washington, DC, USA, 2006.

25.   Jongman, R.G.H. Landscape Ecology in Land Use Planning. In Issues in Landscape Ecology; Wiens, J.A., Moss, M.R., Eds.; Fifth World Congress; International Association for Landscape Ecology: Snowmass Village, CO, USA, 1999; pp. 112–118.

26.   Rees, W.E. Revisiting carrying capacity: Area-based indicators of sustainability. Popul. Environ.**1996**, 17, 195–215.

27.   Wu, J. Landscape ecology, cross-disciplinarity, and sustainability science. Landsc. Ecol.**2006**, 21, 1–4. [Google Scholar] [CrossRef]

28.   Forman 1995, van Lier 1998, [56], Golley and Bellot 1999 cited in [39], p. 66.

29.   Beatley, T. Green Urbanism. Learning from European Cities; Island Press: Washington, DC, USA, 2000.

30.   Musacchio, L.R. Metropolitan landscape ecology. Using translational research to increase sustainability, resilience, and regeneration. Landsc. J.**2008**, 27, 1–8.

31.   Wu, J. Urban sustainability: An inevitable goal of landscape research. Landsc. Ecol.**2010**, 25, 1–4.

32. Musacchio, L.; Wu, J. Collaborative landscape-scale ecological research: Emerging trends in urban and regional ecology. Urban Ecosyst.**2004**, 7, 175–178.

33. Botequilha-Leitão, A. Towards Sustainable Human Habitats. The Role of Landscape Ecology in Urban Planning.In Landscape Ecology in the Mediterranean: Inside and Outside Approaches. In Proceedings of the European IALE Conference, Faro, Portugal, 29 March-2 April 2005; Bunce R.G.H.;, Jongman, Jongman, R.H.G., Eds.; IALE Publication Series: Faro, Portugal, 2005; 3.

34. Botequilha-Leitão, A. Land Use Planning in Portugal: Brief History and Emergent Challenges. The Case of Peri-urban Landscape of Faro (Algarve Region, Portugal). In New Models for Innovative Management and Urban Dynamics; Panagopoulos, T., Ed.; COST publication, European Science Foundation and University of Algarve: Faro, Portugal, 2009; pp. 19–36.

35. Botequilha-Leitão, A.; Cruz, R.; Aguilera, F. Landscape Changes in the Algarve Region, Portugal ('85-'07)—Diagnosis, Prospective and a Proposal for a Green-infrastructure in the Algarve Central Coast. In Proceedings of the International Conference Green-Infrastructures for Biodiversity, Congress Center, Estoril, Portugal, 26 September–1 October 2011; Reis-Machado, J., Ed.;

36. Aguilera, F.; Valenzuela-Montes, L.M.; Botequilha-Leitão, A. The use of landscape metrics in urban patterns analysis. Landsc. Urban Plan.**2011**, 99, 226–238.

37. Aguilera, F.; Botequilha-Leitão, A.; Diáz-Varela, E. Selección de métricas de la ecología del paisaje mediante ACP para la caracterización de los procesos de alteración del paisaje del Algarve (Portugal). Int. Rev. Geogr. Inf. Sci. Technol. (submitted in July 2011).

38. Botequilha-Leitão, A. Sustainable Land Planning. Towards a Planning Framework. Exploring the Role of Spatial Statistics as a Planning Tool. Ph.D. Dissertaiton, High Tecnical Institute, Technical University of Lisbon (Instituto Superior Técnico, UTL), Lisbon, Portugal, 2001.

39. Botequilha-Leitão, A.; Ahern, J. Applying landscape ecological concepts and metrics in sustainable landscape planning. Landsc. Urban Plan.**2002**, 59, 65–93.

40. Ferreira, H.; Botequilha-Leitão, A. Integrating Landscape and Water Resources Planning with Focus on Sustainability. In From Landscape Research to Landscape Planning. Aspects of Integration, Education and Application; Tress, B., Tress, G., Fry, G., Opdam, P., Eds.; Springer: Dordrecht, The Netherlands, 2006; pp. 143–159.

41.   Maturana, H.R.; Varela, F.J. The Tree of Knowledge. The Biological Roots of Human Understanding; Shambhala Publications, Inc.: Boston, MA, USA, 1992.

42.   Capra, F. The Web of Life. A New Scientific Understanding of living SYSTEMS; Anchor Books: New York, NY, USA, 1996.

43.   MacAslan, A. The Concept of Resilience. Understanding Its Origins, Meaning and Utility. The Torrens Resilience Institute: Adelaide, SA, Australia, 2010. Available online: http://torrensresilience.org/ (accessed on 13 October 2011).

44.   Holling, C.S. What barriers? What bridges? In Barriers and Bridges to the Renewal of Ecosystems and Institutions; Gunderson, L.H., Holling, C.S., Light, S.S., Eds.; Columbia University Press: New York, NY, USA, 1995; pp. 3–34.

45.   Gunderson, L.H.; Holling, C.S.; Light, S.S. Barriers Broken and Bridges Built: A Synthesis. In Barriers and Bridges to the Renewal of Ecosystems and Institutions; Gunderson, L.H., Holling, C.S., Light, S.S., Eds.; Columbia University Press: New York, NY, USA, 1995; pp. 489–532.

46.   Forman, R.T.T. Urban Regions. Ecology and Planning. Beyond the City; Cambridge University Press: New York, NY, USA, 2008.

47.   Antrop, M. Landscape change: Plan or chaos? Landsc. Urban Plan.**1998**, 41, 155–161.

48.   Antrop, M. Uncertainty in Planning Metropolitan Landscapes. In Planning Metropolitan Landscapes—Concepts, Demands, Approaches; Tress, G., Tress, B., Harms, B., Smeets, P., van der Valk, A., Eds.; Alterra Green World Research, Wageningen University and Research Centre: Wageningen, The Netherlands, 2004; pp. 12–25, DELTA Series 4.

49.   Naveh, Z. What is holistic landscape ecology? A conceptual introduction. Landsc. Urban Plan.**2000**, 50, 7–26.

50.   Pickett, S.T.A.; Cadenasso, M.L.; Grove, M.; Nilon, C.H.; Pouyat, R.V.; Zipperer, W.C.; Constanza, R. Urban ecological systems: Linking terrestrial ecological, physical, and socioeconomic components of metropolitan areas. Annu. Rev. Ecol. Syst.**2001**, 32, 127–157.

51.   Pickett, S.T.A.; Cadenasso, M.L.; Grove, J.M. Resilient cities: Meaning, models, and metaphor for integrating the ecological, socio-economic, and planning realms. Landsc. Urban Plan.**2004**, 69, 369–384.

52.   Grove, J.M.; Hinson, K.E.; Northrop, R.J. A Social Ecology Approach to Understanding Urban Ecosystems and Landscapes. In Understanding Urban Ecosystems. A New Frontier for Science and Education. In Proceedings of the 8th Cary Conference, Milbrook, NY, USA, 27-29

April 1999; Berkowitz, A.R., Nilon, C.H., Hollweg, K.S., Eds.; Institute of Ecosystems Studies: Milbrook, NY, USA, 2003; pp. 167–186.

53. Alberti, M. The effects of urban patterns on ecosystem function. Int. Reg. Sci. Rev.**2005**, 28, 168–192. [Google Scholar] [CrossRef]

54. Felson, A.J.; Pickett, S.T.A. Designed experiments: New approaches to studying urban ecosystems. Front. Ecol. Environ.**2005**, 3, 549–556.

55. Haberl, H.; Winiwarter, V.; Andersson, K.; Ayres, R.U.; Boone, C.; Castillo, A.; Cunfer, G.; Fischer-Kowalski, M.; Freudenburg, W.R.; Furman, E.; et al. From LTER to LTSER: Conceptualizing the socioeconomic dimension of long-term socioecological research. Ecol. Soc.**2006**, 11, 13. Available online: http://www.ecologyandsociety. org/vol11/iss2/art13/ (accessed on November 2011).

56. Ahern, J. Spatial Concepts, Planning Strategies and Future Scenarios: A Framework Method for Integrating Landscape Ecology and Landscape Planning. In Landscape Ecological Analysis: Issues and Applications; Klopatek, J., Gardner, R., Eds.; Springer-Verlag Inc.: New York, NY, USA, 1999; pp. 175–201.

57. Kato, S.; Ahern, J. Learning by doing": Adaptive planning as a strategy to address uncertainty in planning. J. Environ. Plan. Manag.**2008**, 51, 543–559.

58. Newman, P.W.G.; Kenworthy, J.R. Gasoline consumption and cities. J. Am. Plan. Assoc.**1989**, 55, 24–37. [Google Scholar] [CrossRef]

59. Faludi, A.; van der Valk, A. Rule and Order: Dutch Planning Doctrine in the Twentieth Century; Kluwer Academic Publishers: Dordrecht, The Netherlands, 1994.

60. Batten, D.F. Network cities: Creative urban agglomerations for the 21st century. Urban Stud.**1995**, 32, 313–327.

61. Ribeiro Telles, G. Global Landscape (Paisagem Global). In Challenges for Mediterranean Landscape Ecology: The Future of Cultural Landscapes—Examples from the Alentejo Region. In Proceedings of the 1st National Landscape Ecology Workshop, Montemor-o-Novo, Portugal, 25–28 March 1998; Pinto-Correia, T., Cancela de Abreu, M., Eds.;

62. Ribeiro Telles, G.; Raposo Magalhães, M.; Alfaiate, M.T. Plano Verde de Lisboa. Componente do Plano Director Municipal de Lisboa; Edições Colibri: Lisboa, Portugal, 1997; [in Portuguese].

63. Newman, P.W.G. Sustainability and cities: Extending the metabolism model. Landsc. Urban Plan.**1999**, 44, 219–226.

64. Von Haaren, C. Landscape planning facing the challenge of the development of cultural landscapes. Landsc. Urban Plan.**2002**, 60, 73–80.

65. Kühn, M. From city park to regional park: Landscape in the regional city. Topos**2002**, 39, 65–73.

66. Secchi, B. Urban Scenarios and Policies. In Políticas Urbanas. Tendências, Estratégias e Oportunidades; Portas, N., Domingues, A., Cabral, J., Eds.; Fundação Calouste Gulbenkian: Lisboa, Portugal, 2003; pp. 274–283.

67. Holden, E. Ecological footprints and sustainable urban form. J. Hous. Built Environ.**2004**, 19, 91–109.

68. Catalán, B.; Saurí, D.; Serra, P. Urban sprawl in the Mediterranean? Patterns of growth and change in the Barcelona Metropolitan Region 1993-2000. Landsc. Urban Plan.**2008**, 85, 174–184.

69. Newman, M. The compact city fallacy. J. Plan. Educ. Res.**2005**, 25, 11–26.

70. Kenworthy, J.R. The eco-city: Ten key transport and planning dimensions for sustainable city development. Environ. Urban.**2006**, 18, 18–67.

71. Newman, P.W.G. The environmental impact of cities. Environ. Urban.**2006**, 18, 275–295.

72. Bovaird, T. Emergent strategic management and planning mechanisms in complex adaptive systems. Public Manag. Rev.**2008**, 10, 319–340.

73. Ndubisi, F. Sustainable regionalism. Evolutionary framework and prospects for managing metropolitan landscapes. Landsc. J.**2008**, 27, 51–68.

74. Corner, J. Terra Fluxus. In Landscape Urbanism. A Reader; Waldheim, C., Ed.; Princeton Architectural Press: New York, NY, USA, 2006; pp. 021–033.

75. Corner, J. Recovering Landscape. Essays in Contemporary Landscape Architecture; Corner, J., Ed.; Princeton Architectural Press: New York, NY, USA, 1999; pp. 1–26.

76. Waldheim, C. Landscape Urbanism. A Reader; Princeton University Press: New York, NY, USA, 2006.

77. Mostafavi, M.; Doherty, G. Ecological Urbanism; Lars Müller Publishers: Basel, Switzerland, 2010.

78. Ahern, J. Green Infrastructure for Cities: The Spatial Dimension. In Cities of the Future: Towards Integrated Sustainable Water and Landscape Management; Novotny, V., Brown, P., Eds.; IWA Publishing: London, UK, 2007; pp. 267–283.

79. Ahern, J. From fail- safe to safe-to-fail: Sustainability and resilience in the new urban world. Landsc. Urban Plan.**2011**, 100, 341–343.

80. Steiner, F. Landscape ecological urbanism: Origins and trajectories. Landsc. Urban Plan.**2011**, 100, 333–337.

81. Kates, R.W.; Clark, W.C.; Corell, R.; Hall, J.M.; Jaeger, C.C.; Lowe, I.; McCarthy, J.J.; Schellnhuber, H.J.; Bolin, B.; Dickson, N.M.; et al. Sustainability science. Science **2001**, 292, 641–642. [Google Scholar]

82. Zonneveld, 1988, cited in [38], p. 8.

83. Tippett, J.; Handley, J.F.; Ravetz, J. Meeting the challenges of sustainable development—A conceptual appraisal of a new methodology for participatory ecological planning. Prog. Plan. **2007**, 67, 9–98. [Google Scholar] [CrossRef]

84. Von Bertallanffy, L. General System Theory; Brazzilier: New York, NY, USA, 1968.

85. Steiner, F. Urban human ecology. Urban Ecosyst. **2004**, 7, 179–197.

86. Urban et al. 1987 cited in [52], p. 175.

87. Camazine, S.; Deneubourg, J.L.; Franks, N.R.; Sneyd, J.; Theraulaz, G.; Bonabeau, E. Self-Organization in Biological Systems. Princeton Studies in Complexity. Princeton University Press: Princeton, NJ, USA, 2002.

88. Andersson, E. Urban landscapes and sustainable cities. Ecol. Soc. 2006, 11, p. 34. Available online: http://www.ecologyandsociety.org/vol11/iss1/art34/ (accessed on November 2011).

89. Folke et al. 2004 cited in [88], p. 34; United Nations cited in [80], p. 336

90. Walker, B.; Carpenter, S.; Anderies, J.; Abel, N.; Cumming, G.; Janssen, M.; Lebel, L.; Norberg, J.; Peterson, G.D.; Pritchard, R. Resilience management in social-ecological systems: A working hypothesis for a participatory approach. Conserv. Ecol. 2002, 6, p. 14. Available online: http://www.consecol.org/vol6/iss1/art14 (accessed on November 2011).

91. [32] cited in [51], p. 369

92. Low, B.; Ostrom, E.; Simon, C.; Wilson, J. Redundancy and Diversity: Do They Influence Optimal Management? In Navigating Socio-ecological Systems. Building Resilience for Complexity and Change; Berkes, F., Colding, J., Folke, C., Eds.; Cambridge University Press: Cambridge, UK, 2003; pp. 83–109.

93. Rosenfeld, J.S. Functional redundancy in ecology and conservation. Oikos **2002**, 98, 156–162.

94. Walker 1992 and 1995 cited in [93], p. 156.

95. Quinlan cited in [77], p. 630.

96. Grove, J.M.; Burch, W.R. A social ecology approach and applications of urban ecosystems and landscape analyses: A case study of Baltimore, Maryland. Urban Ecosyst. **1997**, 1, 259–275.

97. Cited in [41], p. 75.

98.   Cited in [41], p. 92, my examples.

99.   Cited in [41], p. 99.

100.   Adapted from [41], p. 197.

101.   Botkin and Keller 1998 cited in [85], p. 189.

102.   Mariotti, H. Autopoiesis, culture and society. Business School São Paulo: SP, Brasil, 1999. Available online: http://www.humbertomariotti.com.br/(accessed on 12 October 2011).

103.   Cited in [49], p. 16–17.

104.   Cited in [72], p. 321.

105.   Cited in [30], p. 3.

106.   Decker and Chase 1997 cited in [39], p. 3.

107.   Hester 1990 cited in [39]

108.   Cited in [72], p. 322.

109.   Price 1995 cited in [22], p. 131.

110.   DeBoer and Dijist 1998 cited in [50], p. 143.

111.   Margules, C. Conservation Planning at the Landscape Scale. In Issues in Landscape Ecology. International Association for Landscape Ecology; Wiens, J.A., Moss, M.R., Eds.; Fifth World Congress: Snowmass Village, CO, USA, 1999; pp. 83–87.

112.   Soromenho-Marques, V. Debate sobre Ecologia e Ideologia (Debate on Ecology and Ideology). In Ecologia e Ideologia; Rebelo, J., Ed.; Livros e Leituras: Lisboa, Portugal, 1999; [in Portuguese].

113.   Kühn 1962 cited in [42], p. 29.

114.   Steinitz 1990, Forman 1995 and Zonneveld 1995 cited in [38], p. 67ff.

115.   For a more thorough review see [24]

116.   Cited in [38], p. 119.

117.   Franklin 1997 cited in [38], p. 119–120.

118.   McHarg, I.L.; Steiner, F.R. To Heal the Earth. Selected writings of Ian L. McHarg; Island Press: Washington, DC, USA, 1998; p. 134, cited in [38].

119.   Cited in [38], p. 46–49.

120.   Beatley and Manning 1997 cited in [29], p. 3.

121.   Bayley, O.; Kim, P.; Mullaney, E.; Calabrese, J.; Walman, L.; Nelson, F.; Yao, X. Will limits of the Earth's resources control human numbers? Environ. Dev. Sustain.**1999**, 1, 19–39.

122.   Campbell, 1997; Youngquist, 1997 cited in Pimentel et al. 1999. See [121]

123.   Breheny, 1992; Banister, 1992; Jenks et al., 1996 cited in [12], p. 202.

124.   Elkin et al., 1991; Fouchier, 1998 cited in [12], p. 202.

125.   Lewis cited in Caves 2005 [11], p. 426–427.

126.   Bunting cited in Caves 2005 [11], p. 127.

127.   Boeri, S. Five Ecological Challenges for the Contemporary City. In Ecological Urbanism; Mostafavi, M., Doherty, G., Eds.; Lars Müller Publishers: Basel, Switzerland, 2010; pp. 444–453.

128.   Burchel et al. 1992 cited in [12], p. 204.

129.   Nogué, J. Territorios sin discurso, paisajes sin imaginarios. Retos y dilemas (Territories without discourse, landscapes without imaginary. Challenges and dilemmas. Abstract in English and in French). Ería **2007**, 73, 373–382.

130.   [67], p. 91.

131.   Suwa cited in Caves 2005 [11], p. 229

132.   Berglund 1991, Toupal 2003 cited in [55], p. 3.

133.   Wascher, D.M.; Pedroli, B. Blueprint for Euroscape 2020. Reframing the future of the European landscape. Policy visions and research support. Landscape Europe, Joint Research Commission, Alterra Greenworld Institute, RECEP-ENELC, the Landscape Research Group: Wageningen, The Netherlands, 2008; pp. 32–33.

134.   [24], p. 179 and 182.

135.   Van Buuren and Kerkstra 1993; Ahern and Kerkstra 1994; cited in [24], p. 182.

136.   Sijmons, D. Landscape: Plans, lectures, essays, and articles produced by H+N+S Landscape Architects; Architectura and Natura Press: Amsterdam, The Netherlands, 2002; (Revised version of Landschap 1998).

137.   Weller cited in [76]

138.   Gauzin-Muller, D. Sustainable Architecture and Urbanism: Concepts, Methodologies, Examples; Birkhauser: Basel, Switzerland, 2002.

139.   Mehrhoff in Caves 2005 [11], p. 443.

140.   Desrochers, P. Eco-Industrial Parks and the Rediscovery of Inter-Firm Recycling Linkages. Mises Institute Working Papers, Mises Institute: Vienna, Austria, 2000. Available online: http://mises.org/literature.aspx?action=search&q=Desrochers (accessed on 21 March 2012).

141.   The Kalundborg case study description hereby included was adapted from information consulted at the following websites. Kalundborg Municipality: Denmark. Available online: http://www.kalundborg.dk/ (accessed on 21 March 2012); [142,143], These were also helpful to complement the information on [29].

142.   Symbiosis Institute. Kalundborg Symbiosis: Kalundborg, Denmark. Available online: www.symbiosis.dk (accessed on 21 March 2012).

143.   Indigo Development. Sustainable Systems, Inc.: Oakland, CA, USA. Available online: www.indigodev.com/Kal.html (accessed on 21 March 2012).

144. Christensen. Proceedings of the Industry & Environment at Indian Institute of Management, Ahmedabad, India, 1999, cited in Rombaut, E. Symbiose industrielle: Le cas de Kalundborg (Danemark). Centre d'Etude, de Recherche et d'Action en Architecture (CERAA), Saint-Gilles, Belgium. Available online: http://www.ceraa.be/uploads/annexes/journeesetudesjedd7/ JEDDVII_orateurs/10_ROMBAUT-pps_en.pdf (accessed on 22 March 2012).

145. Figure source: University of Colorado, CO, USA. Available online: http://www.colorado.edu/AmStudies/lewis/ecology/Kalundborg.gif (accessed on 22 March 2012).

146. Desrochers, P. Eco-industrial parks. The case for private planning. The Independent Review, 2001, V, pp. 345–371. Available online: http://www.independent.org/pdf/tir/tir_05_3_desrochers.pdf (accessed on 22 March 2012).

147. Indigo Development. Developing Agricultural Eco-Industrial Parks in China. An Indigo Industrial Ecology Paper. Indigo Development, Inc.: Santa Rosa, CA, USA, 2005. 2005. Available online: http://www.indigodev.com/AEIPwhitepaper.html (accessed on 27 March 2012).

148. [12], p. 201.

149. [66], p. 276.

150. [42]; see also [113].

151. [46]; see section 3.3.

152. [19], p. 7–8.

153. Skelton in Caves 2005 [11], p. 98.

154. [29], p. 345–348.

155. Hagan in [77], p. 458–467.

156. For a more extensive review see [36].

157. Forman, R.T.T. Mosaico territorial para la región metropolitana de Barcelona. Editorial Gustavo Gili: Barcelona, Spain, 2004.

CHAPTER 5

Simulation Models and GIS Technology in Environmental Planning and Landscape Management

Giuliana Lauro

Department of Industrial and Information Engineering, Second University of Naples, Aversa, Italy

ABSTRACT

Landscape protection that, in the past, has been mainly concerned with its historical, artistic and cultural heritage, follows, nowadays, a systemic methodology that looks at landscape as a high level aggregate of spatial, ecologically different units that interact each other by exchanging energy and materials. Strategic environmental assessment, nowadays, has been adopted in Europe in landscape planning, whose task is to verify the compatibility of territory transformations with respect to their levels of criticality and vulnerability, to evaluate possible future scenarios as consequence of interventions by checking if they are in line with preservation and valorization of environmental. To this aim, we make here a short survey of three different simulation models that can be used as Decision Support System in landscape planning and management. They adopt tools of the Landscape Ecology and are based on GIS (Geographic Information System) technology. The first one consists of a planar graph, the so called ecological graph, whose construction needs the computation of suitable indices of environmental control, proper of Landscape Ecology, such as biodiversity, biological territorial capacity, connectivity. The planar graph, for the considered environmental system, returns a picture of its actual ecological health condition and provides very detailed indications and operational assistance for choosing among possible ecological sustainable interventions. The second one, based on the data used to construct the ecological graph, uses the least-cost path algorithm from GIS technology in order to build an ecological network to prevent and to reduce territorial fragmentation caused by intense processes of urbanisation and industrialisation. At last, an integrated GIS-based approach is developed combining an ecological graph model and a mathematical model based on a nonlinear differential equation of logistic-type with harvesting to perform qualitative predictions on the sustainability of a given territorial plan.

Keywords: Landscape Ecology; Ecological Network; Dynamical System; GIS Technology

1. INTRODUCTION

In recent years, the landscape planning has understood the importance of an "ecological-oriented" analysis of environmental systems more or less affected by degradation and ecological phenomena such as fragmentation and reduction of biodiversity.

To be changed is the very idea of "landscape" which, traditionally identified with that of scenic beauty, has now expanded to indicate all the ways of interactions among nature, environment, land, cultural heritage. The landscape as a "field of knowledge" is the basis for the formulation of the European Landscape Convention, CEP, adopted by the Committee of Ministers of Culture and Environment of the Council of Europe July 19, 2000 and since September 1, 2006, law operating in Italy (Law No. 14 of January 9, 2006). According to this convention "Landscape means a certain portion of territory, as perceived by people, whose character derives from the natural and/or human interrelationships". In addition to defining the term landscape, Convention determines all the rules for the recognition, protection, preservation, management of the landscape. It is the first international treaty with the objective of promoting the protection, management and planning of European landscapes by promoting European cooperation. The Convention is potentially a real conceptual revolution that brings the community to become the primary stakeholders of the evolution of landscapes in which they live, seen as a cultural strategy for the quality of habitat and also relevant from a social and economic point of view. It encourages citizens to take an active part in decision-making processes that affect the landscape at the local and regional levels. The content of the Convention takes into account the whole of the States Parties and covers natural, rural, urban and suburban areas, including land, inland waters and marine waters. Concerns landscapes that might be considered outstanding, and the landscapes of everyday life, even degraded landscapes (Article 2). The methodologies for successful landscape and environmental planning have been severely challenged when concepts like ecosystems preservation and sustainability have been questioned. The actual challenge is to build transparent and flexible decision-making tools to be used in environmental planning and to embrace a broad range of stakeholder needs together with landscape management requirements. Decision-making process needs the use of interdisciplinary models. A modern approach to the study of landscape shows that it should not be viewed as a mere sum of parts but as a system of relations between the different constituent ecosystems and processes that determine its evolution in time. In the language of Landscape Ecology, this means considering the landscape as a system of ecosystems, ecomosaic, organized in a hierarchical structure and interacting with each other through exchanges of energy and matter, in a fragile equilibrium under dynamic disturbances of both natural and anthrop origin ([1-3]). This research, therefore, uses principles and models proposed by the Landscape Ecology to analyze and assess the environmental quality of a landscape through the identification of appropriate indices of control such as biodiversity, "bioenergy", connectivity. We refer to the term "bioenergy" as the energy available in the environment, present in different forms like animals, seeds, plants, ruled by suitable metabolic processes. Energy and material "fluxes", i.e., bioenergy fluxes, through the territory are therefore necessary fundamental processes for biodiversity conservation and capacity of the system to resist at the both natural and anthrop perturbations (resilience). Barriers and surfaces with low permeability to such fluxes hamper the movement of animals, seeds spreading and in general the likelihood of species survival, as they act as external constraints on the natural ecosystem. Modeling these fluxes is therefore necessary to assess the most suitable plan strategies for natural resources conservation management and landscape functionality preservation. A tool in this direction can be represented by the construction of the so called Ecological Graph ([4,5]), as we'll see in the next Section, that furnishes a picture of landscape ecological health condition and provides very detailed indications and operational assistance to guide toward sustainable interventions. Moreover, this model also allows to determine the status of territorial fragmentation and, hence, to verify the need for ecological network, functional to dispersion of animals and plants. As said, the flows of gens and individuals between populations is essential for the survival of those species that are sensitive to the fragmentation of their habitats, therefore, the loss of ecological connectivity, i.e., the difficulty met by organisms in their movement between resource patches, constitutes a challenge for biodiversity conservation. In the 1980s, the idea of developing national ecological networks surfaced more or less simultaneously in several European countries. The ecological network concept not just prioritizes the conservation of core areas as natural or semi-natural values, but also prioritizes the importance of buffering, maintaining and re-establishing ecological connectivity and nature restoration. Depending on the species and spatial scale of interest the characteristics of an ecological network may differ widely, therefore ecological networks can be identified at continental, regional landscape and local scales. To this aim, in Section 3, we'll show a procedure ([6,7]), based on the least-cost path algorithm from GIS technology, to build a local ecological network for the National Park of Cilento. Finally, in Section 4, behind the static frame of these two simulation models, we shall propose a mathematical dynamical model in order to investigate the time evolution of the health condition of a territory, namely, of its bioenergy value.

In fact, changes in bioenergy, due to changes in environmental conditions, may produce territorial modifications toward which individual landscapes will tend to move smoothly (attractors) or may produce, instead, critical thresholds that result in radical changes in the state of the ecological system. In this sense, ecological systems are, in fact, said metastable. The investigation on mestability can be performed by means of the study of the equilibrium solutions of suitable differential equations ([8,9]) that model dynamics of the territory's evolution.

The primary objective of these models is to perform qualitative predictions on the sustainability of the territorial planning finding, possibly, critical values of the parameters characterizing the territory itself. In Section 4 we use a nonlinear ordinary differential equation of Logistic-type with Harvesting ([10]), applied to a brownfield, ex-industrial area in the East side of Naples, subject to a Master Plan in the direction of improving the environment and we show the time-evolution trend of its ecological value and the existence of a critical value of a suitable environmental indicator linked to the geometrical setting of barriers to bioenergy fluxes.

2. ECOLOGICAL GRAPH

In the modern discipline of Landscape Ecology, the landscape is defined as a heterogeneous land composed of interacting ecosystems that exchange energy and matter, and where natural and anthrop events coexist. In the present model, an environmental system is subdivided in a given number of different landscape-units separated from each other by natural or anthrop barriers. The bioenergy content of each unit may be represented by a circle (node) whose diameter is proportional to the magnitude itself. The barriers can have different degrees of permeability to bioenergy's flow ([3]). For example, an highway has almost zero value of permeability and determines, hence, a territorial fragmentation. We can represent the various levels of connection, among the units, by arcs whose width is proportional to the bioenergy flux shared among them. The collection of nodes and arcs is called Ecological Graph ([4]) of the environmental system. It can be drawn by means of a software GIS (ArchView 3.x) by using the information, contained in suitable Shape Files furnished by local government, about land uses, presence and connection of road infrastructures (railroads, highways, government and provincial roads), system of water courses (natural and artificial) and administrative subdivision of the various urban territories within the studied landscape. We show, now, how to construct such a graph relative to National Park of Cilento in Campania Region (Italy). Firstly we choose the subdivision in 18 units ([5]) as shown in Figure 1. Let us note that each landscape-unit, in turn, is composed of different ecotopes, i.e., the smallest ecologically homogeneous distinct features in a landscape mapping, with a proper value of Biological Territorial Capacity, B_{TC}, that is the amount of energy (Mcal/m^2/year) that they need to dissipate in order to maintain their organizational level.

B_{TC} values can be computed on the basis of a standard classification ([3]), as reported in Table 1, once that it is known the kind of ecotopes.

We now define the bioenergy M_j of the landscape-unit j, $j = 1, 2, \cdots 18$ as:

$$M_j = B_j\left(1 + k_j\right)$$

(1)

where B_j is the average value of B_{TC} over all the ecotopes belonging to unit j and $k_j \in [0.1]$ is an environmental index computed as the average between three parameters

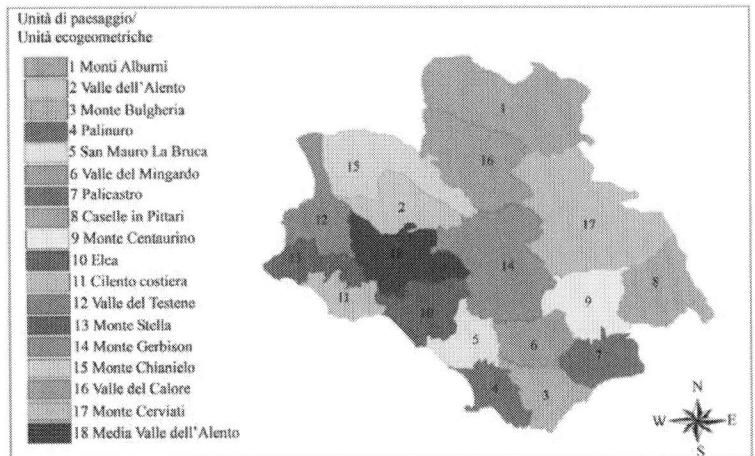

Figure 1. Landscape units for the national park of Cilento.

Table 1. Standard classification of B_{TC} values.

Class	Typology of ecotope	B_{TC} Mcal·m²/year
A (Low)	Prevalence of systems that needs energy (industries, infrastructures, buildings, brownfields, rocky areas).	<0.5
B (Md-Low)	Prevalence of agricultural- technological systems or degraded ecotopes (sowed areas, shed built areas, inculted grassy areas, river corridords).	0.5 - 1.5
C (Medium)	Prevalence of agricultural seminatural systems (sowed areas, orchards, vineyards, hedges to medium resistance).	1.5 - 2.5
D (Md-High)	Prevalence of natural ecotopes (bushes area, pioneer vegetation, rows, poplar areas, reforestations areas, urban green).	2.5 - 3.5
E (High)	Prevalence of natural ecotopes that don't need a supply of energy (woods, mountains areas, damp zones).	3.5 - 5

k_{Fj}, k_{Pj}, k_{Dj}, each with values in [0, 1], as follows:

$$k_{Fj} = 1 - \frac{P_j^C}{P_j}, \quad k_{Pj} = \frac{\sum_{r=1}^{s} p_{rj} L_{rj}}{P_j} \text{ with } P_j = \sum_{r=1}^{s} L_{rj}$$

$$k_{Dj} = \frac{\sum_{i=1}^{5} \frac{n_i}{5} \log_{10} \frac{n_i}{5}}{\log_{10}(1/H)}$$

with P_j perimeter of unit j, P_j^C perimeter of a circle of area A_j, L_{rj} the perimeters of the s portions of P_j which have permeability index p_{rj}, $r = 1, \cdots, s$; n_k is the number of ecotopes of B_{TC} of class k among the whole number H of classes present in the unit j. Note that in our case the maximum number of classes is 5 as shown in Table 1.

The first index, k_{Fj}, is a parameter related to the shape of the patch borders, since their morphology influences strongly the energy exchanges between the patches themselves. Note that the most jagged is the hedge's shape the most favorable are the conditions for hiding and reproducing of wildlife.

The second one, k_{Pj}, again with the purpose of evaluating energy exchanges, takes into account the permeability of the barriers to energy flux, by following some standard classification ([3]) of values of the permeability parameter as reported in Table 2.

Finally, the third parameter K_{Dj} is related to biodiversity, determined by a Shannon entropy value, that takes into account the presence of different ecotopes inside each unit. High values of biodiversity contribute to more stable ecosystems.

Note that, being $\max\{k_j\}=1$, we have that the maximum value of bioenergy, M_{max}, that a given landscape with n units can produce, will be:

$$M_{max} = 2B_{max} \text{ with } B_{max} = \max_{j=1,2,\cdots n}\{B_j\}$$ (2)

The last environmental indicator needed for evaluating the graph is the bioenergy flux, trough the borders of two consecutive units i and j, whose magnitude is proportional to the width of the link between the nodes of the

Table 2. Permeability of barriers.

Typology of barriers	p
Highways, principal net of communication	0.05
urban or secondary roads	0.4
artificial water net	0.4
railroad	0.5
White road	0.7
natural water net	0.85
principal river	1.0

units i and j:

$$F_{ij} = \frac{M_i + M_j}{2} \frac{L_{ij}}{P_i + P_j} p_{ij}$$

(3)

where all the quantities in (3) have already been defined.

For the construction of the ecological graph we have used the software ArcView 3.x of GIS (Geographical Information System).

The shape files, derived from the regional land cover map produced by Campania Region, furnish the information about the land uses, the presence and the connection of road infrastructures (railroads, highways, government and provincial roads), the system of water courses (natural and artificial) and the administrative subdivision of the various urban territories within the studied area.

In Table 3 we have firstly reported the values found, by means of the spreadsheet application of Microsoft Excel, for the bioenergy M_j (normalized to

1), recalling that the diameters of the graph's nodes are proportional to these magnitudes.

Then, in Table 4 we have reported the values of fluxes F_{ij} (see Equation (3)), normalized to one, between consecutive units, recalling that the width of graph's arcs are proportional to these values.

The final result of the construction of the Ecological Graph is shown in Figure 2 that exhibits a picture of the actual state of ecological health of this territory.

From the graph of Figure 2 we get, for example, the information that the energetic content of units 1, 14 and 17 (Monti Alburni, Monte Gerbison e Monte Cerviati) is high, hence they represent the territorial portions of greater ecological value and therefore deserve of more attention as they support the entire environmental system. Note from Figure 3 that units 1 and 17 are Special Protection Areas (SPAs), while in unit 14 there is a Site of Community Importance (SCI). Even though, the flux between 1 and 17 is very weak due to the presence of a highway that is not permeable to energy flow. To strengthen this flow it might be possible to carry out structures that allow wildlife to cross above or below the roadway, as studied by the new discipline of Road Ecology. Moreover, the energy content of unit 9 (Monte Centaurino), even if it is characterized by SCI and SPA areas, is very low, as shown by the small size of its node, for the presence of agricultural areas used for annual crops associated with permanent, it would be appropriate to take action for environmental improvement works (hedges of natural vegetation) aimed to increase the level of biodiversity and thus the overall stability of the system. Unit 6 (Valle del Mingardo) is definitely a part of the territory on which it is advisable to aim for improve the system. It is in fact an area characterized by a low value of bioenergy but by a high number of links (5), characterized however by weak fluxes, even here, this weakness is linked to the widespread presence of agricultural areas on which to intervene in order to increase the biodiversity. We finally note that as decision-making tool we could match 2 graphs, one representing the actual situation, the second one representing the project solution in line with the intents of a Master Plan if any. The comparison among the graphs of the two sceneries allows, in decisional phase, to judge on the sustainability of the intervention on the area.

Table 3. Bioenergy values of the 18 landscape-units.

Landscape-unit	area (mq)	B_j Average Value of B_{TC} over all the ecotopes in the unit j	k_j	Bioenergy $M_j = (1 + k_j)B_j$ normalized to 1
1_Monti Alburni	271795284.779	976415252.511	0.392	0.79
2_Valle dell'Alento	106677569.293	315876438.946	0.481	0.27
3_Monte Bulgheria	90479291.482	308154108.671	0.454	0.26
4_Palinuro	65705173.685	222999377.355	0.466	0.19
5_San Mauro La Bruca	73506910.281	207991143.863	0.536	0.18
6_Valle del Mingardo	88059522.990	256106446.029	0.429	0.21
7_Policastro	83583402.535	223884113.933	0.443	0.19
8_Caselle in Pittari	131787266.802	465534241.171	0.354	0.36
9_Monte Centaurino	117026977.046	315010972.459	0.482	0.27
10_Elea	104082537.911	305804409.005	0.526	0.27
11_Cilento Costiera	60436023.232	180283730.319	0.490	0.16
12_Valle del Testene	113333667.540	272387166.872	0.546	0.24
13_Monte Stella	69489471.921	214733859.953	0.503	0.19
14_Monte Gerbison	240080281.969	875649957.003	0.468	0.74
15_Monte Chianiello	173522652.513	469893200.610	0.513	0.41
16_Valle del Calore	183134923.836	522436272.450	0.512	0.46
17_Monte Cerviati	324094491.338	1223367667.853	0.412	1.00
18_Media Valle dell'Alento	142971752.086	430214999.459	0.531	0.38

Table 4. Bioenergy fluxes between consecutive landscape-units.

Consecutive units	L_{ij}	p_{ij}	$P_i + P_j$	Fluxes values $F_g = \dfrac{M_i + M_j}{2} \dfrac{L_{ij}}{P_i + P_j} p_{ij}$	F_g norm.
1 - 16	19092.000	0.200	157288.685	26088766.816	0.331
1 - 17	9513.000	0.200	182074.021	16128633.135	0.205
2 - 12	3914.000	0.200	130299.592	2670462.353	0.034
2 - 14	7330.000	0.200	141240.674	9100885.786	0.116
2 - 15	28216.000	0.200	139757.782	23793566.474	0.302
2 - 18	19773.000	0.200	129125.858	17247639.944	0.219
3 - 4	14800.000	0.200	111795.429	10256790.053	0.130
3 - 5	3000.000	0.200	107741.349	2136803.108	0.027
3 - 6	10500.000	0.200	101924.128	8384114.553	0.106
3 - 7	7970.000	0.200	106035.225	5794134.282	0.074
4 - 5	11187.000	0.200	102806.136	7033615.419	0.089
5 - 6	6039.000	0.200	92934.835	4454146.045	0.057
5 - 10	15200.000	0.600	104440.840	34324676.081	0.436
5 - 14	8163.000	0.200	131420.783	9971391.297	0.127
6 - 7	7550.000	0.200	91228.711	5701396.261	0.072
6 - 9	10078.000	0.200	103978.820	8072834.087	0.102
6 - 14	8246.000	0.200	125603.562	10844201.535	0.138
7 - 9	12600.000	0.600	108089.916	27624367.021	0.351
8 - 9	12624.000	0.200	118394.474	11701046.248	0.149
8 - 17	9600.000	0.200	153940.225	5943963.589	0.075
9 - 14	8525.000	0.200	142464.768	10488670.454	0.133
9 - 17	12684.000	0.200	156385.776	17798375.337	0.226
10 - 11	6127.000	0.200	105444.358	4272661.374	0.054
10 - 14	4173.000	0.200	318045.318	2299419.999	0.029
10 - 18	17636.000	0.300	124994.751	23814037.862	0.302
11 - 13	22500.000	0.200	109832.078	12114746.575	0.154
11 - 18	2741.000	0.200	580618.501	2919854.799	0.037
12 - 13	16089.000	0.350	130556.205	16043771.238	0.204
12 - 15	11245.000	0.350	151665.537	14687334.610	0.186
12 - 18	9200.000	0.600	141033.613	21131023.445	0.268
13 - 18	4729.000	0.200	129382.471	3586581.485	0.046
14 - 15	3647.000	0.200	162606.619	4478013.621	0.057
14 - 16	7747.000	0.600	153225.182	42264197.828	0.537
14 - 17	18611.000	0.500	178010.519	78760557.246	1.000
14 - 18	14830.000	0.200	151974.695	18973840.167	0.241
15 - 16	20334.000	0.200	151742.291	20108332.455	0.255
16 - 17	13905.000	0.600	167146.191	62825084.073	0.798

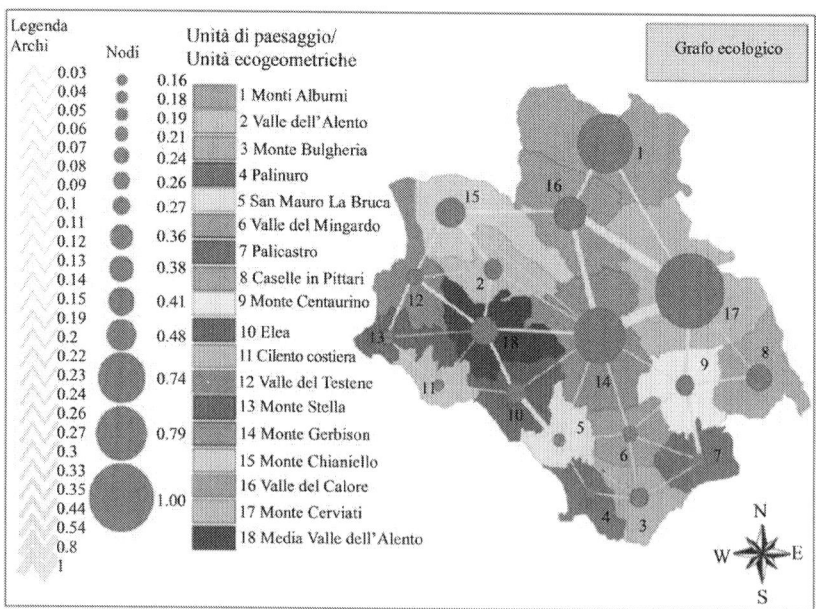

Figure 2. Ecological graph for the national park of Cilento.

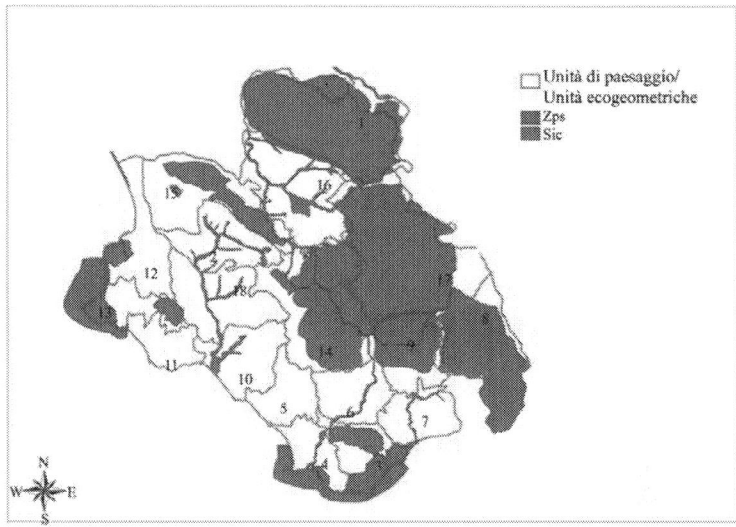

Figure 3. SCI (SIC) and SPAs (ZPS) areas in parco nationale del Cilento.

3. ECOLOGICAL NETWORK

As shown by the ecological graph in Figure 2, the studied area suffers of reduction and fragmentation of natural and semi-natural habitats as outcome of agricultural intensification, infrastructure networks and urbanization, even if it is a place of several SCI (Site of Community Importance) and SPAs (Special Protection Areas), as one can see from Figure 3. Recently, eco-regional planning is playing an increasingly important role on the acknowledgment that it is necessary to integrate, from both ecologically and socio-ecologically points of view, the protected areas in the landscape matrix of the entire territory. Hence the origin of ecological networks, characterized by their emphasis on biodiversity conservation at level of region.

The first step consists in creating a resistance map, that is, a map of resistance of the landscape matrix to the mobility of the selected species. In the language of ecological network the zones with the lowest value of resistance play the role of core areas. Then, the least-cost paths linking such areas, i.e., the potential paths that minimize the cost of mobility, are computed ([6]). They represent the corridors of the ecological network that could be adopted as reference information in environmental evaluation of plans and projects in order to reduce as much as possible territorial fragmentation.

The design of corridors for integration in eco-regional planning demands to know the mobility requirements of certain target animal species with rather wide mobility ranges. In our Study Case we have chosen the Peregrine Falcon that lives on the cliffs of the coastal zone of Cilento and is renowned for its speed, reaching over 325 km/h during is hunting. The Peregrine Falcon requires open spaces in order to hunt, he often hunts over open water, marshes, valleys, fields, and tundra, searching for prey either from a high perch or from the air. While its diet consists almost exclusively of medium-sized birds, the Peregrine will occasionally hunt small mammals, small reptiles, or even insects. Hence, we can say that its living space covers all over the Park.

In order to use the least-cost path algorithm, from GIS technology, we need to relate the specific ecological requirements of the chosen species to land uses of the Cilento Park, through a specific parameter called resistance that measures the degree of environmental opposetion to its spread and colonization.

The areas characterized by low value of resistance (resistance equal to zero to the most suitable area) are considered core areas, i.e., natural areas with ecologically high value, for the potential ecological network specific to the species (they could be protected areas too). By using the GIS software used for the graph's construction we have built the map of resistance ([6]), as shown in Figure 4.

The resistance for the Peregrine Falcon runs from a zero value assigned to bare rocks, cliffs, ponds and meadows, to 60 value for woods, to 90 value for urban areas, rail and road networks. We then apply the PATHMATRIX tool ([7]), an implementation of the least-cost distance algorithm of the GIS software ArcVIEW 3.x, that is able to apply the cost distance algorithm in pair wise fashion among a set of sample locations. PATHMATRIX can also output the length of the least-cost path in geographical distance units. We recall that the least cost path minimizes the sum of resistances along the path (see Figure 5), hence, in our case, it corresponds to the corridor that minimizes the cost of mobility of the target species between the core areas, as shown in Figure 6.

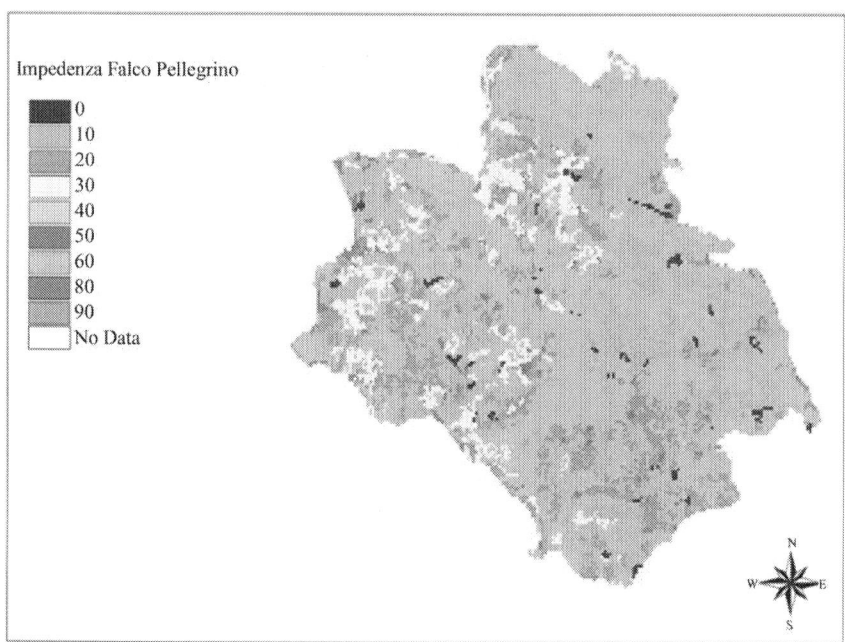

Figure 4. Resistance map for peregrine falcon in the national park of Cilento.

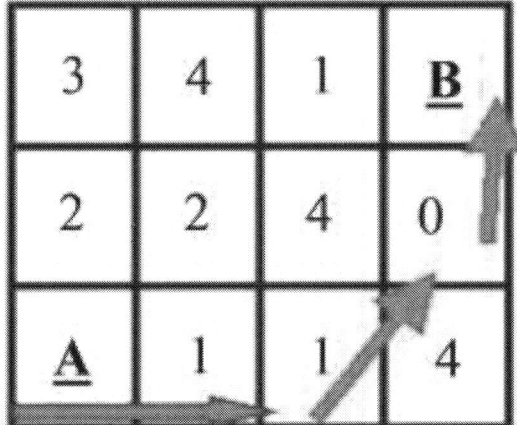

Figure 5. The least cost path minimizes the sum of resistances in going from A to B.

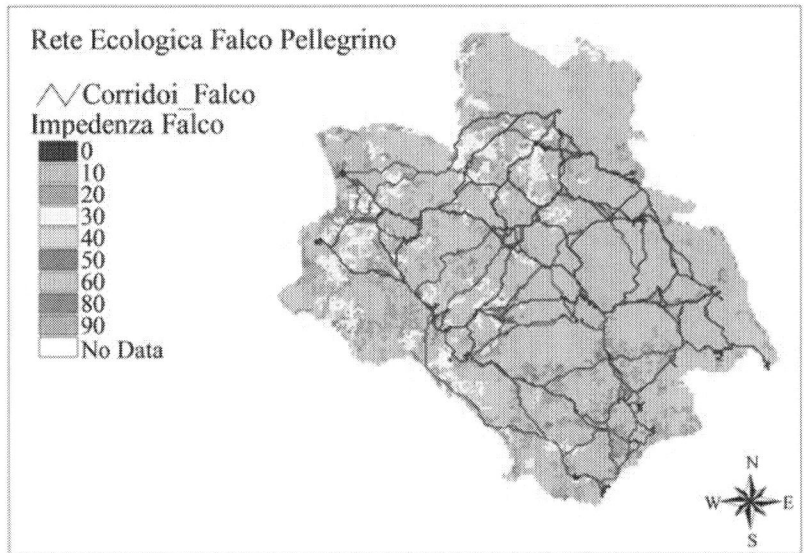

Figure 6. Potential ecological network for peregrine falcon.

The network suggests that along the corridors it would be better to avoid installing equipment in proximity of wetlands, places of wintering of waterfowl and frequented by several species of birds of prey. Also, not installing power lines or wind farms, sources of major impact in the fast flight of Peregrine Falcon. Moreover, rock climbing activity can have negative impact on species whose life is linked to cliffs (nesting, roosting food). Also agricultural activities that impact on the conservation of wildlife could pauperize Falcon's hunting.

These are some of the considerations coming from a rapid analysis of the ecological network that, hence, can be rightly considered as another decision-support tool in environmental planning and landscape management.

4. LOGISTIC EQUATION WITH HARVESTING

As said in the Introduction, the Ecological Graph furnishes the actual state of energy exchange in the territory, hence, it would be interesting to investigate the time evolution of energy, starting from the actual settlement, in order to get information on possible future scenarios and check if the trend is toward a sustainable development of the territory or not. This can be made by studying the equilibrium solution of a suitable differential equation that models dynamics of the territory evolution under an ecological point of view. As said in the Introduction, the term "bioenergy" refers to the energy available in the environment, present in different forms like animals, seeds, plants, ruled by suitable metabolic processes, hence, its value must be limited by some carrying-capacity of the given environment. Then, we propose a simulation model based on a logistic-type differential equation ([8]) like that one approximating the evolution of population over time in presence of limited living resources of the environment. Moreover we add a harvesting term in order to simulate the growth of bioenergy over a landscape in spite of the obstacles coming from territory fragmentation.

Namely, if we denote by M(t) the average value of the bioenergy (see Formula (2)) over all the n Landscape Units constituting the entire system under study:

$$M(t) = \frac{1}{n} \sum_{j=1}^{n} M_j$$

the dynamical simulation model is given by the following nonlinear differential equation

$$M'(t) = cM(t)\left[1 - M(t)/M_{max}\right] - hS_o$$

(4)

where M_{max}, the maximum value of bioenergy the given territory can produce, is given by formula (2) of Section 2 and the connectivity index, c is defined ([8]) by

$$c = \frac{1}{V} \sum_{s=1}^{V} \frac{F_s}{\max_s F_s}$$

with V the number of arcs present in the graph constructed for the given territory, F_s, $s = 1, \cdots, V$, are the values of energy fluxes given by Formula (3).

In Formula (4), the prime indicates the time derivative, t the time variable, the harvesting term, $-hS_o$, is given by the product of h, the ratio between the sum of the impermeable barrier lengths and the total external perimeter of the territory, and S_o, the ratio between the sum of the territory surfaces with low values of B_{TC} and the total surface of the system.

Let us note that, from its definition, the connectivity c represents the territorial ability to spread the bioenergy, furnishing, hence, a measure of territorial fragmentation (the flux F_S through an impermeable barrier is equal to zero). It plays the role of the constant growth rate as in population dynamics.

By using the normalized bioenergy $M(t) = M(t)/M_{max}$, the time evolution equation for M becomes:

$$M'(t) = cM(t)[1 - M(t)] - hS_o \tag{5}$$

In the application of this model to a Brownfield at the East zone of Naples (Figure 7), subject to a Master Plan in the direction of improving the environment, S_o will represent the percentage of edified areas, while h will be the ratio between the sum of the perimeters of edified areas and the total perimeter of the area ([9]).

With a similar procedure as before we can construct the ecological graph for this area. We shall not give the details but, instead, we shall provide the main and significant results coming from the mathematical approach based on the study of the equilibrium solutions of Equation (5).

We outline that the mathematical model basic assumption of Equation (5) is that the time evolution of bioenergy will consist of the balance between two quantities with opposite signs. The first one, positive, describes the bioenergy growth by following a logistic law, driven by the connectivity parameter c; the second one, negative, $-hS_o$, the harvesting term, opposes to bioenergy growth due to the presence of barriers related to the edified areas that hamper the flux of energy.

Note that, once subdivided the territory in units as in Section 2, from the relative ecological graphs, as those in Figures 8 and 9, we see that in the project plan the diameters of the nodes, together with the width and the number of the arcs, are increased in line with the planned environmental improvements.

Figure 7. Brownfield-east zone of Naples.

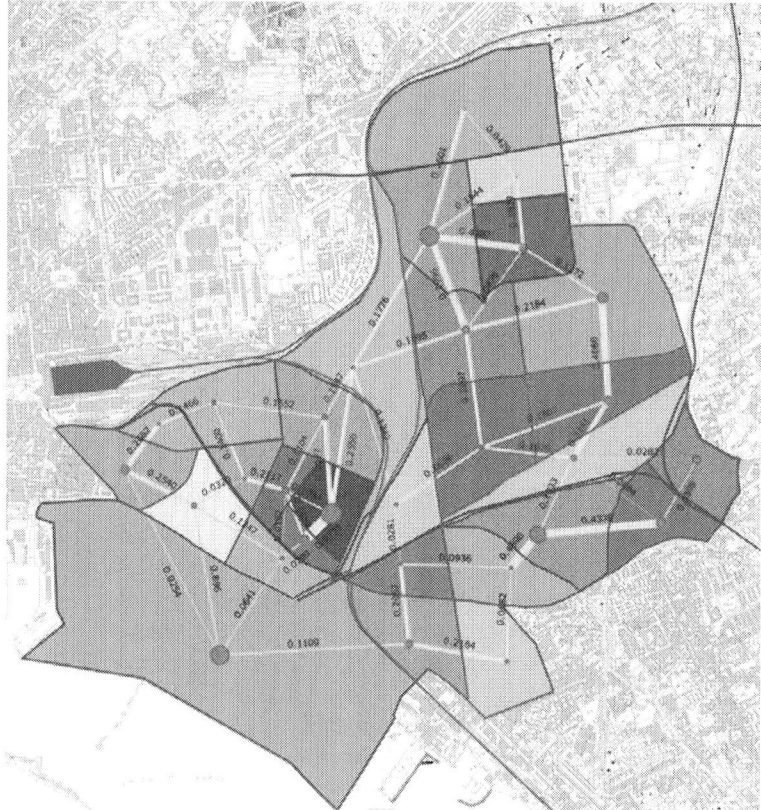

Figure 8. Ecological graph of brownfield actual state.

Figure 9. Master plan project ecological graph.

This happens because between the two situations there will be different typologies of intended use of the ground and barriers that will make changes in the analysis and in the calculation of the environmental indices (the Master Plan foresees in fact, among other, the presence of a Urban Park). As a consequence the initial value of M, M(0), as well as the values of c, h and S_o will change.

If we now turn our attention to the graphs of the solutions ([10]) to the differential Equation (5) of Figure 10, we can see three different possible future sceneries: the first one shows that, starting from the actual state of the area, i.e. with initial value of bioenergy M(0) = 0.048, c = 0.58, h = 0.28, S_o = 0.79, there is a quick trend to the environmental collapse corresponding to M = 0; in the second one, starting from the project plan, M(0) = 0.086, c = 0.65, h = 0.069, S_o = 0.66, the value of the bioenergy grows visibly, due to the environmental improvements given by the interventions in line with the Master Plan and tends to a stable good value; the third one furnishes a critical value of the parameter, h = 0.08, at which the system tends to collapse even if it starts from the project value of bioenergy, showing the important role played by the geometrical configuration of impermeable barriers of the buildings.

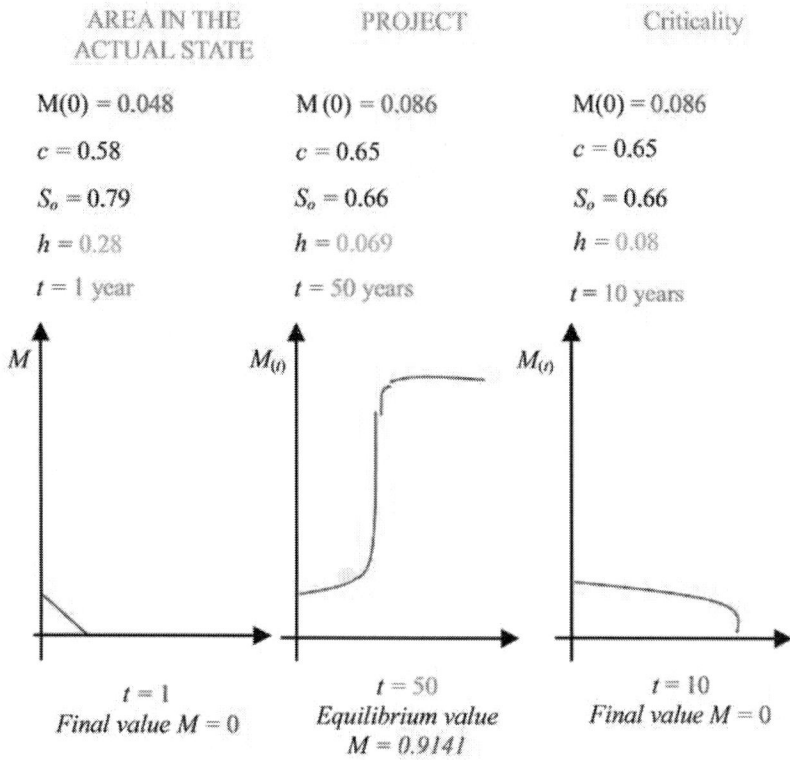

AREA IN THE ACTUAL STATE	PROJECT	Criticality
M(0) = 0.048	M(0) = 0.086	M(0) = 0.086
$c = 0.58$	$c = 0.65$	$c = 0.65$
$S_o = 0.79$	$S_o = 0.66$	$S_o = 0.66$
$h = 0.28$	$h = 0.069$	$h = 0.08$
$t = 1$ year	$t = 50$ years	$t = 10$ years

$t = 1$ Final value $M = 0$	$t = 50$ Equilibrium value $M = 0.9141$	$t = 10$ Final value $M = 0$

Figure 10. Solution graphs.

We would like to stress that the simple simulation models here presented do not pretend to perform quantitative predictions, but to estimate the goodness of a territorial plan, getting an insight to possible criticalities and hence serving as a decision support in sustainable environmental planning and landscape management.

5. CONCLUSIONS

Strategic Environmental Assessment, nowadays, has been adopted in Europe in landscape planning, whose task is to verify the compatibility of territory transformations with respect to their levels of criticality and vulnerability, to evaluate possible future scenarios as consequence of interventions by checking if they are in line with presservation and valorization of environmental quality.

This evaluation must be based, hence, on the knowledge of the mechanisms that rule a territorial transformation. In order to assess the ecological functioning of an environmental system is necessary to single out the energetic contents of the units composing the territory, their connections carrying energy and material fluxes, as well as their breaking points, due to the presence of impermeable barriers, that produce territory fragmentation.

We have presented three different simulation models in the framework of Landscape Ecology, all of them based on GIS technology, which can be used as decision support in environmental planning, such as:

1) **Planar graph**, the so called ecological graph, whose construction needs the computation of suitable indices of environmental control, proper of Landscape Ecology, such as biodiversity, Biological Territorial Capacity, connectivity. The planar graph for the considered environmental system returns a picture of its actual ecological health conditions and provides very detailed indications and operational assistance to guide toward ecological sustainable interventions.

2) **Ecological network**, based on the resistance map of target animal species, that gives useful information to prevent and to reduce territorial fragmentation caused by intense processes of urbanisation and industrialisation

3) A mathematical model based on nonlinear differential equation of logistic-type with harvesting used to investigate the time evolution of the ecological value of a given territory, by starting from a given settlement and looking for the trend to consequent future scenarios, hence, furnishing qualitative predictions on the sustainability of a given territorial plan; a recent implementation of this model can be found in [11,12].

These mathematical and GIS interfaced models can help in understanding environment response and dynamic change in time to correctly manage and preserve natural resources; they can represent a powerful decision support to compare effects and impacts of possible alternative future scenarios.

6. ACKNOWLEDGEMENTS

The author wishes to acknowledge the support given by the Department of Industrial and Information Engineering of the Second University of Naples.

REFERENCES

1. R. T. T. Forman, "Land Mosaics. The Ecology of Landscape and Regions," Cambridge Press, Cambridge, 1995.
2. A. Farina A., "Ecologia del Paesaggio," UTET Libreria, Torino, 2001.
3. V. Ingegnoli, "Landscape Ecology: A Widening Foundation," Springer-Verlag, New York-Berlin, 2002.
4. P. Fabbri, "Paesaggio, Pianificazione, Sostenibilità," Alinea Editrice, Firenze 2003.
5. G. Lauro and R. De Martino, "Environmental assessment of the meta-ecosystem Cilento with the tools of Landscape Ecology," In: C. Gambardella, Ed., Atlante del Cilento, ESI, Napoli, 2009, pp. 533-538.

6.   G. Lauro and R. De Martino, "The Ecological Network as a Tool for the Protection of Coastal Ecosystems. Workshop: The Mediterranean Coastal Monitoring: Issues and Measure Techniques," Livorno, 15-16-17 Giugno 2010, Firenze: CNR-IBIMET, pp. 163-170.

7.   N. Ray, "PATHMATRIX: A Geographical Information System Tool to Compute Effective Distances among Samples," Molecolar Ecology Notes, Wiley Online Library, 2005.

8.   G. Lauro, R. Monaco and G. Servente, "A Model for the Evolution of Bioenergy in an Environmental System," In: T. Ruggeri and M. Sammartino, Eds., Asymptotic Methods in Nonlinear Wave Phenomena, World Scientific, 2007, pp. 96-106.

9.   G. Lauro and M. Musto, "Simulation Models for Environmental Control," Proceedings of Sixth EAAE-ENSHA Construction Teachers' Network Workshop, Mons, 22-24 November 2007, pp. 178-185.

10.   A. D. Bazykin, "Nonlinear Dynamics of Interacting Populations," World Scientific, River Edge, 1998.

11.   F. Gobattoni, R. Pelorosso, G. Lauro, A. Leone and R. Monaco, "A Procedure for Mathematical Analysis of Landscape Evolution and Equilibrium Scenarios Assessment," Landscape and Urban Planning, Vol. 103, No. 3, 2011, pp. 289-302.

12.   F. Gobattoni, G. Lauro, R. Monaco and R. Pelorosso, "Mathematical Models in Landscape Ecology: Stability Analysis and Numerical Tests," Acta Applicandae Mathematicae, Vol. 125, No. 1, 2013, pp. 173-192.

CHAPTER 6

Inventory and Analysis of the Landscape

Murat Özyavuz

¹ Namık Kemal University, Faculty of Fine Arts, Design and Architect, Department of Landscape Architecture, Turkey

1. INTRODUCTION

Landscape planning, often referred to as 'environmental planning' or 'ecological planning', is a way of directing or managing changes in the landscape so that human actions are in tune with nature and environment (Zaucer and Golobic, 2010).

Landscape planning is undergoing change due to new requirements. Its previous main task of controlling spatial uses and the development of nature and the landscape has extended. Implementation of the European requirements for the Natura 2000 network, for the Water Framework Directive (WFD), the Floods Directive as well as the Strategic Environmental Assessment (SEA) can be made considerably easier and can be coordinated with the help of landscape planning. It is ideally suited, for example, as the basis of the Strategic Environmental Assessment or as an extensive information base for river basin planning covering all natural resources. In addition, landscape planning increasingly supports the tasks of providing members of the public with environmental information and their participation in sustainable local community and landscape planning (Haaren, et al., 2008).

Landscape planning can be better used as a versatilely usable information basis for overall spatial planning, impact mitigation regulation or environmental assessments if the information is presented according to the requirements of this planning and these instruments. Spatial planning has always been regarded as an activity that, in seeking solutions for a certain social problem, brings a change into the territory. It primarily looks after social needs, the economical use of the resource and its fertility. Because spatial planning is essentially economic in orientation, it is generally characterized as developmental planning (Zaucer and Golobic, 2010). The basic distinction between spatial planning and landscape planning is that the former is essentially economic and developmental in orientation, while the latter is more concerned with environmental and landscape qualities and thus protective in orientation. It must be noted that landscape planning does not represent a substitute for spatial planning. With developed approaches and methods, it may complement a set of spatial planning approaches and methods and contribute to larger efficiency of bottom-up comprehensive planning (Zaucer and Golobic, 2010). The aim of landscape planning is thus to prevent or at least limit the degradation of the environment to a minimum while increasing, as far as possible, 'creativity' in order to meet the developmental needs. By combining approaches from the natural sciences and the planning disciplines landscape planning has developed a range of different methods and tools for integration of environmental objectives into the process of analysis and the development of planning proposals. Landscape planning approaches and methods are transparent and systematic, which makes them useful for participatory and comprehensive spatial planning (Zaucer and Golobic, 2010).

Landscape planning is currently developing away from rigid planning to a generally accessible and easy to update information base and a basis for action. By using new data processing and transfer technologies, landscape planning can be developed into an information and communication platform, which also communicates data and knowledge about nature and the landscape to the public and makes simple consultation and participation possibilities available via the internet. This development is borne by a new understanding of government action which is characterised by more proximity to citizens and transparency in politics and the administration. Landscape planning therefore supports implementation of Agenda 21 as well as the objectives of the Aarhus Convention and the associated EU Directives and federal laws for the introduction of more democracy in environmental issues. It promotes the enlightenment of members of the public and businesses as well as their commitment to their environment and homeland. Because landscape planning creates fundamentals, competence and incentives for own initiatives, resourcefulness and commitment to the integration of environmental aspects in landscape usage (Haaren, et al., 2008).

Landscape planning is also a way to effectively include the environmental requirements of different sectors into planning process. One of the most valuable approaches is vulnerability analysis, where environmental qualities are assessed from the viewpoint of potential threat resulting from planned actions. It functions as an integrating and conflict-solving tool, since it (Zaucer and Golobic, 2010):

o   includes a whole territory (of a chosen administrative / planning unit),
o   considers a whole range of diverse environmental components,
o   supports active dialog between stakeholders,
o   may embrace all interests (natural, social, economic, and political) and evaluate their consequences and thus supports crosssectoral or comprehensive planning, and
o   supports search for an optimal solution.
o

Landscape planning today should not be viewed as a static plan but as a dynamic, continuously or modularly changeable information and working basis.

Landscape planning is expected to be need- and problem-oriented. Against the background of fast changes in use of nature and landscape, these requirements are becoming increasingly important (Haaren, et al., 2008). Landscape planning, design and management are practised directly or indirectly by many others and in many sectors, including land use planning, agriculture, forestry, nature conservation, amenity land management, and so on, and we include all these in our approach. The term 'landscape' used here is also broad and includes much more than 'the appearance of the area of land which the eye can see at once' (Chambers, 1993). Landscape is an evolving cross-disciplinary area, which draws contributions from art, literature, ecology, geography and much more (Benson and London, 2000). As the legal basis of the various planning and instruments partly name natural resources as sensitive receptors, partly landscape functions, it should be possible to access landscape planning information structured both by natural resources and landscape functions.

Landscape-ecological planning is a specialization within landscape planning that focuses on spatial planning, the organization of uses and relationships of land uses to achieve explicit goals (e.g. habitat improvement, sustainability). While the landscapeecological planning approach is characterized by a focus on the linkage of ecological patterns and processes, it also includes the actions and values of humans, and social and economic dimensions (Hersperger 1994). Finally, landscape-ecological planning adopts the landscape as the principle spatial unit of research and planning recommendations (Ahern, 2006).

Landscape analysis has a significant function in the process of decision making for the future land use, organization of the space, nature protection, and rational use of the nature resources. Basic problems and tasks of the landscape analysis and planning are located in discovering and solving the conflicts among the development of the society and very complex task in nature assignment. The development covers more intensive engagement of space and more intensive land use, organization and arrangement. The features of the land and space with their entire natural and produced substratum are the significant categories for determination of future development. In this context, landscape analysis and planning appear as an activity of primary task to connect developing possibilities and tendencies for certain space (Pecova, 2000). These strategies, in essence, define the planning context with respect to the macro-drivers of change in a given landscape and the strategic nature of the planners' response. Defining these strategies also helps to place the planning activity within a broader context, which is particularly important when planning methods are transferred or adopted for use in different locations, contexts or for different applications (Ahern, 2006). Landscape planning methods can also be classified and understood according to their strategic orientation: protective, defensive, offensive or opportunistic (Ahern 1995). Planning methods can also be understood and classified according to their resource or goal orientation. The abiotic–biotic–cultural (ABC) model is useful to describe the specific goals addressed in planning and the level of integration between these goals (Ahern, 1995) (Figure 1). In this model, abiotic goals include water resources, soil and air quality. Biotic goals focus on biodiversity in general, including individual species and habitat protection and ecological restoration. Cultural goals are human-based and include: transportation, land use, recreation, historic preservation and economic goals (Ahern, 2006).

Figure 1. presents an array of planning types graphically organized within a triangle that represents the ABC model. In this diagram a number of planning sectors or themes are located according to their emphasis and level of integration within the ABC resources. The figure shows that an evolution is occurring towards a more integrated planning perspective as represented by the central circle (Ahern, 2006).

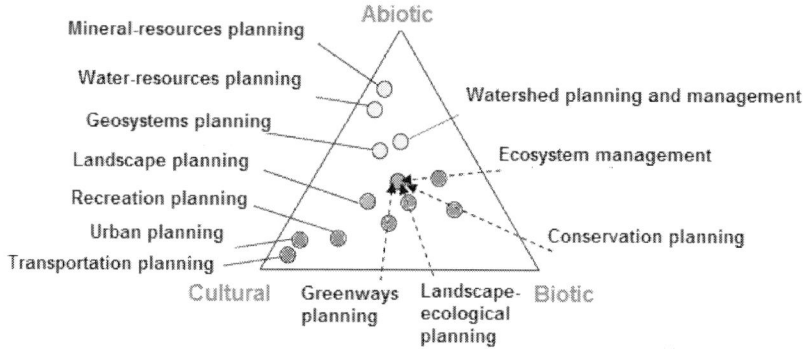

Figure 1. The abiotic, biotic and cultural resource-planning continuum (Ahern, 2006).

The combination of different landscape factors and the interactions between the natural resources are also significant for performance and ability to function. Apart from the description and assessment of the landscape functions (current condition and development potential), statements for specific areas are made on the sensitivity of distinct (sub)landscapes to impacts as well as on the ability to restore their performance and functional capability (Haaren, et al., 2008). This can be achieved by appropriate linking of standard text units in digital landscape plans. It is also advisable not only to be able to select specific individual cartographic areas but also to easily find and collate text statements on special landscape units. This service will assist the administrations or project sponsors in consolidating the respective relevant area descriptions for comments or environmental assessments. The effort spent on these preparations pay off because the information is so conveniently accessible and is easier to integrate in other planning and instruments (Haaren, et al., 2008). In landscape planning the existing condition of nature and the landscape is determined and assessed on the basis of legal and functional objectives and standards, which also include landscape planning objectives at a higher level. To this end the available data and information is collated and, where necessary, it is supplemented and updated by additional surveys. The fundamental information on the soils, geology, bodies of water, air and climate, fauna and flora is used to deduce statements regarding the performance and functions of the individual natural resources and/or the balance of nature and the landscape overall (Haaren, et al., 2008).

2. INVENTORY AND ANALYSIS OF THE LANDSCAPE

2.1. Terrain analysis

Topographical maps of Greek origin Topos (place) and graphics (lines formed figure) is formed from the word, used for lines shows created in the forms. These are the natural and cultural aspects of the land, the horizontal and vertical cases, show a horizontal plane and under a certain scale. Soil maps and other maps are used in making topographic maps as a base map (Figure 2). Soil experts determine the elevation and slope curves, positions the most benefit from relief.. *Physiography* deals with the physical conditions of the surface of the land. The broad physiography of an area can be determined by the knowledge of the physiographic region in which it lies. The important aspects of physiography are elevation and slope. Slope, soils, geology, hydrology, microclimate, plants, and animals may be strongly related to elevation. This means that elevation is an important feature in analyzing landscapes. Topographic contour shape on the map, the location, frequency, taking into account in an area of alluvial fans, colluvial skirt lands, valleys, mountainous-hilly areas, old terraces, alluvial flood plains, river paths, meanders, lagoons and sand dunes landforms such as defined. (Figure) Any contour of the land surface structure identifies the most understandable form. Contour curves, in order to produce a realistic and reliable planning decisions important.

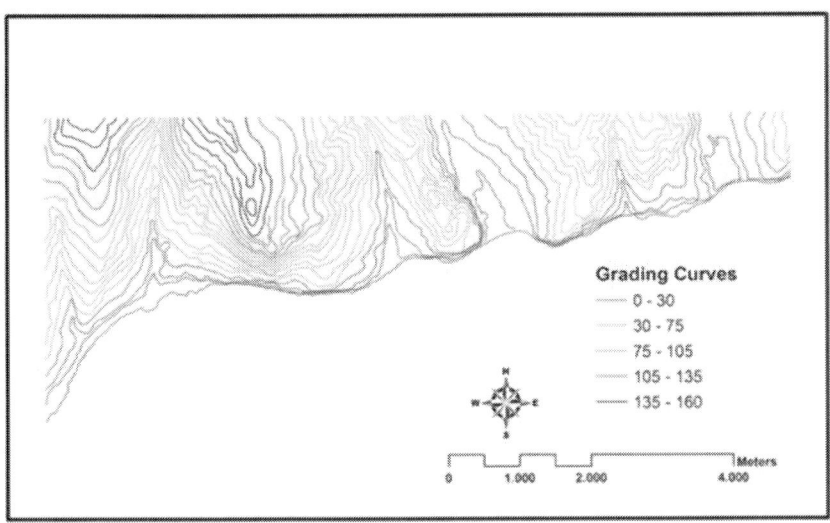

Figure 2. Grading curves for an example area

First of all, check whether the map is up to date. If the current deficiencies detected during the observations, can be added to the map. Or new information obtained from different sources (water sources, wells, dams, ponds, bridges, new uses, etc.) can be placed on the map. Horizontal and vertical cross-section for the map to be more easily interpreted. Thus, the difference in level between the high and low parts of the land surface structure and can be detected (hills, peaks, valleys, valley bottoms, slopes) (Figure 3). This is due to differences in harmful situations, measures to prevent them must be determined. The base terrain, steep slopes, private entities, such as delta, flood deposits, flood areas identified as sensitive areas. The cross sections obtained terrain, concave (depressions, valleys, depressions and so on.) And convex (ridges, hills) allows to perceive how it has developed to take shape. It is also important heights, large and small valleys, flood beds and borders determined and displayed. Accordingly, both precipitation and drainage of surface water streams flow distribution to be determined. Surface water flow intensive - is less dense areas should be considered.

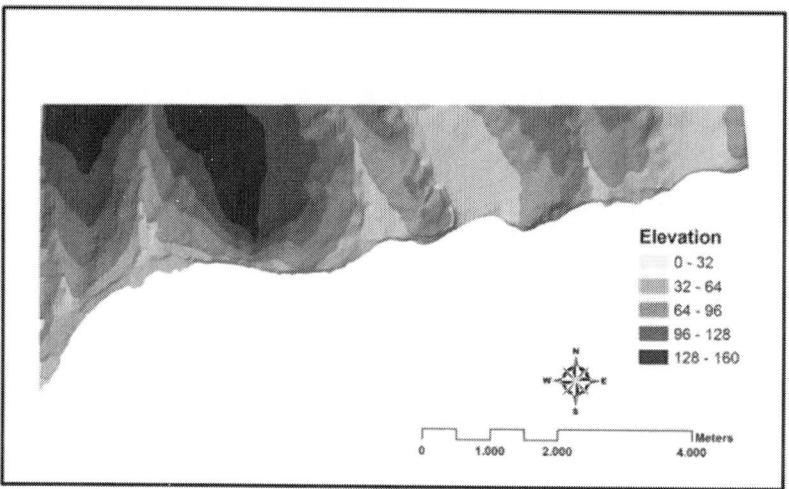

Figure 3. Elevation for an area

According to Kienast, 1993; for many countries topographic maps in the scale of 1:20'000 to 1:30'000 provide reliable monitoring data over the last 100 to 150 yrs (Hodgson and Alexander, 1990). Additionally, there are often rich collections of detailed plans in larger scales (1 :5000 to 1500) that date back to the late 18th and the early 19th century (Di.int, 1990; Miiller, 1990). For the present study the main data sources are two series of topographic maps in the scale of 1:25 '000 dating back to the 1880's (Figure 4). Since topographic maps are perceptions of the environment and often rather a 'text than a mirror of reality' (Harley, 1989), definition of specific landscape elements or habitat types may vary considerably over time and the location of features is often less accurate in earlier map editions compared with today's standards (Hodgson and Alexander, 1990). Some of these misinterpretations are impossible to correct.

However with a strict crossdating and a comparison of the historic maps with data from independent sources, most incompatibilities were eliminated. This was accomplished by consulting the written protocols and field notes of the topographers.

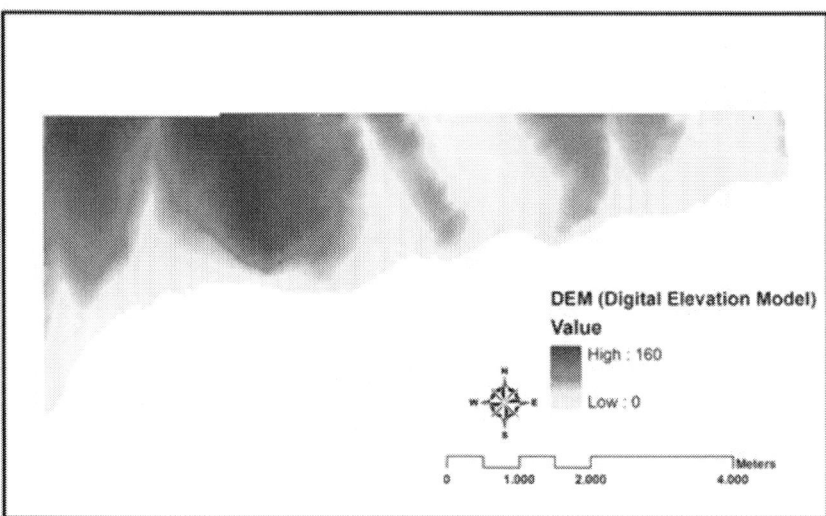

Figure 4. Dijital elevation model (for spatial analyst)

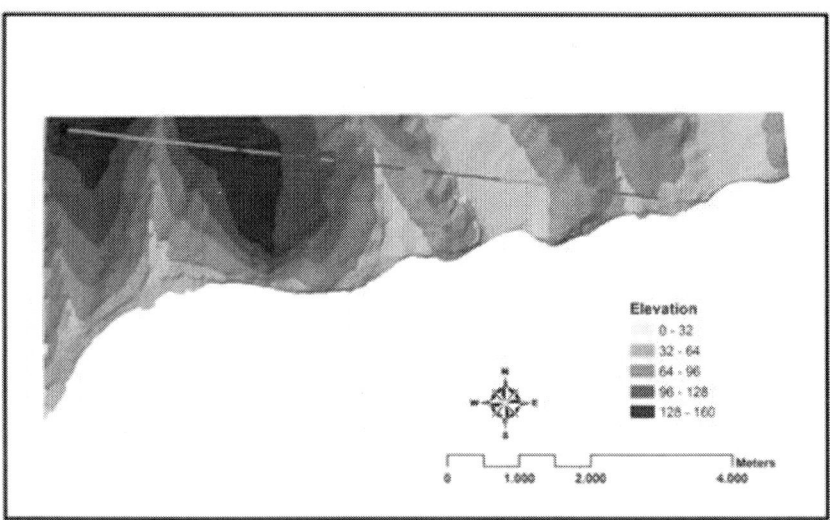

Figure 5. Visibility analysis (observer offset:120 m. and target offset:100 m.)

Elevation maps are easily constructed by selecting intervals from the base maps. Altitudes can be represented by coloring spaces between topographic intervals. Elevation changes are depicted in shades of browns, yellows, or grays with felt markers, colored pencils, crayons, or through the use of computer technology, becoming lighter or darker as elevation increases

In addition, topographic maps allow visibility analysis is made from.Visibility analysis based on viewsheds is one of the most frequently used GIS analysis. This analyst ise used. This analysis is used in many studies of landscape architecture (Figure 5).

Contour curves are some of the features (Karadeniz, 2010):

- All points on a contour curve above are the same height above sea level,
- Each contour curve closes in on itself. Even with the edge of the map, the map will continue and close neighbors,
- A closed contour curve shows the peak or pit.
- Contour curves do not cut each other.
- The slope increases, the contour lines pass as close to each other
- In the cross-sections in order to compare the characteristic profile shapes with each other is possible to make topographic maps. (Figure
- Contour maps of curves that can be obtained; Height groups, slope, aspect maps (except that 3-D images);
- Height Groups
- Curved contour map is obtained by the separation of certain groups heights (Figure 6).

2.2. Slope

Slopes may be subdivided according to steepness and direction. Slope direction is referred to as *aspect,* or orientation (Figure 7). Steepness may be important for such activities as agriculture or the construction of buildings, while the direction of slopes is an important factor for such activities as siting housing for solar energy collection. Slope composition and related *lithology* needs to be determined. Lithology is "the soil and rock material that comprise a slope" (Marsh 1998, 80) or the physical characteristics of sedimentary materials. As with the elevation map, the division of slope categories will depend on the study.On the map using the horizontal and vertical distances are obtained and analyzed in separate groups of different units (Kienast, 1993).

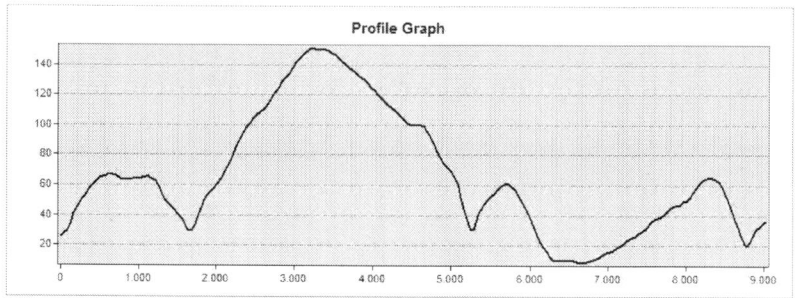

Figure 6. Land profile graph (made by DEM)

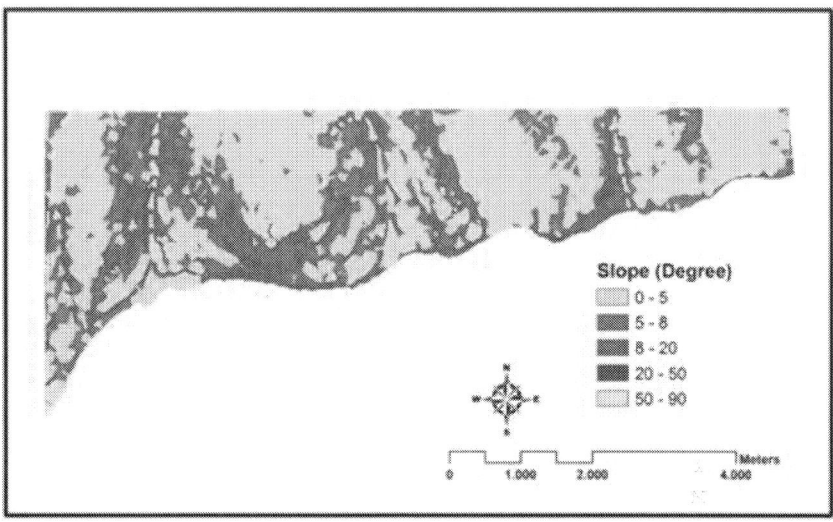

Figure 7. Slope analysis

2.3. Aspect

Aspect map, facing at the surface indicates in which direction. Aspect of a landscape, the climate, especially temperature and rainfall affect the amount of that place. In general, the S, SE, SW, and W aspect "sunny aspects,» is called. N, NE, NW, E aspects, "the shady aspects,» has been described as (Figure 8). The shaded look still water evaporation from the soil temperature is less than would be even less. Therefore, the shady aspects, the same rainfall conditions, sunny look still have a more favorable water economy.

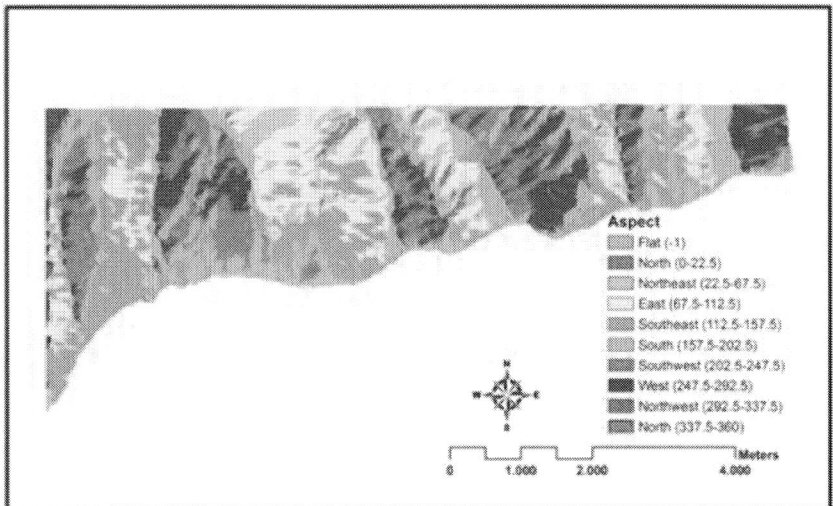

Figure 8. Aspect analysis

2.4. Climate analysis

Climate is the set of meteorological conditions characteristic of an area over a given length of time. It is defined as the study of extremes and long-term means of weather. The regional (or macro) climate is the big picture, the meteorological conditions and patterns over a large area. Macroclimate is affected by physical conditions such as mountains, ocean currents, prevailing winds, and latitude. It in turn affects the formation of the physiographic region through the weathering of the terrain and the amounts of precipitation that fall on the landscape. Climate states the average of the air for a long time in a specific place. Macro, meso and micro are the 3 types of climate. Long-term measurements, temperature, precipitation, air humidity and air movement and extreme values determined for the average for the region, which is characterized by a wide range of climate macroclimate is called. Climate characteristics of macro-states of different characteristics of land in the air near the surface have been described as microclimate. Shape of the face of the land, such as elevation and aspect Relief for small spaces created by the characteristics of the type of climate is called the mesoclimate. This is located between the macro and microclimate. Plant growth and biological analyzes to people about the climate are crucial in terms of comfort, they are temperature and precipitation analysis. Relative humidity, time of sun and cloud cover, wind direction and nature, frost days, and etc. factors have a significant effect on the bioclimatic comfort. Precipitation is the most important factor limiting regeneration of vegetation. The total annual rainfall is 600 mm but not enough for the formation of a local forest as a result of research that indicated

Heat and temperature are very different concepts. Despite the potential value of objects based on the heat, the temperature of this value emerged as a kinetic form. Therefore, the air temperature is not heat.

Bioclimatic limits of comfort with the body temperature of 15 ° C between 37 ° C ambient unless otherwise stated. Temperature of 20-24 C in the summer, optimum indoor temperature of 21 C values of the optimal bioclimatic comfort. At night, the relative humidity of 40-70% and 18-24 C temperature limit values.

Climatic elements in preparing the vegetation ecological environment, climate, type of vegetation and animals living in the rural and urban areas are the limiting factor.

Analysis, planning to put up a field of plant breeding the human race in the comfort of home to learn about the conditions to create a suitable environment is important for people in planning will be done.

Therefore, the climatic comfort of people trying to manifestation of the analysis depending on the season. Bioclimatic comfort level and the amount of which should be of climatic elements allowing research on human beings endure as a medium between the lowest and highest temperatures bioclimatic comfort limits.

Meteorological elements change vertically and horizontally within short distances. Small-scale variations are brought about by changes in slope and orientation of the ground surface; soil type and moisture; variations in rock, vegetation type and height; and human-made features. These different climates found within a small space are grouped together under the general description of *microclimate.* The term *topoclimate* is used when the effects of topographic variations of the land on the microclimate are considered. Generally, topoclimate is an extension of microclimate into the higher layers of the atmosphere and over landscapes, depending on the relief of the land. Therefore, topoclimate can be considered to occupy an intermediate level between macroclimate and microclimate. It is important to understand microclimate and topoclimate for many of the same reasons that macro, or regional, climate is important. These finer layers, however, relate more directly to building and open space design (Kienast, 1993).

2.5. Geology

Geology is the study of the earth. This study involves both what has happened in the past, or geological history, and what is happening on and in the earth today.

This analysis was prepared by institutions engaged in geological studies consists of the interpretation of maps and reports. As a result of this review maps and reports, forms the bedrock of physical and chemical state of the soil, water retention properties of the mass due to the hydrogeological characteristics of sedimentary or igneous efficiency, earthquake status (seismicity), the main rock formations are learned.

The inventory of a place requires an understanding of the geological history and processes of the region. Such understanding can begin with a *geologicalmap,* which is "a graphic representation of the rock units and geological features that are exposed on the surface of the Earth. Accordance with the determination of the geological structure of the area in which such use is put forward. Functional characteristics of a field analysis of geological bedrock exhibits the aesthetic features can be evaluated. Underground long and wide fissures (fault lines) goes on and on. These lines are extremely sensitive to the occurrence of the earthquake areas. These areas should be on residential areas, more use should be included in nature conservation purposes. In addition to showing different types and ages of rocks, most geo- logical maps depict geologic features, such as faults, folds, and volcanoes. The relative timing of events can usually be determined from a geologic map, usually by the application of these principles: superposition (younger layers are above older ones); original horizontality (layers formed from deposition of sediment were originally flat); crosscutting relationships (younger features crosscut older ones; a fault will be younger than the layers or contacts it cuts); and inclusions (a rock unit is younger than the layers from which the inclusions it contains came).

Summary of geologic inventory elements; regional geographic history, depth to bedrock, outcrops, bedrock types and characteristics, cross sections, columnar sections, surficial deposits (regolith): kames, kettles, eskers, moraines, drift and till, mineral resources, major fault lines, earthquake zones, and seismic activity, rock slides and mud slides (Kienast, 1993).

2.6. Geomorphological analysis

Geomorphological structure, depending on the geological structure of different geological times (Era), composed of the structure of the land. Geological times, the base of the valley bottom and the plain plains, alluvial cones, low and high benches, low, medium, and high plateaus, hills, hilly and mountainous areas and etc. geomorphological formations emerge. Generally, plains, hills, high hills and highlands plateaus are around. Maps, brown hills, valleys, yellow and green, and etc.. are shown in colors. Geomorphological structure studies, microclimate, ground water, agriculture, transport, housing and construction issues during the elections helps. For example, the hills are areas that allow private construction. Dense urban developments planned, slope and valleys, parks and woods considered. Relief: mountains, hills, plains, valleys and plateaus of the surface forms of the earth's crust. Soil formation processes such as erosion and drainage efficiency of distribution and orientation plays a significant role in relief.

The following information can be produced as a result of an examination of relief

- the height above sea level
- according to an altitude of objects around the comparison and detection of patterns,
- certain areas, and the average slope inclination
- lands texture local formations (rocks, steep cliffs, sand dunes, karstic, etc.)

Relief, the shape of the land belonging to a landscape on a flat, recessed - ribbed, curved, is a phrase that allows the introduction of phrases such as low or high. Relief characteristics of a landscape, "altitude", "examination", "slope of the land", "shape of the earth» factors are introduced. Relief of a place, climate, vegetation and soil characteristics have a significant impact on.

2.7. Hydrological analysis

According to (Kienast, 1993); Water is essential for all forms of life. It is also a finite resource. Most water in the hydrosphere is salt water (97.20 percent). Water in the polar icecaps and other frozen areas accounts for 2.15 percent of the resource. This means only 0.65 percent of the water in the world is fresh, and its distribution and quality is uneven (Tarjuelo and de Juan 1999). As a result, water is an essential factor to consider in planning Water resources in conjunction with the natural function of the aquatic ecosystem at every stage of life, are needed by all living beings. Hydrological characteristics of the structure is damaged in any way connected to it will affect the life of the living.

The hydrologic cycle expresses the balance of water in its various forms in the air, on land, and in the sea. As the hydrologic cycle and water budget illustrate, *hydrology* deals with the movement of water through the landscape both on the surface and in the ground. *Groundwater* is water that fills all the unblocked pores of materials lying beneath the surface. *Surface water* is water that flows above the ground. Depth to water table, water quality, aquifer yields, direction of movement, and the location of wells are important groundwater factors.

Hydrological survey planned in the field;

- surface drainage analysis
- analysis of groundwater
- surface water analysis
- bottom water analyses can be performed in four different fields.

Types of Utilization of Water Resources (Karadeniz, 2008);

- Drinking water (surface + underground - wells)
- Agricultural activities (surface + underground - wells)
- Industrial activities
- commercial activities
- transportation
- Recreation / tourism activities

2.8. Soils

The skeleton of the geological structure formed by elements of the vitality of plants and animals, which have gained ground, first of all exhaustible not protected, there is no place, no longer works in the life of human nutrition, accommodation, and so on. a substance that has taken over the function of many. Soil, the water and items such as a living source. Soils occupy a unique position in the lithosphere and atmosphere. Sustainable soil use is indispensable for the development of a country. They are a transition zone that links the biotic and abiotic environments. *Soil* is a natural three-dimensional body on the surface of the earth that is capable of supporting plants. Its properties result from the integrated effect of climate and living matter acting upon parent material, as conditioned by relief over periods of time. Many processes are linked within the soil zone, so soils often can reveal more about an area than any other natural factor. The soil survey includes the information necessary:

- To determine the important characteristics of soils
- To classify soils into defined types and other classificational units
- To establish and to plot on maps the boundaries among kinds of soils
- To correlate and to predict the adaptability of soils to various crops, grasses, and trees, their behavior and productivity under different management systems, and the yields of adapted crops under defined sets of management practices
- The principal purposes of the soil survey are:
- To make available all the specific information about each kind of soil that is significant to its use and behavior to those who must decide how to manage it

- To provide descriptions of the mapping units so the survey can be interpreted for land uses that require the fundamental facts about soil

Summary of soils inventory elements: Soil series, permeabilit,texture, profiles, erosion potential, drainage potential, soil associations and catenas, cation and anion exchange, acidity–alkalinity

Soil maps, land capability classes and these classes determine the soil physical, chemical and biological properties of soils, as well as training courses are available. In addition, usage patterns according to these properties are stated and symbols are shown on the map.

Soil landslide, fluidity, displacement characteristics of survey plans or maps are very important in terms of processing uses reach a decision on the erroneous.

Soil, landscape planner is one of the important ecological factors affecting the decision. Should not be the decision of any plan without understanding the properties of the soil as it should be.

Soil, landscape planner is one of the important ecological factors affecting the decision. Different climatic conditions, a wide range of rocks of different plant designs at different elevations formed land, water and air, one of the indispensable elements of life. Soil is a limited natural resource. In Turkey, most of the worlds to represent the soil types have different soil types.

2.9. Vegetation

Plants are important to study for many reasons. They may have economic and medicinal value. They provide habitat for wildlife. They have significant influence on natural events like fires and floods and may reduce the human consequences of such events. Plants are beautiful and contribute to the scenic quality of landscapes. Plants are the source of oxygen, which humans need to survive.

Survey analysis of landscape vegetation is one of the most important works. Flora and vegetation survey done by several methods. The flora of an area of all plant species, surveying, site observations and collected in the form of plants can be identified.

Flora cannot work without knowing the vegetation. Stratification, vitality (naturalness) and the other based on qualitative and quantitative characteristics of the plants are grouped and named all of the vegetation. Vegetation surveys, in the form of detailed investigation, Braun /Blanquet's method and the values determined by the quality and quantity of work done with the vegetation maps we developed this method in many countries.

The vegetation plots will be determined by dividing the area of the study of flora as simple as possible. Sample plots can be created in various sizes. For example, a plot of 20x20 m in size of all trees, shrubs, herbaceous plants is determined by the type and size. Plots more detailed surveys, 4x4 m and 1x1 m in size by dividing the sub-plots of plant species can be detected. These plots are given numerical values are determined by density of plant species.

Vegetation, vegetation can be classified as natural and cultural. In vegetation surveys should be examined in both studies. Vegetation can be determined by aerial photographs. Vegetation remote sensing system is also an important method used in the search. The vegetation surveys, and maps are used in planning the elections of most plant material.

Summary of vegetation inventory elements; Plant associations and communities, vegetative units, species list, species composition and distribution, physiognomic profiles, ecotone and edge profiles, rare, endangered, and threatened species, fire history (Kienast, 1993).

2.10. Fauna

Fauna, terrestrial and aquatic fauna can be studied in two groups. Land fauna of mammals, birds, frogs and reptiles, insects and invertebrates. These populations, habitats, and is determined in terms of activities. Results are reported with a map of a state of wild life. Within the scope of this report, the general condition of the fauna (endemic, rare, endangered, extinction, etc.). Indicated. Broadly, *wildlife* is considered to be animals that are neither human nor domesticated. Insects, fish, amphibians, birds, and mammals are more mobile than plants. While closely linked to vegetative units for food and shelter, wildlife often use different areas to reproduce, eat, and sleep. Like vegetation, wildlife have not been extensively inventoried except where the animals have some commercial value. Because animals are mobile, they are even more difficult to inventory than vegetation. Planners are paying increased attention to wildlife. They observe that in addition to enhancing the quality of life, wildlife protection is important for ethical and moral reasons, for recreational benefits, and for economic and tourist values.

Summary of wildlife inventory elements; Species list, species-habitat matrix, animal populations, habitat value map, habitat of rare, endangered, and threatened species

Animal migration is connected to both the ecological and genetic factors need to be taken into consideration surveying point. Migration event, both vertically and horizontally maintained. Regular vertical migration by plankton, the night side of the sea or lakes go deep into the water again in the morning takes the form of strokes. Horizontal migration of many invertebrate and vertebrate animals conducted by the immigration. Vertebrate animals are more regular migrations. Some animals are effective spread plant species. For example, birds, fruits, taking them to various places after eating the seeds of these plants are to help them develop those areas. In addition, the fauna that need to fertilize plants.

In addition, creating a major impact on hunting fauna. Hunting generating stations was established, hunting periods, must be done regularly. Effects on fauna, there are cultural practices, improper applications, the reductions in fauna, migration, and etc. causes problems.

2.11. Analysis of socio-culturel resources

Landscape architecture, planning activities, as well as natural resources, cultural resources present in the area is important in the determination of planning decisions to be taken for these resources. Planning was done by the people for the people by the people are the most important criterion in planning.

In the selection of a cultural use of natural resources in an area to determine the effects of this field, if necessary, the effects on natural resources, cultural resources, and these effects should be known to be determined.

Surveys, social situation, demographics, migration and urbanization, communication and transportation, education, health and nutritional status, settlement and housing status, issues such as public-state relations under investigation. In surveys, the economic situation, land ownership, party products, credit facilities, cooperatives, marketing and market relations, forms of business power, the income levels of the families studied.

REFERENCES

1.   J Ahern, 1995Greenways as a planning strategyIn: Fabos, J. and Ahern, J. eds. Greenways: the beginning of an international movement. Elsevier, Amsterdam, 131155

2.   J Ahern, 2006Theories, methods and strategies for sustainable landscape planning". In From Landscape Research to landscape Planning: Aspects of Integration, Education and Application. B. Tress, G. Tress, G. Fry, and P. Opdam, Editors. Springer. 119131

3.   J. F Benson, M. H. R London, 2000Landscape are complex systems that require a multiscale approach to fully understand, manage, and predict their behavior. 0-41925-080-8Edition).

4.   C. V Haaren, C Galler, S Ott, 2008Landscape planning: The basis of sustainable landscape development, Federal Agency for Nature Conservation, Hannover.

5.   N Karadeniz, 2008Kaynak Envanter ve Analizi, Ankara Üniversitesi, Peyzaj Mimarlığı Bölümü, Ankara, Türkiye.

6.   F Kienast, 1993Analysis of historic landscape patterns with a Geographical Information System- a methodological outline, Landscape Ecology 82103118

7.   H. S Pecova, 2000Landscape Analysis in Spatial Planning, Landscape Ecology: theory and applications for practical purposes, The Problems of Landscape Ecology Vol VI., 93101

8.   L. B Zaucer, and M Golobic, 2010Landscape planning and vulnerability assessment in the mediterranean, Croatia.

CHAPTER 7

The Role and Heterogeneity of Visual Pollution on the Quality of Urban Landscape Using GIS; Case Study: Historical Garden in City of Maraqeh

Parisa Nami[1*], Parvin Jahanbakhsh[1], Arefe Fathalipour[2]

Department of Geography and Urban Planning, Malekan Branch, Islamic Azad University, Malekan, Iran
[2]Department of Urban Planning, Marand Branch, Islamic Azad University, Marand, Iran

ABSTRACT

The urban landscape heterogeneity has been influenced by the visual, unpleasant and unacceptable face of which there is no charm in it. The present study is formed according to visual pollution and its impact on the appearance and vitality of a city. Preliminary studies and theoretical studies led to the hypothesis with the following themes: 1) There is relationship between the pollution, heterogeneity and visual disturbances urban landscape and urban vitality; 2) Improving the quality, aesthetics and identity of public spaces will increase urban vitality. Therefore, the design process within the theoretical principles, concepts, television and urban landscape, urban art, visual pollution aspects (color, light, and visual symbol) and the quality of urban vitality and explain causal relationships, analytical framework developed and more samples case (garden historic town of Maraqeh[1]) and the area under study were selected using GIS. After designing the questionnaire to evaluate the objectives, assumptions and questions of research and its analysis, the results indicate that the relationship between visual pollution and urban vitalitys, as well as proving hypotheses and vitality after the final analysis, model was introduced based on the components of comfort visual editing and proposed at three levels: micro, middle and macro for objectives, strategies and policies to regulate visual pollution and improve the aesthetic quality of the environment and the promotion of vitality.

Keywords: Urban Landscape, Visual Pollution, Urban Vitality, Quality

1. INTRODUCTION

1.1. Expressing the Problem

In today's world, people are encountered with environmental pollution coming from different dimensions including a visual pollution in cities as far more and deeper wounds on the face has the impact on the urban areas the outcome of the citizens and its outcome is the misidentify of the ravages of unbridled cities. And because the rigorous study has not been done to evaluate the effect of visual pollution, it is thought to have less consequence than other contaminants [1] .

Plenty of space-spirited, cold and lifeless are in our cities that can be seen that do not induce any meaning, concept and performance far from any beauty and utility, as in the evaluation of The Economist in Stock in lively 2012, Tehran as the Iranian capital (132th among 140 cities around the world) is not in a particularly good position [2] .

There is so wrong in defining the structure of urban and public spaces that visual chaos has become a common problem in urban areas and citizens are going there solely because of their daily essential needs in these areas are now thought to be the part of urban life. And "visual comfort, improve the quality of environment, beautification and urban vitality, joy and satisfaction of citizens" comes as the missing link between citizens and urban spaces and often little attention is paid to planning and urban design as one of the important objectives and strategies to be considered [3] .

1.2. The Importance of and Need for Research

"Relaxing, calm space and reduce pollution and regulate urban graphics and visual art" and immunity of citizens of psychological harms of visual disturbances in the environment are the most important productive factors the quality of human life. So the contaminated heterogeneous urban landscape through planning, legislation and full participation in the beautification of the urban visual landscape and creating vitality for peace as a result of the urban environment has become increasingly important as the beauty and vitality of a city to be felt significantly in these places and all ages and social sex with a sense of security and safety can be present in these spaces. See Table 1.

1.3. Project Goals and Research Questions

1) Organizing urban public spaces with an emphasis on reducing visual pollution and visual disturbances resolved to increase the desirability,

attractiveness and satisfaction, and evaluate effective strategies in creating a vibrant urban spaces;

2) Improve the quality of urban aesthetic including visual, safety, environmental and functional quality;

3) Considering the principle of people towards the urban art and design attractive spaces, beautiful and vibrant city.

4) Based on the goals, the following questions have been raised in the context of this study:

5) Is the visual pollution due to lack of charm, quality and vitality in urban areas?

6) Is it possible to increase the vitality using the organization the visual pollution and improve environmental quality?

7) What practical solutions can organize the visual pollution and disturbances in the organization of the urban landscape?

Table 1. Favorable environmental qualities of urban spaces highlighted by experts.

Kevin Lench "good shape of city" 1981			Ian Bentley's "responsive environment" 1985		
Vitality	Meaning	Compatibility	Permeability	Diversity	Readability
Convenience	Supervision & control	6-efficiency	Flexibility	Visual compatibility	Sensory richness
Death and Life of America's great cities 1961			Perspectives for a New American Dream 1994		
1-mixed-use	Emphasize on street	Convenient user	Human scale	Street view	Accountability
Social incorporation	Permeability		Promoting pedestrian-oriented	Outdoor forecast	Mixed-use
Making good cities for people					
			Opportunities for innovation in Training of urban design 1987		
Human scale	Moving walks	General and specific space	Vitality & diversity	Design for implementation	Recreational and cultural environment
1-mixed-use	Visual richness	Urban Management	Architectural environment, historic preservation and urban restoration		
Cleanliness and security	Identity	Dimension of designing urban planning			
Dimension of designing urban planning 1997			Quality urban design 1993		
Environmental sustainability	Quality of urban form	Quality vision	Vitality & diversity	Harmony and diversity	Personalized possibility
Quality urban landscape	Quality public realm	Quality construction form	Controlling development	Richness	Flexibility

1.4. Literature

Many experts have dealt with explaining the promotion of the beauty, vitality, quality of urban space and subject matter. Jicobs in "The Death and Life of America's great cities" [4] , expresses the standards creating diversity and attractiveness contribute to the vitality of the urban environment with emphasis on the social aspect of streets, sidewalks, and parks. Lynch "theory of the City" [5] , in a large-scale study of urban vitality believed that vibrancy along with the meaning, relevance, access, control and discretion, efficiency and justice are functional axes forming a good city. Cullen in "The Selection landscape" [6] , with investigating the effects of aesthetic and sensory experience of urban space and visual appeal (places and content), has considered thee heterogeneity (pollution) in the visual appearance of the floor, walls and street art of visual and believes that structural integrity (identity) to set the location of the city and social investment and a sense of ownership of the space are the living standard of urban space and understanding the vitality. Dandis in "visual literacy principles" [7] , has dealt with to expressing the effects of graphics and urban art and aesthetics, contrast, harmony and balance in urban on space and the citizens understand the meaning and message of visual (visual literacy). Carmona in "public places and spaces" [8] , has investigated visual, aesthetic, cognitive, social, functional aspects of designing a public place. BA, Ashihara in "aesthetic landscape" [9] , emphasizing the place and the people in the urban landscape and has examined the scope of architectural space, composition, landscape, painting memorable spaces and aesthetic elements urban landscape.

2. THEORETICAL FOUNDATIONS

2.1. Of Urban Landscape

According to Kevin Lynch, urban landscape is part of the city that comes to mind is lodged seen and cause for joy. The main factors for constructive are way, node, neighborhoo, edges and tokens [4] .

Urban Landscape is the art of visual and structural integrity of the buildings, streets and places that make the city environment and the art of constructive communication between different parts of the body [10] . See Figure 1.

2.2. Visual Pollution

Visual pollution is "unbridled and uncoordinated diversity of color, form, light and materials and the accumulation of heterogeneous visual elements, ugly, unattractive and man-made space and urban landscape, and is an aesthetic issue and its effects one's ability to enjoy the sights reduced or view it disrupts", which is divided in different dimensions:

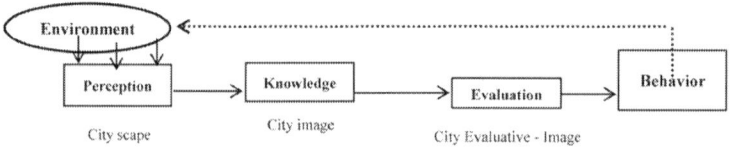

Figure 1. Conceptual model [11] .

Visual pollution, is confusion and visual pollution in urban public space experience, most within the location (street and square) which is the main field of the visual experience and the mental picture of it—and the more the body in the form of street furniture, volumes and elements and ugly extensions, walls and tables, sidewalks, flooring and walls, architecture, facade construction, outdoor advertising and urban, spaces abandoned and dilapidated, buildings and monuments of art, history, religion and business are destroyed.

Light pollution is synonymous with phrases such as Pollution Light Pollution; Luminous shows the negative effects of improper lighting, improper use of light sources and a mismatch between the lighting and the location [12] . The most obvious sign of the improper use of light sources in the urban environment and the lack of compatibility with the space, lack of environmental lighting, lighting disturbing lack of proper guide lighting and lack of lighting.

Color Pollution

Color of the elements affecting the incidence of visual pollution and heterogeneity of urban landscape into a canvas identity and role in increasing feelings and emotions, enhancing beauty and environmental impact is a factor to define the space, enhance readability, sense unity, and identity and create a sense of place [13] . So the color pollution includes "adverse visual landscape due to uneven application of color and human perception of color is the cause of unhappiness and unpleasant visual impact on the urban landscape".

Symbol Pollution

Symbols, monuments, old town and a general indication are the important factors for understanding the city. Clear image, good and certain environment, a person a sense of peace and security and create an accurate picture of the city in mind that if the visual aspect is to strengthen the means it was increased [14] . Symbol pollution includes "any disruption in any of the two elements that concept conception occurs and receiving messages, a clear picture and understanding of the environment it is difficult" such as the existence of vague information, numerous other understandable and relevant to the location.

2.3. Quality

Golkar believes that quality is used to describe the perfection of things and phenomena resulting form (formal or formal quality), performance (quality performance) and meaning (semantic quality) [11] .

2.4. Activity

Vitality, dynamism and attractiveness of urban space in front of the dismal is the reflection of the type of activity that takes place in space. Ian Gehl has divided activities into 3 categories: Essential (mandatory), optional (health) and, social activities. In a nice city, not only essential activities are in good condition but many people's favorite leisure, social activities are done and the city offers quality, tempting and, attractive spaces [6] .

2.5. Vitality and Lively Urban Space Alive

Dr. Golkar assumes the vitality is equal to "Livability and liveliness" that it depends on the achievement of desirable qualities such as urban design, legibility, color belonging, inclusion, flexibility, visual character, learners, the quality of public areas, together with the nature, sense time, energy, climatic comfort, permeability and movement, sensory richness, safety and security, user mix and form efficiency and environmental cleanliness, along with vitality, "the overall quality of urban design" [11] . According to the definitions, quality, activity and vitality of a "lively urban space" include "the place where significant presence of people and their diversity (in terms of age and gender) in a wide range time (night and day) that their activities are mainly in the form of social choice and has two procedures [9] . A procedure is dependent on insight and understanding of their culture and the other is related to the quality, beauty and architecture of the urban spaces that are interrelated".

2.6. City Art or Its Graphic

The result is a series of communications and procedure that offers systematic visual representations of lubricating seen, the quality of forming in the environment and development in the light of the aesthetic and the functional presentation of [8] .

The goals of Urban Graphic include:
- o   -Identity, dignity and readability of urban space-relax and soften the space, reducing the severity of mental suffering caused visual;
- o   -promoting visual culture, visual and coordination with art and urban furniture;
- o   -The development of urban and visual art in the design and implementation of lasting quality in the city, based on scientific principles;
- o   -Organization of color and light, urban furniture, graphics, and visual behavior in the urban environment to suit urban elements and space General.
- o   Bob Jarvis in "urban environment as visual art" believes that the traditions of city art thinking include visual aesthetic traditions, social customs and traditions make the city a place of integration of aesthetics and social tradition, and the guidance of art create urban location, identity, quality public space, ease of mobility, readability, compatibility, variation, qualities of visual and aesthetic experiences [15] .

3. METHODS

Research was done in 7 phases: initial question, exploratory studies, conceptual design research, analytical modeling, observation, analysis, conclusions, taken [16] , which has an "applied purpose" and data is quantitative and a qualitative assessment has been done on them. The study also is a "field study" based on the characteristics of the "solidarity work". The view includes focus groups and direct observation (mapping behavioral and imaging). The sample was estimated by Cochran formula 250, consisting of two groups of experts (systematic sampling) and citizens (random sampling available). Face and content validity of the questionnaire endorsed by prominent scholars and its coefficient Cronbach's alphais 0.853 that its reliability level is very good. After extraction of information through the SPSS, in order to determine the distribution of data, Kolmogorov-Smirnov test was used, to determine the average state variables of the "single-sample T-test" and "T-test", to compare rank variables "Friedman", to examines the relationship between independent variables and the dependent variables of the "analysis of variance" was is used.

4. DETERMINING THE SCOPE OF STUDY

For selecting of the problematic area having highest visual pollution, yet the beauty, charm, and vitality readability in low levels in those spaces is evident, after determining the coordinates of 425 points (public spaces) and scoring intensity maps Statistics Center pollution, over 1700 information code was determined and completed based on the output map GIS most polluted urban environment Maraqeh, accordingly street in the city center, adjacent to New Street Nasir heaven and jam, as well as urban areas with very specific visual pollution Maraqeh is located in the historical context—which often is the destruction, are, in spite of centers and commercial activity, a great deal of visual disturbances in the body, improper walking floor spaces with fences, extensions, open canals, unbridled urban advertising, poor lighting, lack of lighting, street furniture ugly and inefficient, lack of color balance found dead. In summary, an urban environment cool and the lack of appropriate space can be seen that the majority of citizens spend their time due to their essential needs (see Table 2).

5. ANALYSIS

5.1. Direct Conclusion

According to the present status, pollution and visual diversity of urban areas studied range in size (visual, color, light, Symbol) is provided in the form of images and spatial analysis (see Figure 2).

Figure 2. Part of visual pollution in the study area, Source: Authors.

Table 2. Descriptive characteristics of the sample group.

Variable		Frequency	Percent	Variable		Frequency	percent
Sample group	Experts	100	40		Practitioner	122	48.8
	Citizens	150	60		Non-working	20	8.0
Gender	Male	145	58.0	Activity status	Housewife	23	9.2
	Female	105	42.0		Studying	55	22.0
Familiarity with & belonging to Maraqeh	Born in Maraqeh	213	85.0		Income without working	9	3.6
	Residents of Maraqeh h	37	14.8		Other...	21	8.4
Education	High school	17	6.8	Age range	20 years and under	17	6.8
	Diploma	32	12.8		21 - 30	100	40.0
	Technician	37	14.8		31 - 40	92	36.8
	Bachelor	115	46.0		Over 40	41	16.4
	MA or higher	49	19.6		Total	250	100

5.2. Indirect Conclusion

According to the research subject, questionnaire designing has been based on the criteria outlined in the theoretical foundations. Table 2 shows the descriptive characteristics (frequency, percent) of the sample group:

In posing the question of "beauty, calm, charm, vitality and vibrancy of the urban environment and public spaces for citizens" 37.60% announced very high, 39.60% high, 20.80% somewhat, 1.60% low and 0.40% at the very least (Figure 1). In question "the impact of visual disturbances and pollution on the presence in the study area" 12% declared high, high 44.80%, 36% somewhat and finally 7.2% declared that that the visual pollution has a low and very low effect on them (see Figure 3, Figure 4).

In the next step, of the sample was asked to specifies "the reasons priority for their presence in urban areas (study area)". This question is proposed according to Ian Gel activity in determining the vitality and dynamism of urban spaces. Based on the SPSS data output, essential activities (mandatory) is the most important reason of the people for being present in urban public spaces that it would show a lack of attractiveness, vitality within the scope of the study (see Table 3).

In response to the question "how much is the visual pollution effectiveness on beauty, attractiveness, vitality and vibrancy of urban spaces Maraqeh?" the results show that the impact of visual pollution was very high and high on urban space (see Figure 5).

In the question of Table 5 which was mainly on analysis and theoretical foundations of research, the idea of people was questioned on the impact of effective factors related to Quality, beauty, charm, liveliness, vitality, and visual comfort of urban space.

In the next question, criteria related to theoretical foundations were addressed in relation to the characteristics of the urban public space, beautiful, charming, vivacious, and lively and respondents were asked to prioritize according to their importance from one to 10. The variable has the highest priority is the lowest rank (see Table 4).

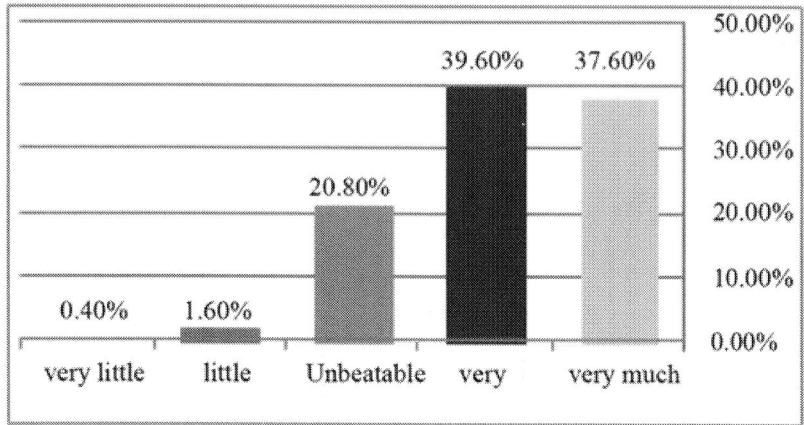

Figure 3. Importance of beauty, relaxation, vitality and vibrancy of urban spaces.

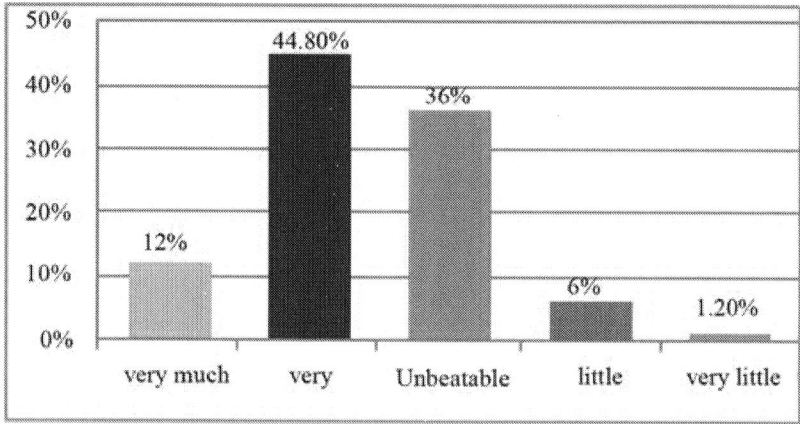

Figure 4. Importance of visual pollution impact on the citizens in urban areas.

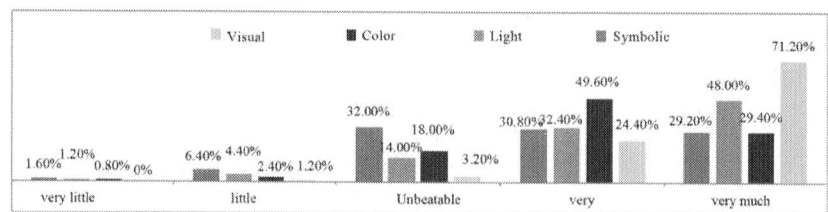

Figure 5. The effect of visual pollution on beauty, vitality of urban spaces.

Table 3. Friedman test to prioritize reasons for participating in the study area.

Varible	Number	Mean	Average Rating	Chi-square value	Degrees of freedom	Significance level
Buying, go to work	250	1.46	1.46			
Travel and leisure	250	2.48	2.48	267,586	4	...
Visiting, meeting, stroll	250	2.77	2.77			
Other goals	250	3.3	3.29			

Statistical Analysis

Before performing the test, to determine the normal distribution of data, the Kolmogorov-Smirnov test was used (see Table 5).

One sample T-test was used to assess the first and second hypotheses. As the significance level is less than 0.05, and the mean is greater than 3, the null hypothesis (H_0) is rejected in both the hypothesis. We conclude that:

(H1) the effect of visual pollution on urban vitality is significantly more than average and the first assumption proved.

Table 4. Distribution of the factors affecting the promotion of quality, beauty, charm, liveliness, vitality, visual comfort and stylized urban space.

Factors	Very low	Low	Somewhat	Much	Very much
Restore and beautify buildings and monuments of historical, artistic, religious, cultural and ...	-	1.6	6.0	34.8	57.6
Proper use of urban furniture, volumes, elements, elements and...	-	0.8	21.2	46.8	31.2
Diversify and strengthen urban green space and vegetation	2.0	1.2	29.2	28.8	38.8
Improvement and beautification of the space walk	-	4.0	.814	38.4	42.8
Ceremony and celebration of cultural, artistic and...	0.8	3.2	34.4	33.2	28.4
Creating an interactive space and enough space to sit and pedestrian walking areas	0.4	2.4	20.8	35.2	41.2
Emphasize architectural identity and organization building facades	1.2	10.8	29.2	30.8	28.0
Shopping, restaurants, entertainment and...	0.8	5.6	21.2	31.6	40.8
Organizing elements and urban outdoor advertising (signs advertising commercial, advertisement, poster, billboard, graffiti, stickers, flyers, leaflets, banners, etc.)	1.2	4.4	22.4	38.0	34.0
The use of lighting and lighting of spaces and street furniture and lighting principles, artwork, historical elements and...	-	3.2	13.6	32.0	51.2
Creating a colorful and standards in the use of color in the city	0.4	4.0	16.0	41.6	38.0
Social security and the safety of pedestrians (cars)	1.2	4.4	20.4	36.8	37.2
Remove extensions unseemly elements (air channels, pipe installation gas, electricity and telephone lines, blinds and awnings shops, studs, etc.)	0.4	4.8	31.2	31.2	32.4
Improvement, organize and beautify abandoned and dilapidated urban spaces.	0.4	5.2	19.2	33.6	41.6
Development, improvement and beautification of public perspectives and legible visual milestones and landmarks of the city	1.2	6.4	27.2	40.0	25.2
User distribution and diversity of individual and group activities	3.2	12.4	28.8	28.8	26.8
Upgrading charm and beautiful squares (flooring, lighting, elements, etc.)	1.2	4.4	16.4	37.2	40.8
Restoration and preservation of old monuments and the creation of artistic and recreational use	-	4.0	17.2	29.2	49.6
Access to the street (permeability) and park location	2.4	7.6	28.8	32.4	28.8
Urban waste management and cleanliness of public spaces	0.8	6.8	21.6	34.8	36.0
Organizing and business use annoying (blacksmith, oil changes, mechanical)	0.8	5.2	23.6	38.4	32.0
Organizing and beautify the city uses and public facilities (toilets, drinking water, kiosks, etc.)	-	5.2	15.6	33.2	46.0

Table 5. Results of Kolmogorov-Smirnov test for normality of the distribution of scores.

Variable	Level of significance	Z statistics Kolmogorov-Smirnoff	Number
Effect of visual pollution and heterogeneity of urban vitality	0.084	2.236	250
Effect of improving the quality and aesthetics and identity of public spaces on urban vitality	0.160	1.124	250
Satisfaction level	0.072	1.879	250

(H1) the effect of improving environmental quality, aesthetics and identity of public spaces on the urban vitality is significantly more than average; the promotion of quality and aesthetics and identity of urban space increases the vitality (see Table 6).

Table 6. Results of t-test one sample for research assumptions.

Variable	Test N. 3						
	Mean difference between the amount of test	Significance level	Degrees of freedom	T	SD	Mean	Number
Effects of pollution and visual diversity with beauty, charm and vitality of urban space	1.19600	0.000	249	36.416	0.51929	4.1960	250
Effect of improving the quality of environment, visual beauty and the vitality of urban identity	1.04291	0.000	249	0.49880	0.3359	0.4294	250

6. SUMMARY

6.1. Results

Quality and vitality has a meaning beyond urban areas that can be investigated at different levels such as environmental quality, sustainability, public welfare, visual, compliance, compatibility, performance, justice, security, inclusion. Based on the findings, vitality can be divided into two levels of macro and micro: the macro level consists of spatial, cultural, economic, aesthetic, social and micro level consists of visual comfort, space and component non-physical (activity) that their subsets of these factors are discussed also in place. For example, cosmetic components include indicators of urban art and visual delight and itself includes a subset of satisfaction, attractiveness and vitality.

Charming spaces, beautiful and lively spaces are those that will attract and satisfy a wide range of subjects (in terms of age and sex) according the diversity of visual and calm environment, which requires the attention to human and the needs of users (citizens) in the creating high quality public spaces.

To designing the proposed model, urban art, urban design human-centered and component visual comfort (color, light, symbol, visual) were considered as a first priority of beauty, vitality, attractiveness and vitality of urban public spaces which leads to the identity, improve the quality of urban environment, and beauty and attractiveness of these spaces which in turn will lead to citizen satisfaction and it is necessary to consider the improving the quality of aesthetic and identity-oriented space with a focus on urban art to be considered as factors affecting the vitality and linked with urban spaces.

6.2. Suggestion

With regard to the subject and importance of beauty and vitality in urban areas, it has been tried to achieve the important objectives of this study be set. The present study suggestions are in three levels of micro, middle, and macro which are comprehensive and besides the logical and fundamental relationship are complementary.

Micro level: at the micro level, the goal is to plan special offers (details) to organize and removing pollution and visual disturbances in the study area in different aspects (visual, color, light, Symbol) and in terms of objectives, policies, strategies.

Middle level: According to the goals of the present study are searched in places and spaces within urban and as the "vitality" is the main feature of sidewalk and place of interaction, leisure, recreation and vitality, shop, relax and sidewalk that attract a wide range of citizens; to enhance the quality of vitality, removing visual pollution, beauty and identity of the area under study, creating a link between urban vitality and components of visual relaxing that can be complementary and visual comfort promote besides the beauty, charm, vitality and quality of urban space and it seems rational.

Macro level: Beautification Master Plan as a unifying and comprehensive urban beautification which includes elements of visual comfort is proposed; a plan that in addition the city's image quality considers quality [17] . while bind cross-sectional and temporary designs of beautification and put them in a logical and consistent with each other, is applicable to the case that the final product is creating the quality, attractive, beautiful, and lively urban spaces that citizens take pleasure in it and find their historical, cultural, and recovered identity of their own city and feel the sense of belonging and vitality in all aspects of urban spaces.

6.3. Conclusion

According to the concepts studied, designing theoretical foundations, analysis and evaluation of the present situation (where are we?), confusion, disparity and visual pollution have a direct influence on the environment and reducing quality environment, charm, satisfaction, joy and vitality of urban areas but as the urban vitality and multidimensional concept is due to various factors and has a multi-dimension meaning, its achievement is dependent on its logical relationship with the parameters of quality of urban design which is the missing link with the citizens and the city triangle with urban management and urban spaces and the components of visual relaxing in city spaces.

REFERENCES

1. Beautification Organization of Tehran (2012) Urban Art. Institute of Shahr Publication, Tehran.

2. Goodey, B. (1993) Two Gentlemen in Verona: The Qualities of Urban Design, Strrewise.

3. Coleman J. (1987) Opportunities for Inviting in Urban Design Education. Australian Planner, 1, 32.

4. Jacobs, J. (1961) The Death & Life of Great American Cities. London Cape.

5. Lynch, K. (2004) Vision City. 6th Edition, Translated by M. Mozayani, Tehran University Press, Tehran.

6. Gehl, J. (2004) Public Spaces, Public Life. Danish Architectural Press, Copenhagen.

7. Dandis, D. (2009) The Principles of Visual Literacy. Translated by M. Sepehr, Soroush Publication, Tehran.

8. Carmona, M. (2009) Public Places and Spaces. Translated by Gharaei F. et al., Tehran Art University Publication, Tehran.

9. Tibaldez, F. (2004) Citizen-Oriented Urban Development, Modernization of Public Areas in Cities and Urban Environments. Translated by Mohammad Ahmadinejad, Khak Publication, Isfahan.

10. Kiwi, R. and Kampenhood, L.V. (2007) Research Methods in the Social Sciences. Translated by Nik Abdul Hussein Gohar, Tutya Publication, Tehran.

11. Golkar, K. (2007) The Concept of Vitality in Urban Design Quality. Journal of the Saffeh, Tehran, 44, 56.

12. Pakzad, J. and Suri, E. (2011) Guide of Lighting Urban Places. Utopia Publication, Tehran.

13. Loshchinsky, N. (2008) Landscape Planting Design: Professional Approach to Garden Design. Translated by M. Kafi, Aeezh Publication, Tehran.

14. Lynch, K. (1997) Theory of Good City. Translated by Sh. Bahraini, Tehran University Press, Tehran.

15. Landry, C. (2000) Uban Vitality: A New Source of Urban Competitiveness. Prince Claus Fund Journal, Archive Issue Urban Vitality—Urban Heroes.

16. Nelessen, A.C. (1994) Visions for a New American Dream. APA Planner Press, Chicago.

17. City of Turlock, California (2010) Beautification Master Plan, Landscape & Signage.

CHAPTER 8

Optimal Electricity Distribution Framework for Public Space: Assessing Renewable Energy Proposals for Freshkills Park, New York City

Kaan Ozgun * , Ian Weir and Debra Cushing

Creative Industries, School of Design, Department of Landscape Architecture, Queensland University of Technology, Gardens Point, 2 George St, GPO Box 2434, Brisbane, QLD, QLD 4001 Australia 4001 Australia

ABSTRACT

Integrating renewable energy into public space is becoming more common as a climate change solution. However, this approach is often guided by the environmental pillar of sustainability, with less focus on the economic and social pillars. The purpose of this paper is to examine this issue in the speculative renewable energy propositions for Freshkills Park in New York City submitted for the 2012 Land Art Generator Initiative (LAGI) competition. This paper first proposes an optimal electricity distribution (OED) framework in and around public spaces based on relevant ecology and energy theory (Odum's fourth and fifth law of thermodynamics). This framework addresses social engagement related to public interaction, and economic engagement related to the estimated quantity of electricity produced, in conjunction with environmental engagement related to the embodied energy required to construct the renewable energy infrastructure. Next, the study uses the OED framework to analyse the top twenty-five projects submitted for the LAGI 2012 competition. The findings reveal an electricity distribution imbalance and suggest a lack of in-depth understanding about sustainable electricity distribution within public space design. The paper concludes with suggestions for future research.

Keywords: renewable energy distribution; public space; sustainability; LAGI; Freshkills Park; New York City; triple bottom line (TBL)

1. INTRODUCTION

A growing body of research suggests energy potential mapping to design more sustainable cities based on local energy potentials at multiple scales [1]. Moreover, the application of renewable energy infrastructure within urban environments is growing rapidly, yet it is still commonly conceived of as an add-on feature, rather than as an integral characteristic of urban space. This underestimation of the potential for renewable energy systems is demonstrated in both the urban design profession and their counterpart policy makers, where the focus is on increasing the environmental sustainability of cities by retrofitting spaces and buildings with so-called "techno-fixes" [2] (p. 24) [3], such as green walls and photovoltaic arrays. Commentators have identified a now common trait, where designers make "crafty attempts to get on the 'eco' bandwagon without linking the project to the messy and unpredictable dynamics of nature" [4] (p. 178). In these cases, the primary design objective is often one of superficial display, rather than genuine and in-depth knowledge of sustainability. Although individual buildings are designed with green infrastructures at ever-increasing rates, landscape architects and urban designers need to investigate the integration of renewable energy within urban open spaces where the contextual issues are more multi-layered than in private domains.

First, a new conception of public space is essential, one that addresses the ever increasing complexity of urban environments. For example, swarm planning theory deals with the increasing complexity and uncertain futures of cities, focusing predominantly on the planning process within a regional scale [5] (pp. 606–609). The theory explains the transformation of spatial land use over time and enables new self-sufficient and resilient developments. Therefore, rather than perpetuating the idea of public space as a static artefact, or end product, the new conception of public space must embrace a more dynamic definition, one that is concerned with connectivity, network flow and multi-functional participatory space [6] (p. 234).

Second, this paper argues that renewable energy can no longer be considered a techno-fix or a mere cosmetic intervention in public space. Instead, designers need to consider renewable energy as an important 'ecological infrastructure' similar to the management of water resources, waste cycling, food production and mass mobility [7] (p. 348). Renewable energy infrastructures can also be fully recognized as complete localized electricity production, consumption and distribution systems when integrated in public spaces. For example, Byrne et al. [8,9] argue for locating "energy-ecology-society relations in a 'commons' space [...]," focusing on techniques and social arrangements that can serve the aims of sustainability and equity. Public space can be a showground for implementing a renewable energy commons approach [10,11]. It can be seen as a bridge that connects mainstream energy with the emerging alternative decentralized energy movements. This approach must complement the rapidly changing renewable energy technologies and their increasing energy generation capacity. Such an approach also exposes social, environmental and economic relationships of renewable energy usage, which brings the accepted triple bottom line (TBL) framework to the foreground. Originated in the 1990s as a medium to integrate sustainability into the business world, the TBL framework operationalizes and implements sustainability into practice [12,13,14] (p. 252). The balance between these three accepted pillars of the TBL becomes a critical aspect to achieve sustainable production, consumption and distribution. Renewable energy-embedded public space designs that encourage direct and indirect consumption and production of electricity can help to increase public engagement, while also educating the public about renewable energy.

In an effort to engage more people with energy in public spaces, the Land Art Generator Initiative (LAGI) is an international enterprise that hosts regular design competitions dealing with renewable energy within urban environments. In comparison to engineering solutions, which often satisfy quantitative metrics of electricity capture, storage and distribution, LAGI exemplifies a qualitative conception of renewable energy within public spaces and uses design competitions to promote its motto, "renewable energy can be beautiful." LAGI's philosophy and innovative approach demonstrates an awareness of the societal issues surrounding the production of energy within public spaces and was honoured as a top sustainable solution at the United Nations Rio+20 conference and published in "Sustainia100" [15].

In 2010, LAGI announced its first international competition to design and construct public art installations for three different locations in the United Arab Emirates. In 2012, LAGI organized a second competition for Freshkills Park (Former Freshkills landfill) in New York City. Most recently, in May, 2014, LAGI held a third design competition for a shipyard site in Copenhagen, Denmark. All competitions advance the same strategic objective to integrate art into the interdisciplinary creative process and re-imagine sustainable design solutions in public domains. Over four years of competitions, LAGI has increasingly sought to address what it means to embed renewable energy into daily public life. The competition recognizes that practitioners of urban design and public art can have agency over the diversity, richness, quality and types of interactions between the user and energy in public spaces. When successful, designs can effectively communicate new information to the community.

This study focuses on the distribution of produced electricity from renewable sources within a public space context. It introduces an optimal [16] electricity distribution (OED) framework for public space design that organizes potential relationships of local electricity production, consumption and distribution by adapting ecologist Howard T. Odum's theories about energy flow and hierarchy in nature. It then uses the OED framework to assess the top 25 LAGI 2012 proposals. The paper concludes with a discussion of the results and implications of using the OED framework to assess and design new conceptions of energy-embedded public space. Areas of future research are also explored.

2. LINKING PUBLIC SPACE AND RENEWABLE ENERGY: THE OPTIMAL ELECTRICITY DISTRIBUTION FRAMEWORK

"Environmental sustainability", a concept stemming from sustainable development, is defined as social and economic development that is also environmentally responsible [17] (p. 6). Renewable energy has since become associated with sustainable development, enabling projects to have less environmental impact and much greater energy capacity compared to fossil fuels and nuclear energy, while being self-sufficient, locally based and less dependent on national energy networks [18] (p. 172). This conception of renewable energy acknowledges its agency over the economic and social dimensions of sustainable development, including, but not limited to, new jobs, by producing ones' own power facilities, avoiding infrastructure costs (transmission, transport, distribution), promoting decentralized new economic relationships, increasing productivity by having fewer conversion steps and spreading ownership [19] (pp. 75–76). Of particular interest to designers and policy makers, the social aspect of renewable energy needs to be emphasized within the context of well-designed and well-used public space.

To enable this shift, this study developed the OED framework to effectively distribute on-site-produced electricity into public space. The framework requires an understanding of the economic-social-environmental TBL relationships of the produced electricity within a public space context. The European commission's report on sustainable cities argues that the environmental function is achievable if only the economic and social components are also in line [20] (p. 2). That is, a balance between all three is required for a truly sustainable distribution of produced electricity in public spaces.

Similarly, the renowned ecologist Howard T. Odum made significant contributions to ecosystems ecology and incorporated thermodynamics law into ecology. One of his provisional ideas [21,22], "Tripartite Altruism," is useful to landscape and environmental design because it identifies an energy/nature equation. For example, this self-regulatory feedback system is applied in permaculture, a holistic gardening practice that works with nature, not against it [23] (p. 15). Rabbits exemplify the "Tripartite Altruism" theory. "They eat grass to live, grow and reproduce. Their manure fertilizes the grass that feed[s] them, and they 'sacrifice' weak rabbits to predators to help keep the population fit and in balance" [24] (p. 73). According to "Tripartite Altruism", approximately one-third of the energy in an organism or a mature complex system [25] is used for self-maintenance and/or energy storage, one-third is for lower order operations and one-third is contributed upward to higher-order system controllers [23]. The following diagram (Figure 1) conceptualizes an optimal distribution of produced electricity from renewable resources embedded in public open spaces, representing the optimum balance between TBL components.

PUBLIC SPACE OPTIMAL ELECTRICITY DISTRIBUTION FRAMEWORK

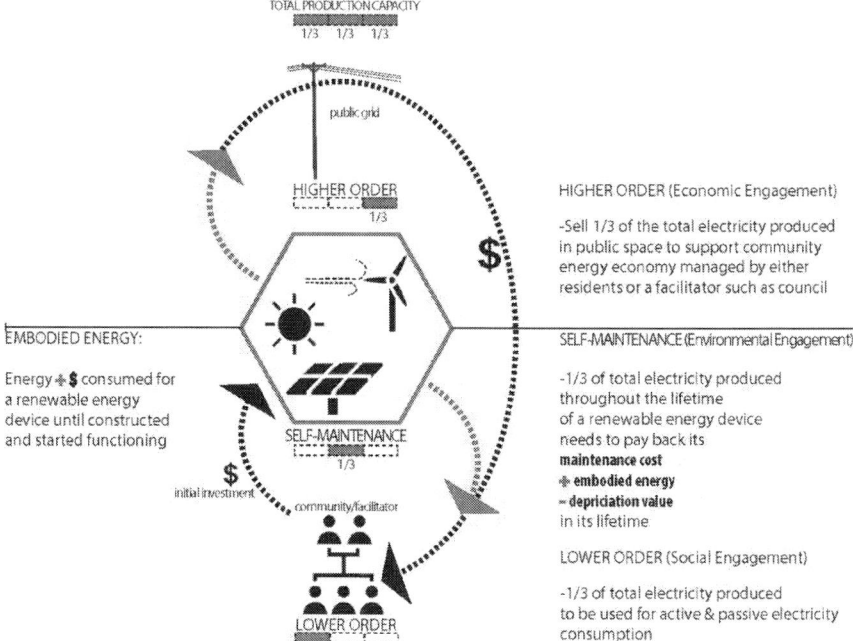

Figure 1. Public Space Optimal Electricity Distribution Framework. This figure was initially published in the Journal of Landscape Architecture, Taylor & Francis Ltd. [26].

The OED framework illustrated in the diagram simplifies Odum's provisional energy/nature equation, designating one-third of the on-site-produced electricity to be used for active and passive engagement, representing "social engagement." One-third of the on-site-produced electricity can be sold to the public grid to create a local energy economy, representing "economic engagement." The remaining one-third of the on-site-produced electricity can be used for self-maintenance, representing "environmental engagement."

2.1. The OED Framework Lower Order: Social Engagement with Renewable Energy in a Public Space

Generating social engagement by on-site-produced electricity from renewable sources is rooted in the innate nature of public space. Public space is a social place where people communicate, interact and engage with their surroundings. For example, Miller [27] (p. 204) argues that "Public spaces do not exist as static physical entities but are constellations of ideas, actions, and environments." The social aspect of public spaces can best be described by Amidon [4] (p. 178), who states that "New public space designs need to arouse desire in the public to participate, to cultivate and to advocate." Unlike embedding renewable energy into a building, designers need to complement the evolutionary and dynamic nature of a public space when embedding renewable energy. Accordingly, North [28] (p. 15) argues, "While a building begins to erode once built, a landscape continuously evolves." Lefebvre contends that the spaces of the modern city have to provide not only consumable material goods for its dwellers, but also evoke the need for creative activity and information [29] (p. 18). Similarly, Gehl states that public spaces provide a source of information about the social world outside, as well as a source of inspiration for action [30] (p. 21). Public space can, therefore, be seen as an education and information agent, through which renewable energy can be introduced to a community.

Odum particularly focused on useful information as concentrated energy and as one possible product of the energy cycle in self-organized systems. "Concentrated energy" has an important role in the energy hierarchy, because it monitors, controls and provides feedback to higher and lower orders constantly. In this instance, an ecologically well-designed public space can play a similar role by interacting with its users, as well as its immediate vicinity and the city's greater energy grid. Similarly, Abel describes useful information as [31] (p. 85), "[f]undamentally a product of the self-organization of systems, wherein its function is to remember successful configurations of cells, organisms, ecosystems, and [...] human adaptations."

This paper stresses public spaces as an education and information agent to encourage a sustainable lifestyle and increase general environmental awareness in an effort to maximize energy efficiency in the broader community. A growing body of literature indicates urban environments as complex systems [5,32]. When conceptualized as a self-organized system, public space can be considered a platform to create useful information for a community, which can thus promote greater uptake of sustainable energy across multiple domains in society. This claim is grounded in the "maximum power principle", which is considered the fourth law of thermodynamics [33,34]. According to this law, in the self-organizational process, systems develop parts, processes and interactions that maximize efficiency and production [35] (p. 71) [36].

For the purpose of this paper, interactions with renewable energy are identified as active and passive. Active social interaction with on-site-produced electricity includes activities that promote direct consumption of electricity, including educational, performance or recreational based activities, such as electric car charging points, personal device charging utilities and wireless services. Active interaction also refers to direct electricity production from users' movements, such as capturing energy from the downforce of footsteps via piezoelectric generators.

Passive social interaction with renewable energy refers to activities that have an indirect relationship with electricity consumption. Passive modes are characterized by activities involving artful play and the interpretation of renewable energy systems, including information centres, interactive energy toys, interpretive energy screens and media displays. Simply put, the on-site produced electricity needs to be consumed internally without any external output. For example, a public space user consumes the on-site-produced electricity for way-finding through the site using the embedded interpretive energy screen.

Active and passive interactions are imperative for the generation of shared knowledge, because they directly connect users with their environment and economics [37,38,39] in the public space, both literally and symbolically. For optimal electricity distribution, active and passive social engagement with renewable energy must achieve a combined total of one-third of the electricity production capacity. This comprises the 'lower order usage' in the devised OED framework. The two interaction modes demonstrate the necessity for an integrated approach to renewable energy and public space, to not only achieve meaningful and measureable sustainability, but to also communicate the reciprocal relationship between society and energy. To achieve this, designs must employ best practice principles of interpretation and sense of place into the design.

This paper argues that such enhancements in our interactions with energy correlate with the observed tendencies of self-organized mature ecosystems. For example, the fifth law of thermodynamics states that, 'system processes maximize power by interacting abundant energy forms with ones of small quantity, but a larger amplification ability' [40] (p. 122). Therefore, the more ecologically-sustainable public space is one that responds to the fifth law by engaging with renewable energy, through both active and passive interactions. The greater the number of active and passive interactions that exist between renewable energy and public space users, the greater the likelihood that the public space will influence society's sustainable energy lifestyle.

2.2. The OED Framework Higher Order: Economic Engagement with Renewable Energy in a Public Space

In Odum's "Tripartite Altruism", another one-third is assigned to "economic engagement", where electricity distribution contributes to the local energy economy. Applied to the context of public space, produced electricity could be sold to the utility grid and used to support the community renewable energy economy managed by either local residents or a facilitator, such as a local council. The initial investment cost to accomplish this can be subsidized by the community or the facilitator. There is an expanding body of literature about sustainable energy transition that points to a shift from centralized autocratic energy economies, towards decentralized modes of electricity production that bring new socio-economic relationships to cities [41,42,43,44].

To understand the potential for a decentralized energy economy based on public spaces, it is useful to refer to 'system size' in ecology, which is the spatial extent or physical boundary of a system. The system size measurement of the energy capacity of conventional public spaces would include an assessment of the total energy demand supplied from the main energy grid. A public space also contains, but is not limited to: users; hard landscapes, such as paved floors, stairs, ramps and street furniture; soft landscapes, such as grass and other plant material; infrastructure; the continuous information and matter flow; and the built structures within and around it. Thus, an energy system in a public space has many components, not unlike ecosystems composed of a community of organisms and chemical cycles [45] (p. 523).

The concept of system size simply frames the energy demand and supply relationship. When a conventional system requires more energy to sustain its demand, an external energy supply feeds the system. System size becomes more significant because of energy availability that is dependent on the produced electricity from renewables. Both the quantity and quality of available energy in the system determines the optimum system size [46]. As current research [1] on potential energy mapping underpins the importance of the local energy potentials for sustainable city design and planning, environmental designers also have to consider the optimum system size of each energy resource [47] (pp. 33–34). A public space as an optimum system may be achievable by considering both the quality and quantity of on-site-produced electricity. Energy quality refers to the emergy concept, which is discussed in the next section.

2.3. The OED Framework Self-Maintenance: Environmental Engagement with Renewable Energy in a Public Space

To complete Odum's "Altruistic Tripartite", the final one-third of the produced on-site energy is designated for environmental engagement. This engagement refers to the electricity utilized for "self-maintenance" of the public space and to recoup its maintenance cost and embodied energy of the renewable energy devices [48,49,50]. Embodied energy is also directly related to the 'emergy' concept. Emergy represents energy memory emphasized by the prefix (em) in emergy and defined as the history, the time and the processes involved up to the present state of a system [51] (p. 33). Odum quantifies 'energy quality' in an urban environment and defines it via the emergy concept [52,53]. This parallels the fifth law of thermodynamics, which states that information generally has the highest energy quality and the densest form of the emergy/energy ratio, as highlighted in Table 1 [39] (p. 88).

Table 1. Exemplars show the emergy/energy ratio; a higher number means a higher quality of work [36] (p. 69).

ITEM	Solar Emcalories per calorie *
Sunlight energy	1
Wind energy	1500
Organic matter, wood, soil	4400
Potential of elevated rainwater	10,000
Chemical energy of rainwater	18,000
Mechanical energy	20,000
Large river energy	40,000
Fossil fuels	50,000

Table 1. *Cont.*

ITEM	Solar Emcalories per calorie *
Food	100,000
Electric Power	170,000
Protein foods	1,000,000
Human services	100,000,000
Information	1×10^{11}
Species formation	1×10^{15}

* calories of solar energy previously transformed directly and indirectly to produce one calorie of energy of the type listed. Source: Odum 1996 [35].

The depreciation value of a renewable energy device in its lifetime can be calculated based on existing data from energy payback time (EPT) and embodied energy values, subtracted from the production value. Applied to the public space context, this would include the basic energy demands, such as lighting. This type of electricity consumption is similar to that which occurs in a normal household, including the energy need of appliances. By grouping consumption modes, we can monitor, control and create better sustainable outcomes.

According to Odum, it is beneficial to have a large amount of electricity production, as long as enough storage is available for the lower and higher order interactions to exist in the system. Odum states, 'With increasing scale of available energy (the production capacity of renewable energy in public space), storages increase, depreciation decreases and pulses are stronger but less frequent' [39] (p. 63). This definition depicts the behaviour of mature complex ecosystems [39] (p. 54). From a public space point of view, a larger amount of electricity produced from renewables means that more interaction and storage will be required to use the produced electricity sustainably.

The application of Odum's "Tripartite Altruism" to the urban space context establishes the OED framework, through which speculative and built projects can be assessed. The next section describes how this study used the OED framework to assess competition entries for the LAGI 2012 competition, set in Freshkills Park, NYC.

3. METHODS: USING THE OED FRAMEWORK FOR ASSESSMENT

Out of the 250 entries submitted in LAGI's 2012 competition, 65 projects were selected and published in the book, Regenerative Infrastructures of Freshkills Park, NYC [54]. To better understand current design thinking about renewable energy embedded into public space, the study used the first 25 entries, including four place-winning and twenty-one shortlisted schemes, for content analysis. These schemes were selected for LAGI 2012 by experts from a multidisciplinary jury and a selection committee [54] (p.29).

For the purposes of the study, the authors overlaid the OED framework with LAGI's judging criteria. Three out of the seven judging criteria directly aligned with the framework:

- The annual electricity production capacity (economic engagement);
- How the proposal engages with the public (social engagement); and
- The embodied energy required to construct the renewable energy infrastructure (environmental engagement) [54] (p. 30).

The other four judging criteria are not directly related to renewable energy usage and were, therefore, excluded. The authors determined how the projects responded to the three judging criteria using thematic content analysis of images and text in the Regenerative Infrastructures [54,55] book and also LAGI's official website [56]. Thematic content analysis focuses on the occurrence and meanings of keywords and concepts in texts to generate themes, employing either a predefined analytical structure or an interactive structure [57] (p. 83). The authors employed NVivo software to thematically code the collected data based on the three criteria.

Competition submissions, active on the official LAGI website at the time of data collection, communicate their designs through A4 pages with project descriptions, as well as four A1 panels with graphics and text. The published content in the book is a refined version of the original A1 panel submitted through the website. The amount of information published differs, depending on the jury's selection order and editing. While the four place-winning projects have six pages of content published, shortlisted projects have four pages.

This assessment addresses the social, environmental and economic engagement with on-site-produced electricity identified in the devised OED framework. To quantify this, we created a quality impact assessment scoring scale from one to three to align with the OED framework. The analysis aims to quantify the quality of each project's energy interventions: a score of one for no/low quality, a score of two for medium quality and a score of three for high quality. Entries obtaining higher scores were perceived as more responsive to renewable energy distribution.

First, the study assessed the social engagement (lower order) aspects of an entry and determined the extent of public engagement that it was likely to generate by using on-site-produced electricity from renewable sources. For example, if an entry does not consider any engagement, or the assessment outcome is unknown, the entry scores a one. If an entry considers either active or passive engagement, it scores a two. If an entry considers both active and passive engagement, it scores a three.

Next, the study investigated the economic engagement of renewable energy (higher order usage). For example, if an entry designates none of its on-site electricity production to be sold to the local grid or if this is unknown, it scores a one. If an entry considers all on-site-produced electricity from renewables to be sold to the local grid, without any maintained for self-maintenance described below, it scores a two. If the on-site-produced electricity is to be partially sold to the local grid, an entry scores a three.

Finally, the study assessed the environmental engagement (self-maintenance) aspects of the entries, including embodied energy, using a portion of the produced electricity for maintaining the renewable energy installation, energy storage, general public space maintenance and other primary electricity needs of services within the space. If an entry does not appear to respond to any of these aspects or the situation is unknown, it scores a one. An entry that partially considers these factors scores a two. If an entry considers most or all of these, it scores a three.

In summary, the content data were analysed against the framework and the three LAGI judging criteria relevant to renewable energy usage. The next section discusses the findings from this assessment.

4. FINDINGS

The following table (Table 2) illustrates the quality impact level (scores from one to three) of each competition entry, displaying their individual, average and total scores using the OED framework. The order of projects in the table follows the same order in the competition book, Regenerative Infrastructures of Freshkills Park, NYC. Although not explicitly clarified by LAGI organizers, the order of shortlisted projects in the publication somewhat indicated the LAGI jury's order of preference. The embedded text under Table 2 is a brief summary of the methods in Section 3.

The content analysis of the four place-winning entries (labelled Entries 1–4 in Figure 2) revealed that the designs focused on economic engagement first (higher order), with social engagement (lower order) and environmental engagement (self-maintenance) considered as secondary (Figure 2). Similarly, the shortlisted entries scored higher for economic engagement (higher order), with environmental engagement and social engagement secondary (Figure 3).

Figure 2 shows the assessment quality impact level (scores from one to three) of four place-winning design entries based on social (orange), environmental (green) and economic (blue) engagement. The results showed that the four place-winning design entries did not score overwhelmingly higher than the shortlisted projects, indicating that they do not necessarily promote the most ideal electricity distribution for a public space context according to the OED framework. Instead, the average score for the place-winning entries was 6.50 out of nine, which is slightly higher than the average score for the shortlisted entries of 6.40 out of nine.

For example, one of the top scoring projects was Entry 22, Solar bloom, which scored nine out of nine. The project addressed the OED framework criteria fully. Entry 22 integrated a sterling-based solar dish engine into a sculptural installation. The installation generates 35,500 MWh of electricity annually and can power 3087 houses every day. While visitors can directly engage with the produced electricity through charging outlets as active engagement, they can also engage indirectly through LED lighting that demonstrates the systems efficacy through visual means and refers to passive engagement with produced electricity. Thus, the project scored a three,

Table 2. Distribution assessment for the LAGI 2012 renewable energy proposals.

Distribution Assessment for the Lagi 2012 Renewable Energy Proposals							
Quality Impact Level			Annual Capacity	Social	Environmental	Economic	
1	2	3	MWh	Lower Order	Self-Maintenance	Higher Order	
Four winning entries							Total
Entry 1-scene-sensor			5500	3	2	3	8
Entry 2-fresh hills			238	1	2	3	6
Entry 3-pivot			1200	1	1	2	4
Entry 4-99 red balloons			14,000	3	2	3	8
(4 entries) Total				8	7	11	26
(4 entries) Average				2	1.75	2.75	6.50
Twenty-one shortlisted entries							
Entry 5-solar loop			10,000	1	2	3	6
Entry 6-power play			100	2	1	2	5
Entry 7-in between scapes of light			4800	2	1	3	6
Entry 8-inefficiency can be beautiful			672	2	1	3	6
Entry 9-field of energy			13,000	2	2	3	7
Entry 10-flightaic			1,000	1	2	3	6
Entry 11-biofuel armature			60,000	1	1	2	4
Entry 12-robo zoo			10	2	1	1	4
Entry 13-flirt			72,000	3	2	3	8
Entry 14-solar cairn			1000	1	2	1	4
Entry 15-electric meadow			unknown	1	3	1	5
Entry 16-art-wind-energy unit			145	1	3	3	7
Entry 17-blossommings			520	3	3	3	9
Entry 18-heliofield			15,000	2	2	3	7
Entry 19-beauty of recycling			3600	2	2	3	7
Entry 20-cloudfield			5910	2	2	3	7
Entry 21-fresh clouds			65,000	2	2	3	7
Entry 22-solar bloom			35,500	3	3	3	9
Entry 23-tree			1700	2	2	3	7
Entry 24-nawt balloons			30,500	1	2	3	6
Entry 25-currents			28,470	2	2	3	7
(25 entries) Total				46	48	66	160
(25 entries) Average				1.84	1.92	2.64	6.40

Table 2. *Cont.*

Distribution Assessment for the Lagi 2012 Renewable Energy Proposals

Economic Engagement (Higher Order)

(1) None/Unknown of the electricity produced to be sold to the local grid

(2) All on-site electricity produced to be sold to the local grid

(3) On-site produced electricity to be partially sold to the local grid

Environmental Engagement (Self Maintenance) *

(1) None/Unknown

(2) Only considers partially

(3) Considers majority/all

Social Engagement (Lower Order) ***

(1) None/Unknown **

(2) Active or passive engagement through direct electricity consumption or production ****

(3) Active and passive engagement through direct electricity consumption or production [†]

* Electricity demand of permanent functions such as lighting, heating, energy storage and other primary electricity needs of services of public spaces Energy demand of maintaining the energy device/installation Embodied energy consideration; ** No engagement through direct electricity consumption/production; *** Educational, informative, event and recreational use; **** For example Piezoelectric generator used to generate power from people movement. [†] Personal device, event, electric car recharge in the car park, wireless.

addressing economic and social engagement. Lastly, the project is also responsive to environmental engagement because the dish engine is made of an eco-friendly resin that is 40 percent recycled content and 100 percent recyclable. The installation is modular and complies with the LEED (Leadership in Energy and Environmental Design) green building practice to reduce its environmental impact. The project also includes energy storage units. Thus, the project considered the majority of environmental criteria and scored a three for environmental engagement.

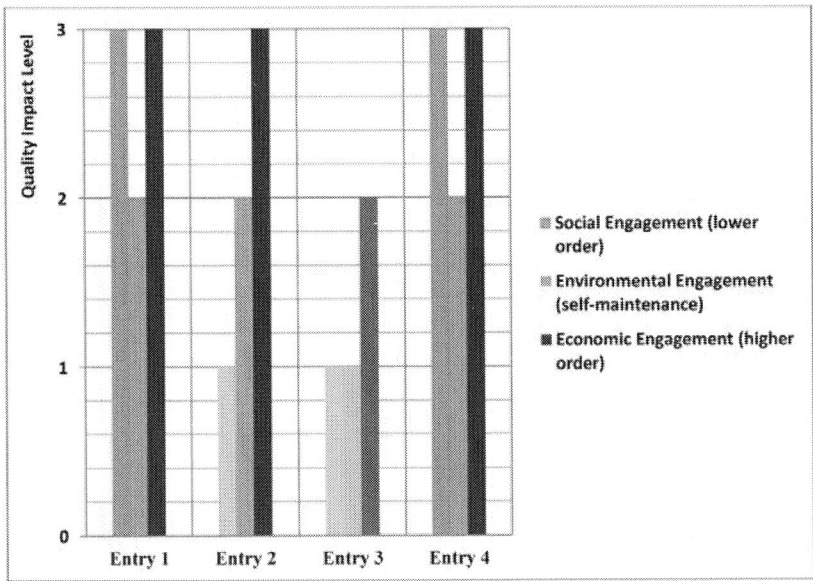

Figure 2. Optimal electricity distribution assessment of the four LAGI place-winning entries.

Entry 1, Scene-sensor, scored eight out of nine, using piezoelectric generators for electricity production through people movements and wind power. According to the OED framework, Entry 1 addressed active social engagement through direct electricity production from footsteps, whereas no data were provided concerning the direct on-site electricity consumption. Entry 1 also addressed the passive engagement with the produced electricity through wind mapping and LED lighting performance integrated into the installation. Therefore, Entry 1 scored a three by addressing active and passive engagement through direct electricity consumption or production. From an environmental engagement perspective, only minor data were found with regards to lighting. This enabled Entry 1 to score a two; since other factors underpinned in the OED framework, including embodied energy, energy storage, and other primary electricity needs, were not stated anywhere in the project description. Lastly, at an economic engagement level, Entry 1 produced electricity (5500 MWh annually) for 1200 households while using part of the electricity for LED lighting performance, therefore scoring a three.

One of the lower scoring projects according to the OED framework was Entry 12, Robozoo. This entry produced 10 MWh of electricity annually through solar ivy, a novel solar energy generating system inspired by ivy leaves. However, no data were found in the project submission content about selling the on-site-produced electricity to the city grid. The project proposed a mechanical ecosystem with electricity producers (flora) and electricity consumers (fauna). The visitors can engage with this ecosystem by harvesting the batteries from electricity producers and integrating them into the mechanical creatures. This refers to passive engagement with electricity, and no data were found concerning active engagement with electricity. Therefore, Entry 12 scored a two out of three for social engagement. The project also scored a one from environmental engagement, since no data were identified.

High annual clean electricity production capacity requires more environmental engagement (self-maintenance) and social engagement (lower order) to create an optimal distribution, according to the OED framework. Out of twenty-five entries assessed, ten entries produced over 10,000 MWh of electricity annually.

The findings showed that the total assessment scores for these entries were also higher than the entries producing less than 10,000 MWh (Table 3). The table displays the annual energy capacity of twenty-four entries [58]. While ten out of twenty-four have more than a 10,000-MWh annual capacity, the other fourteen have less than 10,000 MWh. This result aligns with the theory reasoning that high production capacity entries not only produce more electricity, but also sell energy to the public grid, generating more income.

Table 3. Distribution assessment of LAGI winning entries with their annual electricity production capacity.

Annual Capacity		Social	Environmental	Economic	Total
10 entries	>10,000 MWh	2	2	2.90	6.90
14 entries	<10,000 MWh	1.78	1.78	2.57	6.13

However, it is important to note that entries with the highest production capacity did not necessarily score highest using the proposed framework. For example, Entries 20 and 25 were compared, and both scored seven out of nine (Figure 3). Entry 20 produced 5910 MWh of electricity, and Entry 25 produced 28,470 MWh of electricity, nearly six-times more. Therefore, Entry 25 required innovations with a greater intended social and environmental engagement impact, in order to balance the higher electricity production. Entry 20 promoted passive engagement through direct electricity consumption for music and theatre events, but did not promote active engagement; whereas, Entry 25 promoted only active engagement and provided electric car plug-ins from electricity produced on-site. Thus, both entries scored a two out of three under the social engagement criterion. However, since Entry 20 provided these interactions with less electricity production capacity, it is actually more energy responsive and sustainable according to the OED framework.

The findings from this study demonstrate a discrepancy between sophisticated designs as chosen by the LAGI jury and their approach to sustainable distribution of on-site-produced electricity (indicated by their resulting OED assessment in Figure 3).

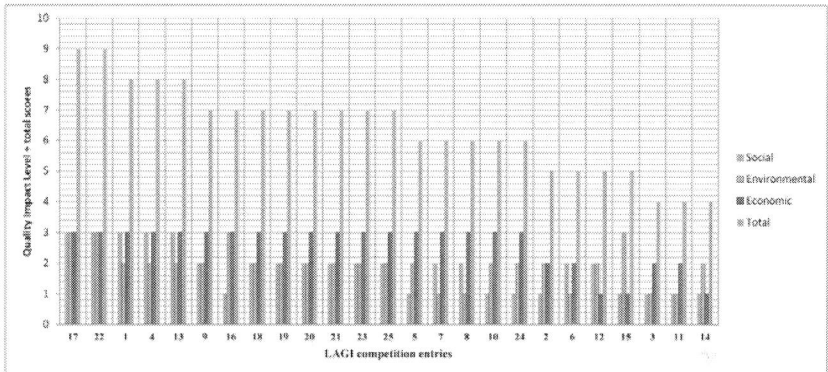

Figure 3. The graph shows entries ranked according to the optimal electricity distribution (OED) framework assessment from highest to lowest score. Entry numbers in bold black represent LAGI competition ranking order. For example, Entry 1 refers to LAGI's first place winner project, and Entry 25 is the very last shortlisted project.

The next section, therefore, discusses the implications of these findings and the significance of the proposed OED framework from the perspective of current design thinking about renewable energy-embedded public spaces.

5. DISCUSSION

This study set out with the aim of assessing cutting-edge design propositions that integrate clean electricity production into public space. The assessment of twenty-five LAGI 2012 competition entries using the proposed OED framework described in this paper revealed that the primary focus was on economic engagement with on-site clean electricity production. A secondary focus was on environmental and social engagement.

In addition, the four winning LAGI entries did not score highest in the OED assessment. This suggests a lack of association between cutting edge design propositions and the science of sustainability, with respect to the optimal distribution of produced electricity from renewable sources. The findings also show that although predefined themes relevant to renewable energy usage were included in the judging criteria list, competition entries did not address them specifically. Likewise, LAGI's assessment criteria are perhaps not precise enough to reveal the relationship between sophisticated designs and their genuine sustainability. This could be attributed to LAGI's highly artistic and conceptual emphasis, which prompts designers and artists to focus heavily on the aesthetic attributes of their entries, rather than sustainable energy production and distribution.

A further reason might be the lack of a well-defined design framework that effectively addresses renewable energy usage within the public space context. LAGI's judging criteria includes three types of engagement; however, the criteria are not specific and, therefore, remain secondary. Instead of embedding the three types of engagement (economic, environmental and social) into the criteria, the LAGI enterprise could potentially provide this information to designers as foundational public space sustainability knowledge with respect to electricity distribution.

In addition, ecologically-sophisticated public space designs need to address energy more deliberately. Initiatives similar to LAGI are imperative to advancing the uptake of these concepts in the broader society. While LAGI is primarily an art initiative and, therefore, focuses on the aesthetics of renewable energy, our developed OED framework seeks to expand the relationships and interactions between public space users and renewable energy. This includes the production of electricity from on-site renewable sources and its effective and optimal distribution with respect to three different types of public space-specific engagement: environmental, social and economic. This could be beneficial to LAGI for the continued evolution of their art/science/urban design framework and to leverage LAGI's artistic approach to advance sustainable energy transition. Considering the current conjecture about sustainable energy transition, LAGI's role in promoting renewable energy is indispensable.

The next section concludes with the implications of using the devised OED framework as a method of assessing and designing energy-embedded public spaces, the limitations of this study and recommendations for future research.

6. CONCLUSIONS

Both the findings and the developed OED framework contribute to the sustainable design and assessment of public spaces. The framework, when used as a design tool, enables designers to engage with sustainability throughout the design phases, rather than after the project has been completed, which is what commonly happens. Rather than perceiving renewable energy as a 'techno-fix' addendum to the existing public space designs, this paper introduced a novel path to treat renewable energy-embedded public space as micro-scale ecological infrastructure. This infrastructure would potentially establish new social, cultural, economic and environmental relationships between the city environment and its dwellers, complementing the sustainable energy transition and the increasing number of urban production activities. Likewise, when conceived of as a method of assessment, the devised OED framework can potentially be integrated into the existing public space sustainability assessment tools [59,60], which only assess renewable energy as an indicator of environmental sustainability and often downplay the social and economic aspects of local electricity production. Thus, the method employed in this study will serve as a starting point for future research to advance an effective assessment tool.

6.1. Limitations

The OED framework specifically focuses on clean electricity distribution in public spaces in relation to the economic, social and environmental dimensions of engagement. Therefore, one limitation is the lack of recognition of the aesthetic dimension of design. Each public space design contains site- and designer-specific features, such as site characteristics, aesthetic sensibilities, historically- and culturally-significant features, the financial context and budget and universal access. Yet, the LAGI 2012 competition entrants are speculative, without real-life political, financial and logistical constraints. Although the proposed OED framework accepts and works with this diversity and assumes designers will accommodate these opportunities and constraints as necessary, further research is needed to apply the OED framework to built projects.

An additional limitation includes the limited detail available for each LAGI 2012 entry. LAGI's entries are conceptual, and therefore, the energy-relevant data are limited. For example, the available data for each entry do not provide an exact quantity of energy designated for social, environmental and economic engagement. Therefore, for the purposes of this study, entries were only analysed to understand if their energy interventions aligned with the devised OED framework.

6.2. Future Research

The theories contributing to the OED framework of this study provide several implications for future research. From a landscape architecture and environmental design perspective, the extant research focuses on energy-conscious planning and design within a regional scale, often neglecting the micro-scale. The devised OED framework for renewable energy-embedded public space fills this gap.

Scholars of energy-conscious design and planning focus predominately on the first and second law of thermodynamics [61,62], yet this study integrates the fourth and fifth law into energy-conscious design. This expanded theoretical framework has the potential to connect society, energy and information at a micro-urban scale, specifically in public space. Despite the criticisms of Odum's approach to information by conventional ecologists and information theorists, systems ecologists and emergy scholars have started to integrate emergy research into cultural and societal studies [31,63]. Additional research possibilities exist to apply emergy analysis to public spaces.

Sustainable energy transition can only be achieved with the right policies and tools. This transition can occur when renewable energy in public spaces is regarded as an embedded and context-specific feature of public space, rather than as an add-on or techno-fix to conventional spaces. Such rethinking presents opportunities for new urban perspectives regarding planning policies, new levels and modes of community participation and engagement, place-making strategies, entrepreneurship and management of clean electricity-producing public spaces. With the increasing number of production activities in cities, public spaces offer great opportunities to share renewable energy knowledge and to educate the public in order to facilitate a quicker transition to sustainability. Any policy or framework that identifies the relationships between renewable energy and urban environments, considering the social, economic and environmental perspective simultaneously, supports this transition. This research clearly demonstrates the need for further discussion on the aesthetics of renewable energy technology when electricity production and its emerging TBL relationships come into focus.

ACKNOWLEDGMENTS

Many thanks to Christian Long and two other anonymous reviewers for their constructive comments on an earlier version of this paper. The authors would also like to acknowledge the financial support of Queensland University of Technology for funding this publication.

AUTHOR CONTRIBUTIONS

Kaan Ozgun conceived of and designed the OED framework as part of his Ph.D. research project. Kaan Ozgun collected and analysed the data, and wrote the first draft. Ian Weir revised the initial version and co-wrote the second draft. Debra Cushing helped to develop the study and revised the paper. All three authors read and approved the final manuscript.

REFERENCES AND NOTES

1. Van den Dobbelsteen, A.; Jansen, S.; van Timmeren, A.; Roggema, R. Energy potential mapping—A systematic approach to sustainable regional planning based on climate change, local potentials and exergy. In proceedings of CIP World Building Conference, Cape Town, South Africa, 14–17 May 2007; 2007; pp. 2450–2460.

2. Huesemann, M.; Huesemann, J. Techno-Fix : Why Technology won't Save Us or the Environment; New Society Publishers: New York, NY, USA, 2011.

3. Huesemanns [2] (p. 24) argue in their book Techno-fix that 'science and technology, as currently practiced, cannot solve the many serious problems we face and a paradigm shift is needed to reorient science and technology in a more socially responsible and environmentally sustainable direction.' This paper, therefore, used the term "techno-fix" to indicate the research statement and the need to have a counterpart design solution.

4.   Amidon, J. Big nature. In Design Ecologies : Essays on the Nature of Design; Tilder, L., Blostein, B., Amidon, J., Eds.; Princeton Architectural Press: New York, NY, USA, 2009; p. 255.

5.   Roggema, R.; van den Dobbelsteen, A. Swarm planning for climate change: An alternative pathway for resilience. Build. Res. Inf.**2012**, 40, 606–624.

6.   Wall, A. Programming the urban surface. In Recovering Landscape: Essays in Contemporary Landscape Architecture; Corner, J., Ed.; Princeton Architectural Press: New York, NY, USA, 1999; pp. 233–249.

7.   Belanger, P. Redefining infrastructure. In Ecological Urbanism; Mostafavi, M.D., Gareth, D., Eds.; Lars Muller Publisher: Baden, Germany, 2010; pp. 332–349.

8.   Byrne, J.; Martinez, C.; Ruggero, C. Relocating energy in the social commons: Ideas for a sustainable energy utility. Bull. Sci. Technol. Soc.**2009**, 29, 81–94.

9.   Eizenberg, E. Actually existing commons: Three moments of space of community gardens in new york city. Antipode**2012**, 44, 764–782.

10. 'The commons is a way of thinking and operating in the world, a way of organizing social relations and resources; existing commons should not be seen as a "return" of some noble but possibly archaic ideal but as a springboard for critiquing contemporary social relations and as the production of new spatiality, initiating the transformation of some fundamental aspects of everyday life, social practices and organization, and thinking' [9] (764–782).

11. Energy commons is not a new approach, and some countries, like Denmark and Germany, have been experiencing sustainable energy transition starting as a grassroots, community-based initiative supported by local governmental policies and cooperative small-scale private decentralised ownership [42].

12. Elkington, J. Cannibals with Forks: The Triple Bottom Line of 21st Century Business; New Society Publishers: Gabriola Island, BC, Canada; Stony Creek, CT, Canada, 1998.

13. McDonough, W.; Braungart, M. Design for the triple top line: New tools for sustainable commerce. Corp. Environ. Strategy**2002**, 9, 251–258.

14. This paper adopts the TBL framework not only to substantiate Odum's provisional idea 'Tripartite Altruism', but also to explicitly reveal the relationships of economic, social and environmental objectives of the produced clean electricity that exist, but are commonly neglected by public space designers.

15. Alslund-Lanthén, E.; Riiskjær, J.; Gerdes, J. Sustainia100. In A Guide to 100 Sustainable Solutions; Eika, C., Alslund, E., Eds.; Sustainia Publisher: Copenhagen, Denmark, 2012.

16. For the purpose of this paper, optimal refers to distributing produced electricity for social, economic and environmental purposes within a public space context. The definition of optimal in this paper was not used as a proven quantitative formula, but an approximation to the ideal design of electricity distribution for creating ecologically-sustainable public spaces.

17. Moldan, B.; Janoušková, S.; Hák, T. How to understand and measure environmental sustainability: Indicators and targets. Ecol. Indic.**2012**, 17, 4–13.

18. Dincer, I. Renewable energy and sustainable development: A crucial review. Renew. Sustain. Energy Rev.**2000**, 4, 157–175.

19. Scheer, H. Energy Autonomy: The Economic, Social and Technological Case for Renewable Energy; Earthscan: London, UK; Sterling, VA, USA, 2007.

20. Rostami, R.; Khoshnava, S.M.; Lamit, H. Heritage contribution in sustainable city. IOP Conference Series Earth Environ. Sci.**2014**.

21. Odum, M. Personal communication, Texas A&M University: San Marcos, TX, USA, 2014.

22. Tripartite Altruism was a provisional idea in the 1980s, which Odum refined in the 1990s with the 'emergy' concept [21].

23. Holmgren, D.; Services., H.D. Permaculture: Principles & Pathways beyond Sustainability; Holmgren Design Services: Hepburn, Austrilia, 2002.

24. Yeang, K. Ecodesign : A Manual for Ecological Design; Wiley: London, UK, 2006; p. 499.

25. One of the key lessons ecology can teach is that as the system's biomass increases and the system moves towards becoming self-organizing, more recycling loops and complex interactions are needed to prevent it from collapsing. In emulating ecosystems, we must design our human-built environment to contain more recycling loops and interactions [24] (pp. 47–48).

26. Ozgun, K.; Cushing, D.; Buys, L. Renewable energy distribution in public spaces: Analyzing the case of ballast point park in sydney, using a triple bottom line approach. J. Landsc. Archit.**2015**, 2. in press.

27. Miller, K.F. Designs on the Public : The Private Lives of New York's Public Spaces; University of Minnesota Press: Minneapolis, MN, USA, 2007.

28. North, A. Community evolution through public space. Available online:

http://www.researchgate.net/publication/266289676_LANDSCAPE_F
RAMEWORKS_COMMUNITY_EVOLUTION_THROUGH_PUBLI
C_SPACE (accessed on 17 February 2015).

29. Mitchell, D. The Right to the City: Social Justice and the Fight for Public Space; Guilford Press: New York, NY, USA, 2003.

30. Gehl, J. Life between Buildings: Using Public Space; Island Press: Washington, DC, USA, 2011.

31. Abel, T. Emergy evaluation of DNA and culture in 'information cycles'. Ecol. Model.**2013**, 251, 85–98.

32. Portugali, J. Self-Organization and the City; Springer: New York, NY, USA, 1999.

33. Sciubba, E. What did lotka really say? A critical reassessment of the "maximum power principle". Ecol. Model.**2011**, 222, 1347–1353. [

34. Valyi cited in [33]; so far, no publications can be considered as evidence for the applicability of the 'maximum power principle'; however, it should be noted that the results may be interpreted under a different paradigm.

35. Odum, H.T. Environmental Accounting; Wiley: Hoboken, NJ, USA, 1996.

36. Odum, H.T.; Odum, E.C. A Prosperous Way Down: Principles and policies; University Press of Colorado: Boulder, CO, USA, 2008.

37. Shuman, M. Going Local: Creating Self-Reliant Communities in a Global Age; Free Press: New York, NY, USA, 1998.

38. Economics is the science of efficiency dealing with production, consumption and distribution [37]. 'Efficiency is the traditional measure used to represent energy transformations. It is the percentage of input energies that is output energy' [39] (p.64).

39. Odum, H.T. Environment, Power, and Society for the Twenty-First Century: The hierarchy of Energy; Columbia University Press: New York, NY, USA, 2007.

40. Tilley, D.R. Howard T. Odum's contribution to the laws of energy. Ecol. Model.**2004**, 178, 121–125.

41. Hauber, J.; Ruppert-Winkel, C. Moving towards energy self-sufficiency based on renewables: Comparative case studies on the emergence of regional processes of socio-technical change in germany. Sustainability**2012**, 4, 491–530.

42. Wächter, P.; Ornetzeder, M.; Rohracher, H.; Schreuer, A.; Knoflacher, M. Towards a sustainable spatial organization of the energy system: Backcasting experiences from austria. Sustainability**2012**, 4, 193–209.

43. Van Timmeren, A.; Zwetsloot, J.; Brezet, H.; Silvester, S. Sustainable urban regeneration based on energy balance. Sustainability**2012**, 4, 1488–1509.

44. Droege, P. 100 Per Cent Renewable : Energy Autonomy in Action; Earthscan: London, UK, 2009.

45. Stremke, S.; Koh, J. Ecological concepts and strategies with relevance to energy-conscious spatial planning and design. Environ. Plan. B **2010**, 37, 518–532.

46. Odum, H.T.; Odum, E.C. Energy Basis for Man and Nature; McGraw-Hill Book Company: New York, USA, 1976.

47. Stremke, S. Designing sustainable energy landscapes: Concepts, principles and procedures. Ph.D. Thesi, Wageningen University, Wageningen, The Netherlands, 2010.

48. Alsema, E.A.; Fthenakis, V.M. Photovoltaics energy payback times, greenhouse gas emissions and external costs: 2004–early 2005 status. Prog. Photovolt. **2006**, 14, 275–280.

49. For example, energy pay back times of photovoltaics, 1–7 years, depending on the mode of technology. Another research's findings concerning energy pay back times of solar, geothermal, wind wave and tidal power is an average of three years [48,49,50].

50. Roberts, F. Energy accounting of alternative energy sources. Appl. Energy **1980**, 6, 1–20.

51. Bastianoni, S.; Marchettini, N. Emergy/exergy ratio as a measure of the level of organization of systems. Ecol. Model. **1997**, 99, 33–40.

52. Odum, H.T. Self-organization, transformity, and information. Science **1988**, 242, 1132–1138.

53. 'It is a measure of value in the sense of what has been contributed. Self-organizing systems use stores and flows for purposes commensurate with what was required for their formation. To do otherwise is to waste resources, making products without as much effect as alternative designs. Therefore, the higher emergy use there is, the more real work is done, the higher is the standard of living, the more money can buy' [36,373839] (pp. 6–139).

54. Monoian, E.; Ferry, R. LAGI 2012 design guidelines. In Regenerative Infrastructures : Freshkills Park, nyc: Land Art Generator Initiative; Klein, C., Ed.; Prestel: Munich, Germany; London, UK; New York, NY, USA, 2013; p. 30.

55. The other four judging criteria included 'adherence to the design brief and submission requirement; the integration of the work into the surrounding environment, landscape and the draft master plan of Freshkills Park; the sensitivity of the work to the environment, to local and regional ecosystems and to the integrity of the landfill cap and underground infrastructure; the originality and social relevance of the concept [54] (p. 30).'

56. Land art generator initiative. Available online: http://www.landartgenerator.org/winners2012.html (accessed on 21 January 2012).

57. Carley, K. Coding choices for textual analysis: A comparison of content analysis and map analysis. Soc. Methodol.**1993**.

58. Twenty-four entries were taken into consideration, since Entry 15's annual energy capacity data were not clear.

59. Sustainable SITES Initiative. Available online: http://www.sustainablesites.org/education?q=about/faqs (accessed on 26 March 2015).

60. The most common ones include 'BREEAM' (Building Research Establishment Environmental Assessment Method) in the U.K., 'LEED' (Leadership in Energy and Environmental Design) in the USA and 'Greenstar' in Australia. These assessment methods have become an industry standard for sustainable architecture and have later guided sustainable landscape architecture. Recently developed after 'LEED' by the American Society of Landscape Architects (ASLA) in conjunction with the Lady Bird Johnson Wildflower Centre at The University of Texas at Austin and the United States Botanic Garden, 'The Sustainable SITES Initiative' (SITES) primarily focuses on the ecosystem services and aims to encourage more sustainable land development and management practices. The SITES creates 'guidelines and performance benchmarks for sustainable design, construction and maintenance in landscape architecture projects' [59].

61. Dincer, I.; Rosen, M.A. Thermodynamic Fundamentals. In Exergy; Elsevier: Amsterdam, The Netherlands, 2007; pp. 1–22.

62. According to the first law of thermodynamics, energy cannot be destroyed or produced and can only be transformed and conserved. The second law deals with this transformation and states that the work capacity (exergy) of energy becomes extinct while disorder (entropy) occurs [61].

63. Abel, T. Culture in cycles: Considering H.T. Odum's 'information cycle'. Int. J. General Systems**2013**, 43, 44–74.

CHAPTER 9

Use of Watersheds Boundariesin the Landscape Planning

Aybike Ayfer Karadağ

[1] *Düzce University, Faculty of Forestry, Department of Landscape Architecture, Turkey*

1. INTRODUCTION

Interactions between human society, biosphere, atmosphere, and hydrosphere have increased extensively, sometimes for the welfare of mankind and environment, but frequently for their man. These interactions are characterized by increasing complexity, diversity, use, and misuse of natural resources, the latter permanently decreasing. And this holds true for any scale in space and time, from global to local and from long-term to short term.On the regional and local scale the interactions between society, hydrosphere, and biosphere are relevant (Kaden, 2003) and these interactions determine the future of the landscape.

Landscape is complex and far-reaching. People have strong ties to landscapes and use them in various ways. Thus landscape is interweaves with climate change and ecology, development, economics, politics, and culture (Bastian *et al.,* 2006; Jones, *et al.,* 2007). Landscape changes as a result of these relationships that human-nature interaction. The changes in landscape were brought up idea of planning for sustainable use, conservation and management. But landscape character and structure make difficult landscape planning decisions. Therefore it must be understood primarily "landscape" to successful landscape planning.

Two different approaches have emerged to defining landscape, when the definitions of landscape are evaluated. According to the first approach, landscape is ecological units. In this context Forman (1995) defined landscape as a mosaic where the mix of local ecosystems or land uses is repeated in similar form over a kilometers-wide area. A landscape manifests an ecological unity thought its area. Within a landscape several attributes tend to be similar and repeated across the whole area, including geologic land forms, soil types, vegetation types, local faunas, natural disturbance regimes, land uses, and human aggregation pattern. Thus a repeated cluster of spatial elements characterizes a landscape. Burel and Baudry (2003) argue that landscape is a level of organization of ecological systems that is higher than the ecosystem level (Farina *et al.*, 2005). It is characterized essentially by its heterogeneity and its dynamics, partly governed by human activities. It exists independently of perception. Landscape is considered mainly a mosaic of geographical entities in which organisms deal with the spatial arrangement of these entities determined by complex dynamics (Farina *et al.*, 2005). Landscape is geographic unit at second approach. Geography, where the landscape plays a central role and may be considered a fundamental unit, is of particular importance in the attempt to delineate a clear, scientifically useful concept of landscape. The definitions in geography essentially focus on the dynamic relationship between natural landforms or physiographic and human cultural groups (Forman and Godron, 1986). Landscape refers to a common perceivable part of the earth's surface. Landscape became a core topic of geography, in particular regional geography. It was seen as a unique synthesis between the natural and cultural characteristics of a region (Mander and Antrop, 2003). As Zonneveld (1979) stated, landscape is part of the spaces on the earth's surface, consisting of a complex of systems, formed by the activity of rock, water, air, plants, animals and man, and that by its physiognomy forms a recognizable entity (Forman and Godron, 1986). The European Landscape Convention defines landscape as "an area, as perceived by people, whose character is the result of the action and interaction of natural and/or human factors" (Anonymous, 2000). In this context Turner *et al.*(2002) indicated landscape as an area that is spatially heterogeneous in at least one factor of interest. Opdam *et al.* (2006) defined landscape as a "geographical unit characterized by a specific pattern of ecosystem types, formed by interaction of geographical, ecological and human-induced forces."

Regardless of how landscape is defined, landscape can be characterized by *structure, function, and change* (Kurum and Şahin, 2000). *Structure,* the spatial relationships among the distinctive ecosystems or elements present-more specifically, the distribution of energy, materials, and species in relation to the sizes, shapes, numbers, kinds, and configurations of the ecosystems. Landscape structure is generally defined in terms of "composition" and "configuration". Dunning *et al.* (1992), these are, respectively, the kinds of patches present in the landscape and the amount of each, and the spatial relationships among them as indices of landscape structure, landscape metrics can be used to describe the composition and spatial arrangement of a landscape (Forman and Godron, 1986). They can be applied at different levels to describe single landscape elements by such features as size, shape, number or for whole landscapes by describing the arrangement of landscape elements and the diversity of landscape (Forman and Godron, 1986; Waltz, 2011). Forman and Godron (1986) defined *landscape function* from a systems-theoretical point of view as "the interactions among the spatial elements, that is, the flows of energy, materials, and species among the component ecosystems". Leser (1997) emphasized that it is necessary to analyze functions and functional interactions between landscape factors and landscape components in order to understand the relationships within the system (Bastian, *et al.* 2006). *Change,* the alteration in the structure and function of the ecological mosaic over time (Baker, 1989).

Landscape change, because they are the perceivable expression of dynamic interactions between the physical and material environment and natural and cultural forces. In addition, Consequently, landscapes differ from place to place and different landscape types can be recognized as well as different landscape regions (Mander and Antrop, 2003). In this context three main factors can be identified in determining landscape: physical, biological and anthropic. Their interaction are continuously composing the landscape in such a way that we can distinguish between a spatial and a temporal aspect of this composition. The spatial landscape variety consists in the present interrelation of these three factors in a certain place (Kerkstra *et al.,* as cited in Makhzoumi, 1973 and Pungetti, 1999). In addition history and ecology are essential factors in the structuring and understanding of landscapes. No reference is made to "special" landscapes such as "spectacular" or "ordinary" ones, to rural, industrial or urban ones; all landscapes should be considered equally (Antrop, 2005).

Landscape ecology provides understanding of change of landscapes. In addition landscape ecology provides a strong conceptual and theoretical basis for management and planning at the landscape level by contributing to a better understanding of the structure and function (Uzun, 2003; Ivits *et al.*, 2005). Landscape ecology, a subdiscipline of ecology, is the study of how landscape structure affects the abundance and distribution of organisms. Landscape ecology is the study of the pattern[1] - and interaction between ecosystems within a region of interest, and the way the interactions affect ecological processes, especially the unique effects of spatial heterogeneity on these interactions (Clark, 2010). Landscape pattern consists of three elements: patches[2] - , corridors[3] - and a matrix[4] - . In addition, landscape ecology involves the application of these principles in the formulation and solving of real-world problems (Forman, 1995).

Landscape planning has come up in the process of understanding, maintainable usage and preservation of the landscape that changed as a result of the relationship and interaction between the man and the nature (Bastian *et al.,* 2006; Jones, *et al.,* 2007). Landscape planning is the key planning instrument for nature conservation. The basis for the concept of planning is formed by the idea of ''balancing the needs and the sources by complying with rational priorities in the long term to reach certain goals with scarce resources'' (Keleş, 2004). From upper scale to subscale, the planning includes physical environment, socio-cultural life, history on economic and political issues, decisions concerning today and future (Uzun, *et al.,* 2012). Also, social and physical are grouped as executive and object planning (Zaimoğlu, 2003). At that point, landscape planning is evaluated as the subtopic of physical planning (Köseoğlu, 1982) and accepted as the basis for it (Zaimoğlu, 2003).

The European Landscape Convention defines landscape planning as "strong forward-looking action to enhance, restore or create landscape" (Anonymous, 2000). Landscape planning is an activity that analyses, plans and localises landscape and environmental characteristics, resources and values (Dökmeci, 1996). Steiner (1999) used the term of landscape plan to emphasize that such as plans should incorporate natural and social consideration. Uzun *et al.,* (2012) stated that landscape planning had two basic approaches which were ''depending on a certain territory'' and ''directed to problem solving''. The landscape planning studies that depend on a certain territory are examples of the planning studies concerning an area having a developmental potential. It contains the approaches concerning the formation of criteria about determining the territorial usage (agriculture, recreation, etc.) in the process of development of a newly developing region or a sub-region. The landscape planning studies directed to problem solving have the aim of solving the present problems in the landscape planning and the problems concerning the planned usage. Choosing places for industry, settlement, highway route, etc. and landscape renovations are examples for these planning. In addition Uzun and Gültekin (2012) emphasized landscape planning which is one of the fields of study that creates a balance between natural sciences and engineering sciences in the best possible way is also important for natural resource management. One of the main purposes is a balanced planning of people and nature, instead of people oriented planning. In landscape planning, the approaches in which landscape functions are analyzed and the structure and change of landscape is presented have been supported by ecology and landscape ecology sciences.

L. McHarg (1920-2000), the pioneer of the environmental movement, revealed that natural sciences should be evaluated in solving the problems, by focusing on the natural life processes and their determinative effects on area usage plans (Şahin, 2009a). In this context, putting preservation and usage balance forward, examining the ecological features, analysing the usages and accordingly the ecological relationships, and after these examinations, defining the actions and forming an environment which people will take the most benefit of, but will be less threat for other animals are emphasised in the landscape planning (Uzun et al., 2012). At the same time landscape planning provides a coordinated information basis for all natural resources, which enables us to rapidly obtain an overview of the nature and landscape situation within the planning areas; fragmented changes to individual parts of nature and the landscape can be assessed with respect to their effect on the whole existing condition; planning and nature conservation experts in the administration can use this as basis for quick and uncomplicated comments. The complex interaction of all the factors affecting the balance of nature such as soil, water, air and climate, plants, and animals, as well as diversity, characteristic features and beauty and the recreational value of nature and landscape as well as the effects of existing and foreseeable land usages, are analysed and assessed within the landscape planning. As a result, extensive basic information about nature and the landscape is available for the whole area. The spatial objectives, measures and requirements developed in the landscape on the protection, maintenance and development of nature and the landscape (Anonymous, 2008).

In the basis of a successful landscape planning lies understanding and knowing the landscape. In this context, landscape structure, landscape processes and the changes in landscape were effective items. Uzun et al. (2012) states that the structure and functions of landscape are evaluated, landscape processes are analysed and landscape ecology based approaches are put forward in the recent landscape planning studies. In this context landscape ecology have also attached great significance to the issue of scale, and the "landscape units" is more widely canvassed as a framework for analyzing inter-relationship and delivering joined-up policy within a comprehensible and identifiable space (Selman, 2006).

In this study, the concept of boundary in landscape planning is emphasized. Additionally, the use of the ArcHydro Model was described to delineate watershed boundaries.

2. EXPLORING THE BOUNDARIES IN LANDSCAPE PLANNING

Natural systems are usually considered parts of hierarchies—an ordering from biggest to smallest (or vice-versa). For our purposes (planning, management, etc.), ecological hierarchy will be discussed from the largest to the smallest scale.

Scale is the dimension of an object or process. It can be described as resolution and range, which indicate in how much detail the object or process has been understood (Du-ning and Xiu-Zhen, 1999). Scale is a key issue in planning. Due to the interdependencies of ecosystems, a planning approach is need that examines a site in its broader context. Scale is related to three dimensions (Selman, 2006).

- A spatial dimension: -the mostly cited component of landscape scale, based on both a rational and intuitive recognitions of distinct physical units.
- A temporal dimension: -implying a continuum from the earliest human use of a landscape into the sustainable use by future generation.
- A modification dimension: -from intensely urbanized areas, through farmland and other types of natural use, to pristine or wilderness areas, with some areas processing such intense degrees of alteration that the landscape requires human assistance to accelerate the recovery of its "regenerative" properties.

The concept of scale can allow to the analysis on the level of different hierarchal system that can be related to each other and it can be related to the hierarchy theory. Allen and Star (1982) stated that the hierarchy theory was developed as a study outline for analysis of complex systems or situations which became organized in certain types. The systems that become organised hierarchically can be divided into functional components. These elements' structure, function and characteristics related to time and space can be formed in scale or on different levels. There is no basic hierarchy in the hierarchy theory. Its focus level can change according to considered events (Hersperger, 1994 as cited in Uzun, 2009). The hierarchical theory is a useful instrument for exploring numerous patterns and processes through various scales in space and time. Considering complexity as an attribute that is intrinsic to a landscape, the hierarchy paradigm explains how the various components located on certain scales enter into contact with other ones that are visible on different scales of resolution. The hierarchical theory views a system as a component in a larger system that consists of subsystems (Allen and Starr, 1982; O'Neill et al., 1986; Allen and Hoekstra, 1992 as cited in Farina, 2001).

The concept of boundary is a spatial expression of the scale and it can be expressed in different ways with hierarchy theory. Such as the biosphere or planet is boundary and is subdivided into continents (and oceans) within hierarchical theory. Continents are subdivided into regions, region into landscapes, and landscapes into local ecosystems or land uses. Region is a broad geographical area with a common macroclimate and sphere of human activity and interest. This concept links the physical environment of macroclimate, major soil groups, and biomes, with the human dimensions of politics, social structure, culture, and consciousness, expressed in the idea of regionalism (Forman, 1995). A region therefore almost always contains a number of landscapes (Forman and Godron, 1986; Du-ning and Xiu-Zhen, 1999). In addition the region is composed of patches, corridors and a matrix that vary widely in size and shape. In this case the spatial elements are whole landscapes. Unlike the recurring landscape elements in a landscape, a region does not exhibit a pattern of repeated landscape. Usually the distribution of landscape simply mirrors the typically coarse-grained, geomorphic land surface. Thus, most regions are coarse grained or variable-grained with group of small landscapes. In short, the spatial pattern or arrangement of landscape in a region is just as important functionally as the pattern of continents on the globe, local ecosystems in a landscape (Forman, 1995).

Landscape is a dynamic and hierarchical setting. Landscape comprises so many hierarchically constructed ecosystems from a single molecule to the whole Earth and even the limitless emptiness called the space (Selman, 2006). Considering complexity as an attribute that is intrinsic to a landscape, the hierarchy paradigm explains how the various components located on certain scales enter into contact with other ones that are visible on different scales of resolution (Farina, 2001). Every ecosystem has its own boundaries yet is in relation with other ecosystems through the flow of energy and data which ensure the continuity of the system. A system is theoretically in balance when the inputs and outputs required for its functions within its natural boundaries are equal. Therefore, the assessments in defining the capability, capacity and sensitivity of the area for any human activity should be performed within the natural boundaries (Şahin, 2009b). For instance, bioregionalists have argued that "nature" defines its own integral systems and that, historically, sustainability in human systems has been a consequence of close alignment between socio-economic practices and environmental capacity. This leads to arguments, discussed more fully below, that natural, rather than political, boundaries could form the basis of many planning and management choices (Selman, 2006).

A landscape can vary in size from a few centimeters to tens of kilometers. The heterogeneity might be expressed as physically identifiable structures. At any rate, the degree of heterogeneity varies according to the spatial arrangement of the single component parts. Landscapes do not exist in isolation. Landscapes are nested within larger landscapes that are nested within larger landscapes, and so on. In other words, each landscape has a context or regional setting, regardless of scale and how the landscape is defined. The landscape context may constrain processes operating within the landscape. Landscapes are "open" systems; energy, materials, and organisms move into and out of the landscape. This is especially true in practice, where landscapes are often somewhat arbitrarily delineated. That broad-scale processes act to constrain or influence finer-scale phenomena is one of the key principles of hierarchy theory and "supply-side" ecology. The importance of the landscape context is dependent on the phenomenon of interest, but typically varies as a function of the "openness" of the landscape. The "openness" of the landscape depends not only on the phenomenon under consideration, but on the basis used for delineating the landscape boundary. For example, from a geomorphological or hydrological perspective, the watershed forms a natural landscape, and a landscape defined in this manner might be considered relatively "closed". Of course, energy and materials flow out of this landscape and the landscape context influences the input of energy and materials by affecting climate and so forth, but the system is nevertheless relatively closed. Conversely, from the perspective of a bird population, topographic boundaries may have little ecological relevance, and the landscape defined on the basis of watershed boundaries might be considered a relatively "open" system (Farina, 2001).

Landscape has different hierarchical systems. The classification of a landscape as one goes from lower to increasingly higher levels in the hierarchy: ecotope (the basic unit in a landscape consisting of biotic and abiotic elements); microchore (the spatial distribution of ecotopes); mesochore (the environmental system composed of a group of microchores); macrochore (a mosaic of landscapes); and megachore (a group of geographical elements covering several kilometers). A system exists independently of its components and is generally able to organize itself and to transmit information; in other words, it is able to exist as a cybernetic system. A landscape exhibits its own type of complexity, and in order to understand it fully it is necessary to focus on a certain organizational level. There are innumerable hierarchical levels and thus an equal number of systems that are nested inside them in one way or another. The behavior of a given subsystem conditions nearby systems both above and below it. The speed with which the processes unfold and thus the scales in time are specific to each level. When going from one level to another, it is therefore necessary to adjust the resolution (Farina, 2001). In the most variants of the landscape, researchers refer to something framed at the human scale. However, this is revised upwards to reveal patterns from satellites, and downwards to reveal mosaics related to the life-spaces of meso- and micro- organism. McPherson and DeStefano (2003), writing from an ecological perspective, identify landscape studies as being those undertaken at quite an extensive spatial scale: less extensive than the "biome" or biosphere", but larger than the ecosystem, community, population, organism or cell (Selman, 2006).

Landscape ecological concepts and applied metric are likely to be useful to addresses the spatial dimension of sustainable planning. The landscape ecological aspect of spatial scale has received so much attention in the literature. Landscape ecology is the study of the interactions between the temporal and spatial aspects of a landscape and its flora, fauna, and cultural components in so far as this impact on ecosystem properties. However, the subject also incorporates the study of water movements, particularly insofar as these impact on ecosystem properties. An understanding of ecological and hydrological pattern and processes not only reveals the complex web of natural interdependencies, but also enroll economic and social systems at these strongly modify the energy and materials inputs into landscape (Selman, 2006). In this context, watershed boundaries, for having well-defined edges make up a fundamental unit for landscape planning (Makhdoum, 2008).

A widely advocated approach to landscape planning is to steward resources on the basis of biogeographic units: that is, segments of the earth's surface defined, not on the basis of traditional political and administrative boundaries. Selman (2006) stated that landscape planning has three main reasons for the popularity of biogeographic units. First, natural systems, as watershed, often form logical units for many resources management decisions, and focusing on an integrative landscape unit may help reduce fragmentation of environmental processes and of policy delivery. Second, neither wildlife species nor hydrological systems recognize administrative boundaries, and their natural geographical range and extend must be taken into account in spatial planning, or even serve as its framework. Finally, people develop particular attachments to landscape on the basis of both physical and cultural factors, and so may possibly identify with distinctive biogeographic space more than with, say, local government districts (Selman, 2006).

Graff, 1993; Metzger and Muller, 1996; Şahin, 1996; Tangtham, 1996; Farrina 2006; Uzun, 2003; Selman, 2006; Bulley et al., 2007; Karadağ, 2007; Şahin, 2007; Makhdoum, 2008; Şahin, 2009b; Uzun, 2003 and Uzun et al., 2012 have drawn attention to the information of watershed in landscape planning. Watersheds can be considered as landscapes. It seems useful to study landscapes by applying the scale of watersheds, which can be considered as multifunctional units in which flows of water and the transfer of nutrients are distinctive processes (Farina, 2001).

3. WATERSHED

Water effects on the environmental and on life in all forms in distribution and circulation of waters (O'Callaghan, 1996). Surface flow, travel of water which is called hydrological circuit and feeding of ground waters, form the basis for ecological processes. The flow of water not only provides a unique ecological feature, but also forms geographically unique areas/spaces.

Surface flow, travel of water which is called hydrological circuit and feeding of ground waters affect landscape from different aspects. Surface flow of water and feeding of the ground waters are related to water period of landscape. Water period depends on permeability values (Uzun and Gültekin, 2013). Hydrological circuit is the process of evaporation and condensation of surface waters with the effects of climactic factors (Karadağ, 2007).

A watershed is the area drained by a river or stream and its tributaries. Generally many watersheds are included in a landscape, and a landscape boundary may or not correspond to the boundaries of watershed (Forman and Godron, 1986). A watershed is a landscape surface area that surrounds and drains into a common waterbody such as a lake, small stream or river basin system (Anonymous, 2012a). Davenport, et al. (2012) defined as watershed is an area of land that drains into a lake or river. As rainwater and melting snow run downhill, they carry sediment and other materials into streams, lakes, wetlands and groundwater. According to United States Environmental Protection Agency (EPA), watershed is the area of land where all of the water that is under it or drains off of it goes into the same place (Anonymous, 2012b). A watershed is a catchment basin that is bound by topographic features, such as ridge tops (Anonymous, 2012c). In addition watershed defined as a physiographic landscape (Şahin, 2007) and units of hydrologically independent areas (McHarg, 1991). In addition a functioning natural unit with interacting biotic and abiotic components in a system whose boundaries is determined by the cycles and flux of energy, materials and organisms. It is valid to describe different ecosystems with different, overlapping sets of boundaries in the same geographic area (e.g. forest ecosystems, watershed ecosystems and wetland ecosystems). A watershed is one of many types of ecosystems (O' Keefe *et al.,* 2012).

A large numbers of terms are very frequently and loosely used to classify watershed in different sizes (micro, small, and large). "Small watersheds are those where the overland flow is the main contributor to peak runoff / flow and channel characteristic do not affect the overland flow". "Large watersheds are those give peak flows are greatly influenced by channel characteristics and basin storage". Watershed classified according to drainage systems; main river watersheds, watersheds and sub-watersheds (micro watersheds). River watersheds are the areas which all the flows on the ground (river, lake, etc.) flow into the sea through a single river mouth, an estuary or delta from a certain point on the water route. Watersheds are defined as multiple territorial areas which feed a certain water resource (river watershed). However, sub-watersheds (micro watersheds) are defined as catchment areas concerning drainage lines in various sizes which feed watersheds and river watersheds (Karadağ, 2007).

Hydrological systems have along with ecological units, long been viewed as a natural basis for division of the earth's surface. Thus the "watershed" or "catchment" has often been proposed as the most appropriate division for landscape planning. Key reasons have been: its relative self-containment in terms of flows of water, other materials and energy; its relationship to geomorphic processes and the consequent recognisability of landform characterizing individual catchments; and the importance of water, often in short or excess supply, to human settlements. Increasingly, landscape ecologists also recognize the importance of water catchments in influencing the nature and functionality of ecosystem, through their role not only in supplying moisture but also moving chemical nutrients along rivers and though ground and soil water (Selman, 2006).

Watershed classification provides a means for generalizing or grouping watersheds by characteristics such as ecological properties, or land use patterns, so that they can be managed, treated, or compared efficiently. Classification can be based on a number of attributes related to natural or anthropogenic differences in watersheds. Natural features include climate, physiography, soils, nutrient productivity, watershed size and connectivity to other aquatic ecosystems. Anthropogenic features are primarily related to land use and include land-use types (urban, agriculture, forest), the degree of hydrologic disturbance and imperviousness, water withdrawals, water quality, in stream habitat conditions, and riparian integrity (Page, *et al.,* 1999).

Climate, hydrology, and geomorphology are physical template to shape forces of ecosystems. The three elements of the physical template and other factors also interact significantly in determining the structure and composition of a watershed and its biotic communities. As a result of different combinations of these formative processes, different types of watersheds are created (O' Keefe *et al.,* 2012). Besides watersheds are continually changing and evolving. Some changes are natural, or are accelerated by human activities. A watershed contains information about all the things happening and lands use history within it (Anonymous, 2012d). Because of that watersheds are frequently used to study and manage environmental resources because hydrologic boundaries define the flow of contaminants and other stressors (O' Keefe *et al.,* 2012).

Each part of a watershed is unique, even though the characteristics of any watershed are similar. All watersheds flow from headwaters to outlets, eventually ending in an ocean. As the water flows, it passes through many parts. And like the parts of a puzzle, if one happens to be damaged, the result affects the whole picture (Anonymous, 2012d). The watersheds are complex ecosystems in which land use, surficial geology, climate, and topography are interrelated with biological components such as vegetation communities (Page, *et al.,* 1999). Weekes (2009) believe that headwater stream flow patterns are homogenous when they have similar climate, bedrock type and hardness, topographical range, drainage area, soils and vegetation). In addition his investigations strongly support that meso-scale geomorphic processes and structures are first order drivers of hydrologic regimes. Geomorphic processes are a part of landscape function. Landscape ecology and catchment hydrology, both disciplines deal with patterns and processes as well as their interactions and functional implications (Schroder, 2006).

A watershed has three primary functions. First, it captures water from the atmosphere. Ideally, all moisture received from the atmosphere, whether in liquid or solid form, has the maximum opportunity to enter the ground where it falls. The water infiltrates the soil and percolates downward. Several factors affect the infiltration rate, including soil type, topography, climate, and vegetative cover. Percolation is also aided by the activity of burrowing animals, insects, and earthworms. Second, a watershed stores rainwater once it filters through the soil. Once the watershed's soils are saturated, water will either percolate deeper, or runoff the surface. This can result in freshwater aquifers and springs. The type and amount of vegetation, and the plant community structure, can greatly affect the storage capacity in any one watershed. The root mass associated with healthy vegetative cover keeps soil more permeable and allows the moisture to percolate deep into the soil for storage. Vegetation in the riparian zone affects both the quantity and quality of water moving through the soil. Water moves through the soil to seeps and springs, and is ultimately released into streams, rivers, and the ocean. Slow release rates are preferable to rapid release rates, which result in short and severe peaks in stream flow. Storm events which generate large amounts of run-off can lead to flooding, soil erosion and siltation of streams (Anonymous, 2012b). This situation, as Schroder (2006) stated, forms the interaction between the landscape and watershed.

Implementing a watershed approach has environmental, financial, social and administrative benefits. As well as its potential for considerable impact on the environment, this type of approach can result in cost savings by building upon the financial resources, knowledge and the willingness of interested people in the watershed to take action. An action plan that focuses on solutions evolves from those knowing the local issues and opportunities. This can help to enhance local and regional economic viability in ways that are environmentally sound and consistent with defined watershed objectives (Anonymous, 2012b).

4. HOW TO DELINEATE WATERSHED USING THE ARCHYDRO MODEL

The advantage of Geographic Information Systems (GIS) technology lies in its data synthesis, the geography simulation, and spatial analysis ability. Spatial analytical techniques, geographical analysis and modeling methods are therefore required to analyses data and to facilitate the decision process at all levels within an urban regional context. GIS approach is very efficient as a tool to facilitate the decision-making process (Laurini, 2001). GIS has emerged as a significant support tool for managingand analyzing water resources using digital elevation models (DEM) of land surface terrain.

Various methods are used in determining the river basin boundaries. The traditional methods are determining and drawing the boundaries of drainage divides, peaks, stream beds on the topographical maps by hand. However, the modern methods are determining the boundaries by digitising and analysing the contour lines developed by GIS.

Arc Hydro is a geospatial and temporal data model for water resources designed to operate within ArcGIS (Maidment, 2003). Arc Hydro is a geographical data model that describes hydrological systems. A data model is a set of conceptsexpressed in a data structure; the data model describes a simplification of reality using tables and relationships within a database. Geographic data models use database structures to describe the world or part of it usingGIS technology. The ArcHydro data model is a conceptualization of surface water systems and describes features such as river networks, watersheds and channels. The data model can be the basis for a "hydrologic information system" which is a synthesis of geospatial and temporal data supporting hydrologic analysis and modeling. Arc Hydro integrates geospatial and temporal information into a defined structure. Based on this structure analysis and modeling tools can be applied. The data model provides a common characterization and understanding of the hydrological system and this description canbe utilized by multiple models, analysis tools and decision support systems all referring to the same common structure (Kovar and Nachtnebel, 1996; Strassberg *et al.,*2011).

This study is going to demonstrate the use of the ArcHydroModel to determine watershed boundaries of a small stream (Köprü stream) in the Central Mediterranean Basin. The hydrologic modeling involves delineating streams network and watersheds, and getting some basic watershed properties such as area, flow length, stream network density, etc. Traditionally this was (and still is!) being done manually by using topographic/contour maps. But in ArcHydro

Model analysis is performed by using DEM (Ayhan *et al.*, 2012). DEM generation from topographic maps that derived from a 10 meter DEM from the General Command of Mapping (Turkey).

Watershed and drainage systems can define generally with 4 stages and 11 analysis in ArcHydro module. At the first stage of the analysis, ''DEM reconditioning'' and ''fill sink'' analysis, which are confirmation and preparation processes for the given analysis, are carried out. At the second stage, ''Flow direction, Flow Accumulation, Stream Definition'' and ''Stream Segmentation'' analysis, by which evaluations concerning surface flow are made, are carried out. At the third stage, ''Catchment Grid Delineation'' and ''Catchment Polygon Processing'' analysis, by which catchment areas are determined, are carried out. At the last stage, "Drainage Line Processing", "Drainage Point Processing" and "Batch Watershed Delineation" analysis, by which watershed boundaries are defined by evaluating drainage systems according to surface flow and catchment areas, are carried out. But first of all, Archydro tools must be downloaded to the computer to start the analysis. Archydro tool 1.3 is downloaded because of ArcMap 9.3 is used in this study.

First stage of Archydro Model is Terrain Preprocessing. Arc Hydro Terrain Preprocessing should be performed in sequential order. All of the preprocessing must be completed before Watershed Processing functions can be used. DEM reconditioning and filling sinks might not be required depending on the quality of the initial DEM. DEM reconditioning involves modifying the elevation data to be more consistent with the input vector stream network. This implies an assumption that the stream network data are more reliable than the DEM data, so you need to use knowledge of the accuracy and reliability of the data sources when deciding whether to do DEM reconditioning. By doing the DEM reconditioning you can increase the degree of agreement between stream networks delineated from the DEM and the input vector stream networks (Mervade *et al.,* 2009; Ayhan *et al.*, 2012; Mervade, 2012).

DEM Reconditioning: This function modifies a DEM by imposing linear features onto it (burning/fencing). The function needs as input a raw dem and a linear feature class (like the river network) that both have to be present in the map document (Mervade *et al.,* 2009; Ayhan *et al.*, 2012; Mervade, 2012). This function is located on Terrain Preprocessing on the ArcHydro Toolbar (*Terrain Preprocessing*→DEM Manipulation→DEM Reconditioning) (Mervade *et al.,* 2009; Ayhan *et al.*, 2012; Mervade, 2012).

Select the appropriate Raw DEM (köprü_dem) and AGREE stream feature (köprü_str). Set the Agree parameters as shown. You should reduce the Sharp drop/raise parameter to 10 from its default 1000. The output is a reconditioned Agree DEM (default name Agree DEM). A personal geodatabase with the same name as your ArcMap document has also been created as shown in the following ArcCatalog view (Figure 1.)(Mervade *et al.,* 2009; Ayhan *et al.*, 2012; Mervade, 2012).

Fill Sinks: This function fills the sinks in a grid. If cells with higher elevation surround a cell, the water is trapped in that cell and cannot flow. The Fill Sinks function modifies the elevation value to eliminate these problems.The model readjusts the height value with this stage to solve the problem. Therefore, the drainage networks' being asunder is prevented. This function is located on

Terrain   Preprocessing   on   the   ArcHydrotoolbar   (Terrain
Preprocessing→DataManipulation→Fill Sinks) (Mervade *et al.*, 2009; Ayhan *et al.*, 2012; Mervade, 2012).

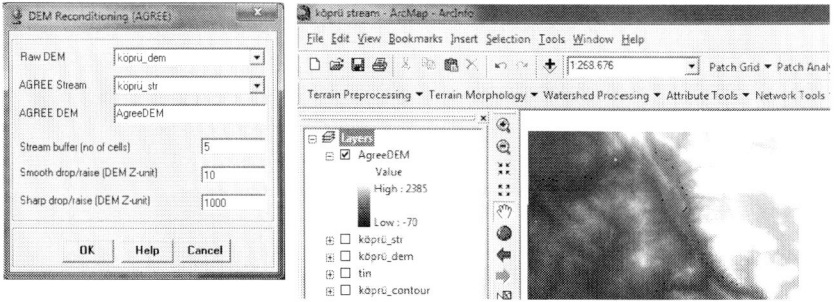

Figure 1.DEM Reconditioning menu and AgreeDEMlayer.

Confirm that the input for DEM is AgreeDEM. The output is the Hydro DEM layer, named by default Fil. This default name can be overwritten. Leave the other options unchanged. The Fil layer is added to the map, when the process completed (Figure 2.) (Mervade *et al.*, 2009; Ayha n*et al.*, 2012; Mervade, 2012).

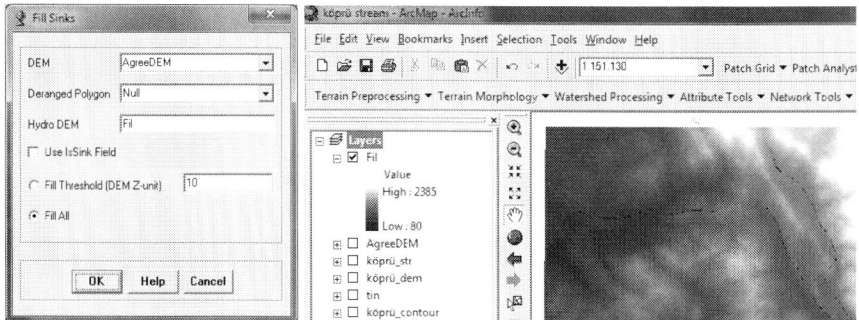

Figure 2. Fill Sinks menuand Fillayer.

Flow direction: This function computes the flow direction for a given grid. Each grid has a value of height and water flow will be towards the lowest one, by comparing the height values of 8 grids. The flow direction is defined as ''8 directional flow model'' in the computer environment. Digital values, which are developed depending on the directions, are used to show the flow direction of the grid in the module. This function is located on Terrain Preprocessing on the ArcHydro toolbar (Djokic 2008, Mervade *et al.*, 2009; Ayhan *et al.*, 2012; Mervade, 2012).

Confirm that the input for Hydro DEM is Fil. The output is the Flow Direction Grid, named by default Fdr. This default name can be overwritten.

The flow direction grid Fdr is added to the map, when the process completed (Figure 3.) (Mervade *et al.,* 2009; Ayhan *et al.,* 2012; Mervade, 2012).

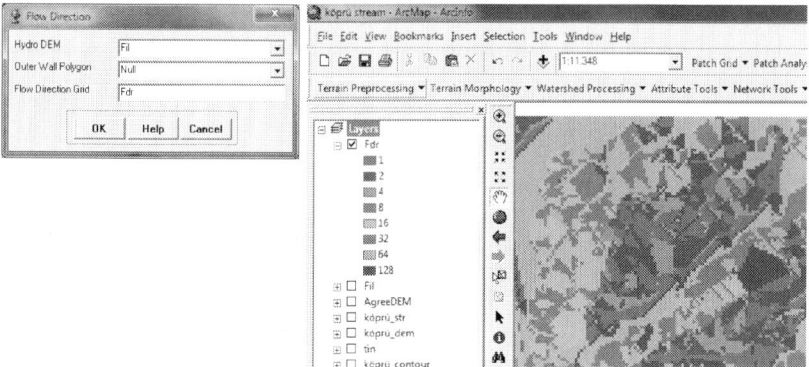

Figure 3. Flow direction menu and Fdrlayer.

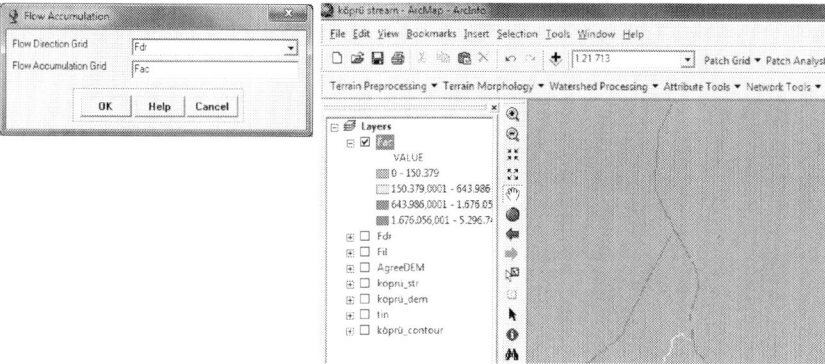

Figure 4. Flow Accumulation menu and Faclayer.

Flow Accumulation: This is the stage in which the cells taking place in the catchment area of each cell are calculated. The water gathered in the lowest grade is calculated, by assuming that each cell has 1 unit of water. The system defines the value of the cells having no flow as zero, and cells in which water gathers are defined in the number of cells having flow. The flow calculation is carried out by taking 8 cells as basis. This function is located on Terrain Preprocessing on the ArcHydrotoolbar (Mervade *et al.,* 2009; Ayhan *et al.,* 2012; Mervade, 2012).

Confirm that the input of the Flow Direction Grid is Fdr. The output is the Flow Accumulation Grid having a default name of Fac that can be overwritten. The flow direction grid Fac is added to the map, when the process completed. Adjust the symbology of the Flow Accumulation layer Fac to a multiplicatively increasing scale to illustrate the increase of flow accumulation as one descends into the grid flow network (Mervade *et al.,* 2009; Ayhan *et al.,* 2012; Mervade, 2012).

Zoom-in to a stream network junction to see how the symbology changes from light to dark color as the number of upstream cells draining to a stream increase from upstream to downstream. If you click at any point along the stream network on Fac grid using the identify button you can find the area draining to that point by multiplying the Fac number by the area of each cell (cell size x cell size which is 30.89 x 30.89 in this case) (Figure 4.) (Mervade *et al.,* 2009; Ayhan *et al.*, 2012; Mervade, 2012).

Stream Definition: This function computes a stream grid which contains a value of "1" for all the cells in the input flow accumulation grid that have a value greater than the given threshold. All other cells in the Stream Grid contain no data. This function is located on Terrain Preprocessing on the ArcHydro toolbar (Mervade *et al.,* 2009; Ayhan *et al.*, 2012; Mervade, 2012).

Confirm that the input for the Flow Accumulation Grid is "Fac". The output is the Stream Grid. "Str" is its default name that can be overwritten. The stream grid Str is added to the map, when the process completed (Figure 5.) (Mervade *et al.,* 2009; Ayhan *et al.*, 2012; Mervade, 2012).

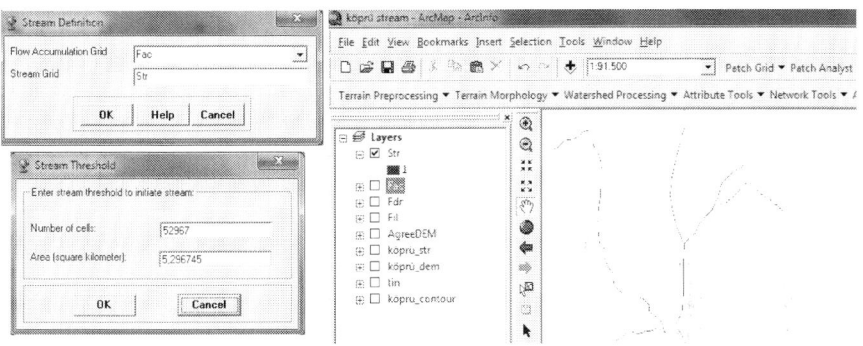

Figure 5. Stream Defination menu and Str layer.

Stream Segmentation: This function creates a grid of stream segments that have a unique identification. Either a segment may be a head segment, or it may be defined as a segment between two segment junctions. All the cells in a particular segment have the same grid code that is specific to that segment. This function is located on Terrain Preprocessing on the ArcHydro toolbar (Mervade *et al.,* 2009; Ayhan *et al.*, 2012; Mervade, 2012).

Confirm that Fdr and Str are the inputs for the Flow Direction Grid and the Stream Grid respectively. Unless you are using your sinks for inclusion in the stream network delineation, the sink watershed grid and sink link grid inputs are Null. The output is the stream link grid, with the default name StrLnk that can be overwritten. The link grid StrLnk is added to the map, when the process completed (Figure 6.) (Mervade *et al.,* 2009; Ayhan *et al.*, 2012; Mervade, 2012).

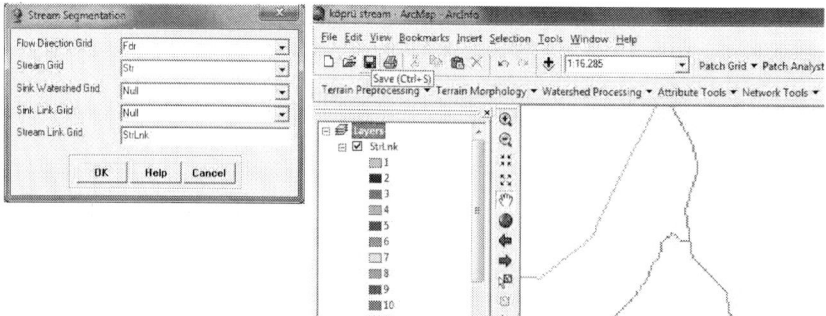

Figure 6. Stream Segmentation menu and StrLnk layer.

Catchment Grid Delination: This function creates a grid in which each cell carries a value (grid code) indicating to which catchment the cell belongs. The value corresponds to the value carried by the stream segment that drains that area, defined in the stream segment link grid. This function is located on Terrain Preprocessing on the ArcHydro toolbar (Mervade *et al.*, 2009; Ayhan *et al.*, 2012; Mervade, 2012).

Confirm that the input to the Flow Direction Grid and Link Grid are Fdr and Lnk respectively. The output is the Catchment Grid layer. Cat is its default name that can be overwritten by the user. The link grid StrLnk is added to the map, when the process completed. The Catchment grid Cat is added to the map, when the process completed. In addition study case will have 70 catchment (Figure 7.) (Mervade *et al.*, 2009; Ayhan *et al.*, 2012; Mervade, 2012).

Catchment Polygon Processing:This function converts a catchment grid into a catchment polygon feature. This function is located on Terrain Preprocessing on the ArcHydro toolbar (Mervade *et al.*, 2009; Ayhan *et al.*, 2012; Mervade, 2012).

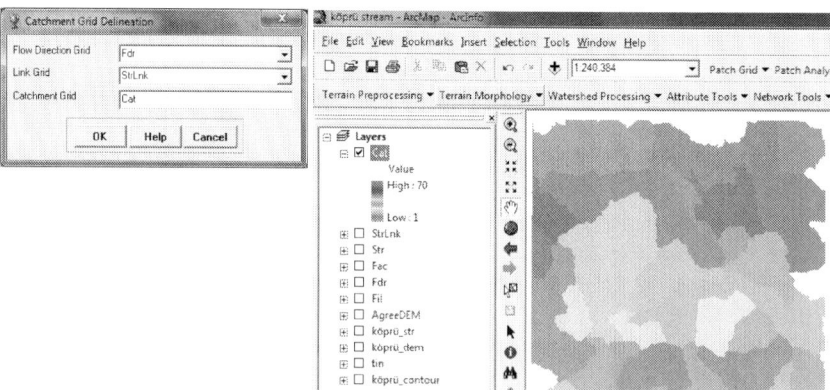

Figure 7. Catchment grid menu and Cat layer.

Confirm that the input to the CatchmentGrid is Cat. The output is the Catchment polygon feature class, having the default name Catchment that can be overwritten. The polygon feature class Catchment is added to the map, when the process completed. In addition there are important information (HydroID assigned, Length and Area attributes of catchment) in attribute table of Catchment (Figure 8.) (Mervade *et al.,* 2009; Ayhan *et al.*, 2012; Mervade, 2012).

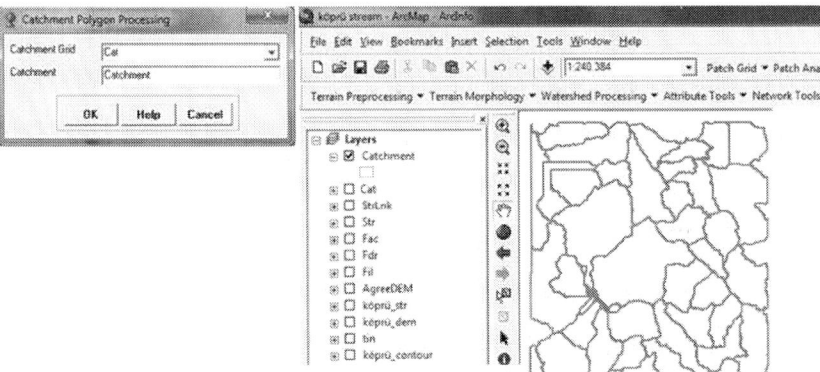

Figure 8. Catchment polygon processing menu and Catchment layer.

Drainage Line Processing: This function converts the input Stream Link grid into a Drainage Line feature class. Each line in the feature class carries the identifier of the catchment in which it resides. This function is located on Terrain Preprocessing on the ArcHydro toolbar (Mervade *et al.,* 2009; Ayhan *et al.*, 2012; Mervade, 2012).

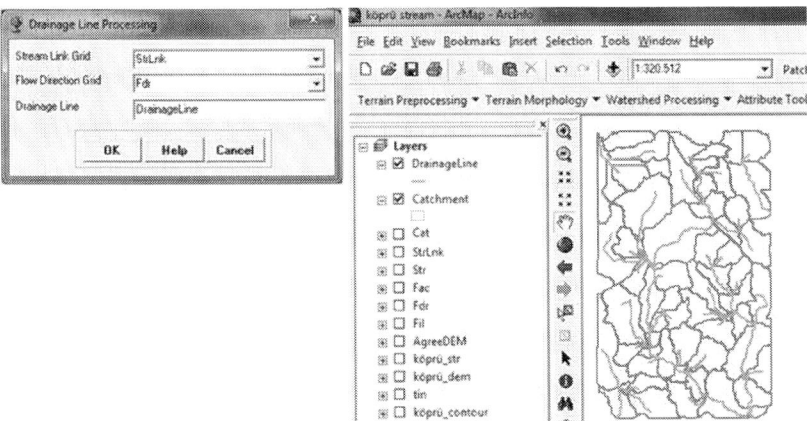

Figure 9. Drainage line processing menu and Drainageline layer.

Confirm that the input to Link Grid is Lnk and to Flow Direction Grid Fdr. The output Drainage Line has the default name DrainageLine that can be overwritten.The linear feature class DrainageLine is added to the map, when the process completed (Figure 9.) (Mervade *et al.,* 2009; Ayhan *et al.,* 2012; Mervade, 2012).

Drainage Point Processing: This function allows generating the drainage points associated to the catchments. This function is located on Terrain Preprocessing on the ArcHydro tools(Mervade *et al.,* 2009; Ayhan *et al.,* 2012; Mervade, 2012).

Confirm that the inputs are as below. The output is Drainage Point with the default name DrainagePoint that can be overwritten. Upon completion of the process, the point feature class "DrainagePoint" is added to the map (Figure 10.) (Mervade *et al.,* 2009; Ayhan *et al.,* 2012; Mervade, 2012).

Watershed Processing: Arc Hydro toolbar also provides an extensive set of tools for delineating watersheds and subwatersheds. These tools rely on the datasets derived during terrain processing.

Batch watershed delineation function delineates the watershed upstream of each point in an input Batch Point feature class. Batch Point Generation can be used to determine the outlet of the watershed. Arrange your display so that Fac, Catchment and DrainageLine datasets are visible. Zoom-in near the outlet of the Köprü stream watershed (Figure 11.). The display should look similar to the figure shown below and be zoomed in sufficiently so you can see and click on individual grid cells. Our goal is to create an outlet point on the flow accumulation path indicated by Fac grid where the flow leaves the Köprü stream watershed (Mervade, 2012).

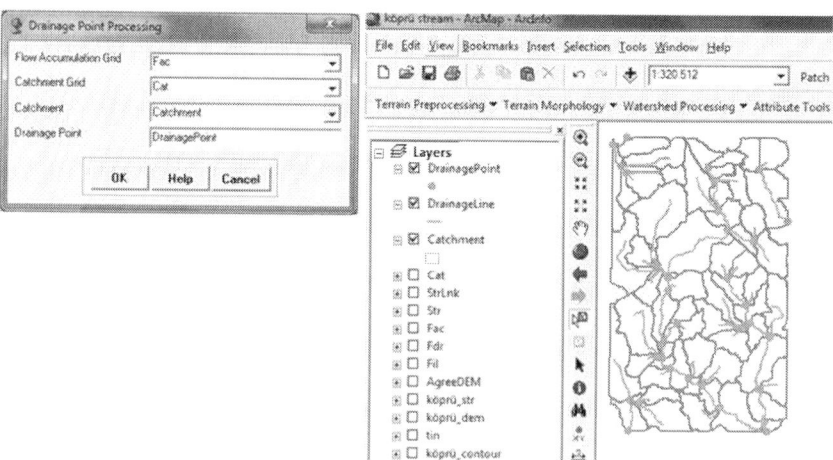

Figure 10. Point processing menu and Drainage point layer.

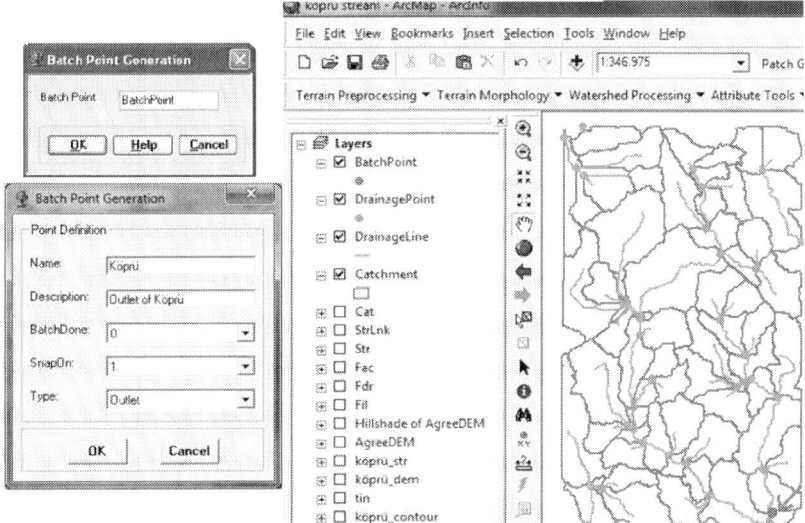

Figure 11. Batch Point generation

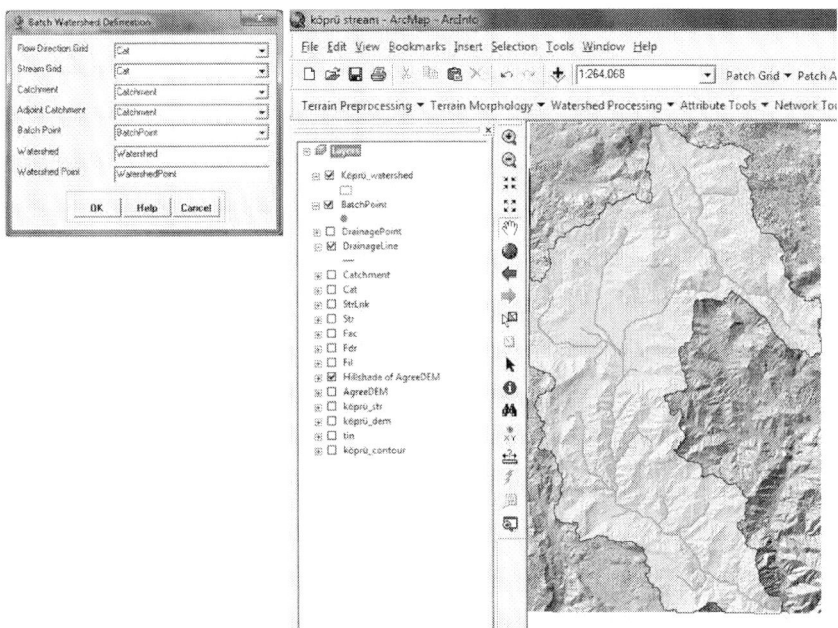

Figure 12. Batch watershed delineation and Köprü stream watershed

Batch watershed delineation function delineates the watershed. This function is located on Watershed Processing on the ArcHydro Toolbar. Confirm that Fdr is the input to Flow Direction Grid, Str to Stream Grid, Catchment to Catchment, AdjointCatchment to AdjointCatchment, and BatchPoint to Batch Point. For output, the Watershed Point is WatershedPoint, and Watershed is Watershed (Figure 12.). WatershedPoint and Watershed are default names that can be overwritten (Mervade *et al.,* 2009; Ayhan *et al.*, 2012; Mervade, 2012).

You can see that area and length, if you open the attribute table of Köprü_watershed. In addition you will see that these two are related through HydroID–the DrainID of WatershedPoint is equal to the HydroID of the watershed, when you open the attribute table of catchment and DranaigePoint. At the same time you can learn length of drainage line from attributes table of DranaigeLine (Mervade *et al.,* 2009; Mervade, 2012).

5. CONCLUSION

The future of our present societies is determined by environmental, social, economic and political situations and the problems and the solutions concerning these issues. Landscape, which can be defined as the interaction space or product of natural and cultural processes, puts its relation with future at this point. The concept of future brought about the concept of planning, which is evaluated in different topics, and transformed landscape planning into a part of the future. Therefore, ecological, aesthetic and economic importance of landscape has become a topic for many researchers and the need for landscape planning is emphasised. At last, vital importance of landscape and need for its being planned were transferred to a legal text when European Landscape Convention was signed in 20 October 2000.

The aims of European Landscape Convention are to promote landscape protection, management and planning, and to organise European co-operation on landscape issues. The Convention applies to the entire territory of the Parties and covers natural, rural, urban and peri-urban areas. It includes land, inland water and marine areas. It concerns landscapes that might be considered outstanding as well as every day or degraded landscapes. Landscape planning means strong forward-looking action to enhance, restore or create landscapes in the Convention (Anonymous, 2000). The convention intended to plan landscape with a "comprehensive, pedant, holistic, coordinated, participant, rationalist" approach. Also, with the expression "transboundary landscape", which took place in 9th clause of the convention and which emphasised local and regional cooperation, brought about the concept of "planning boundary".

The planning, which is defined as balancing the needs and the resources in the long run by complying with the reasonable priorities to reach certain aims with limited resources (Keleş, 2004), is a versatile activity from upper scale to subscale and a body of decisions related to past, present and future integrating social, economic, political, physical, anthropogenic and technical elements, as Alipour (1996) stated (Uzun *et. al.* 2012). This general definition of planning requires landscape planning to be made in different scales and accordingly in certain boundaries. Also, as Uzun *et. al.* (2012) stated, the boundaries of the study area are the first stage of planning and are very important in clarifying the goal. The data gathering, which enables the planning to be carried out systematically and defines success (Mcharg, 1967), depends on the boundaries of the planning area. Ultimately, the management process of realising the plan will be integrated with the administrative structuring within the boundaries. All these put the importance of the question ''What should be the boundary of landscape planning?'' forward.

When determining the boundaries of landscape, the fact that landscape is "a space in which natural, socio-cultural and economical life come together" should not be ignored. This situation emphasises that the boundary of landscape shouldn't just describe the natural areas (eco-zone, ecoregion, habitat, etc.) or administrative spaces. Therefore, the boundary will be integrated with the body of the landscape. Within this context of approach, there are various consistent points of view about the boundary of landscape. Meijerink (1985) considered that watersheds were the best units in which the interactions of human and natural resources, and the geographical distribution of their consequences could be observed and modeled (Metzger and Muller, 1996). Gregersen et al. (1987), said watersheds can use as a physical-biological and a socioeconomic-political units for planning of natural resources (Graff, 1993). According to Farrina (2006) watersheds are examples of the hierarchical organization of the landscape. River watershed is composed of sub-watershed, each of which is composed of smaller-order watershed. The upper and lower limits of this hierarchy are not definitive but it is possible to move in both directions, including smaller and larger basins. Tangtham (1996) and Karadağ (2007) lay stress on watershed classification is thus anticipated as a useful tool for management and planning of natural resources. Selman (2006) emphasized the importance of watershed boundaries in landscape ecology. Makhdoum (2008) indicated that the mapping unite (or land unit) is freely derived from watershed, land system, land form units and ecosystems, at different scale level. He accepted watershed as one of mapping units in land ecology. Bulley et al. (2007) point out that watershed provides an important spatial framework to develop a classification system. Şahin (2007) and Şahin (2009) suggests that watershed can be descriptive and administrative units for landscape planning. According to EPA watershed is an example of hierarchical system in nature (Anonymous 2012a). Efe and Aydın (2009), indicated that the provincial boundaries which constitute the framework of the administrative organization where planning is currently authorized do not coincide with the natural boundaries. They suggest redefining the provincial boundaries compatible with watershed for the protection of the nature.

Actually watershed clarifies the complexity of boundary in the landscape. When we consider the importance of water in life of living things, its effect on establishing, developing and even collapsing civilisations, it is clear that watershed will be effective boundaries in landscape planning; because water turns into the interaction space of natural and cultural life while forming socio-cultural and economical life by its presence. This situation enables a watershed to turn into not only a natural boundary, but also a boundary that effects a human's life. Also, the landscape changes as nature reshapes with human life. The changing landscape gains a new character. This character is not only a product of the change of the natural structure caused by the human presence, but also can be expressed with a hierarchal system from a local scale to upper scales. Thus, the watershed supports the scale approach in the planning with its hierarchal structure (main river basin, basin, subbasin, microbasin). Along with that, the main river basins that go beyond the national boundaries will be able to easily define its collaborators in the transnational landscapes.

FOOT NOTES

1) Pattern refers to its spatial arranges of ecosystems and their type, number, size, shape, and relative relationship over the landscape (Forman, 1995).
2) Patch is a wide relatively homogeneous area that differs from its surroundings. Patches have familiar attributes, such as large or small, rounded or elongated, and straight or convoluted boundaries (Forman, 1995).
3) Corridors, as strips that differ from their surroundings, permeate the land (Forman, 1995).
4) Matrix is the background ecological system of a landscape (Forman, 1995).

REFERENCES

1.   Anonymous2000The European Landscape Convention. Council of European, 51p.
2.   Anonymous2008Landscape Planning the Basis of Sustainable Landscape Development. Federal Agency for Nature Conservation. Gebr Klingenberg Buchkunst Leipzing Gmbh Printed, 52p.
3.   Anonymous, . Watershed Planning http://www.conservewy.com (Accessed on: 10.11.2012)
4.   Anonymous, . What is a Watershed. United States Environmental Protection Agency. www. http://www. water.epa.gov Anonymous, 2012a. (Accessed on: 2.12.2012)
5.   c Anonymous, . An Introduction to Urban Watersheds. http://www.gdrc.org/uem/water/watershed/introduction.html (Accessed on: 12.01.2012)
6.   d Anonymous, . Getting to Know. http://www.agric.gov.ab.ca (Accessed on: 12.12.2012).
7.   M. Antrop, 2005From Holistic Landscape Synthesis to Transdisciplinary Landscape ManagementIn: Tress, B., Tress, G. Fry, G. En Opdam, P. From Landscape Research to Landscape Planning: Aspect of Integration, Education and Aplication. Wageningen, Frontis, 2750

8. N. G. Ayhan, K. Seyrek, and A. H. Sargin, 2012Application of Geographic Information Systems in Hydrology and Water Resources Management.Printed Course Note. ISLEM Group of Companies, Ankara, Turkey, 80p.

9. W. L. Baker, 1989A Review of Models of Landscape Change. Landscape Ecology, 2(2), 111133

10. O. Bastian, R. Krönert, and Z. Lipsky, 2006Landscape Diagnosis on Di?erent Space and Time Scales- A Challenge for Landscape Planning. Landscape Ecology, 21359374

11. B. Booth, 2000Using ArcGIS 3D Analyst GIS by ESRI.Environmental Systems Research Institute, Inc, USA, 212p.

12. H. N. N. Bulley, J. W. Merchant, D. B. Marx, J. C. Holz, and A. A. H. Holz, 2007A. Gis-based, Approach to Watershed Classification for Nebraska Reservoirs. Journal of The American Water Resources Association605620

13. CaO'CallaghanJ. R. 1996Land Use: The Interaction of Economics, Ecology and Hydrology.Chapman & Hall Publisher, London, UK, 197p.

14. W. Clark, 2010Principles of Landscape Ecology. Nature Education Knowledge, 3(10), 34p.

15. T Davenport, ., K Minsch, ., C Novak, . S Olsenholler, ., F Sagona, ., E Sprunger, . and J Warren, . 2012. Getting to Know Your Local Watershed. http://www.ctic.purdue.edu

16. D. Dijokic, 2008Comprehensive Terrain Preprocessing Using Arc Hydro Tools. ESRI, New York, USA.

17. V. H. Dökmeci, 1996Sustainable and Landscape Planning. Unprinted Master Seminar. University of Ankara, Graduate School of Natural and Applied Sciences, Department of Landscape Architecture, Ankara, Turkey.

18. X. Du-ning, and L. Xiu-zhen, 1999Core Concepts of Landscape Ecology. Journal of Environmental Sciences, 11(2), 131135

19. M. Efe, and B. A. Aydin, 2009Change ability of Planning Based on Administrative Boundaries and Proposal Boundaries of Provincial Based on Basin. Aegean Geographical Journal, Izmir, Turkey, 18(1-2), 7384

20. A. Farina, 2001Landscapes and Their Ecological Components. The Living World, 4435448

21. A. Farina, J. Bogaert, and I. Scipani, 2005Cognitive Landscape and Information: New Perspectives to Investigate the Ecological Complexity.Bio Systems79235240

22. A. Farina, 2006Principles and Methods in Landscape Ecology. Kluwer Academic Publishers, USA, 436p.

23. R. T. T. Forman, and M. Gordon, 1986Landscape Ecology.Wiley and Sons, New York, USA, 620p.

24. R. T. T. Forman, 1995Land Mosaic. The Ecology of Landscapes and Regions.Cambridge University Press, UK, 632p.

25. J. Graff, 1993Soil Conservation and Sustainable Land Use. A Royal Institute Series, Amsterdam, 191p.

26. E. Ivits, B. Koch, T. Blaschke, M. Jochum, and P. Adler, 2005Landscape Structure Assessment with Image Grey-Values and Object-Based Classification at Three Spatial ResolutionsInternational Journal of Remote Sensingp.2975-2993.

27. M. Jones, J. Howar, K. R. Olwing, J. Primdahl, and I. S. Herlin, 2007Multiple Interfaces of the European Landscape. NorwegianJournal of Geography, 61(4), 207216

28. S. Kaden, 2003GIS in Water-Related Environmental Planning and Management: Problems and SolutionsApplication of Geographic Information Systems in Hydrology and Water Resources Conference, IAHS Publisher, Vietnam, 211385396

29. A. A. Karadag, 2007Development of Participatory Watershed Management Model: The Case of Kovada Lake. Ph. D. Thesis. University of Ankara, Graduate School of Natural and Applied Sciences, Department of Landscape Architecture, Ankara, Turkey, 254p.

30. R. Keles, 2012Urbanization Policy. Imge Kitapevi Press, 8Ankara, Turkey, 703p.

31. K. Kovar, and H. P. Nachtnebel, 1996Application of Geographic Information Systems in Hydrology and Water Resources ManagementThe International Association of Hydrological Sciences, 235p.

32. M. Köseoglu, 1982Landscape Evaluation Methods. University of Ege Faculty of Agriculture. Ofset Publisher, Izmir, Turkey, 430p.

33. E. Kurum, and S. Sahin, 2000Determination of High Landscape Value in Watershed Management Areas: Analysis of Soil-Vegetation Protection. Our Environment and Landscape Architecture in the 2000's Symposium. University of Ankara, Turkey, 7581

34. R. Laurini, 2001Information Systems for Urban Planning.Taylor & Francis, London and New York, 368p.

35. D. R. Maidment, 2002Arc Hydro: GIS for Water Resources,ESRI Press, Redlands, 201p.

36. M. F. Makhdoum, 2008Landscape Ecology or Environmental Studies (Land Ecology) (European Versus Anglo- Saxon Schools of Thought). J. Int. Environmental Application & Science, 3 (3), 147160

37. J. Makhzoumi, and G. Pungetti, 1999Ecological Landscape Design and Planning: The Mediterranean Library. E&FN Spon Publisher, New York, USA, 330p.

38. Ü. Mander, and M. Antrop, 2003Multifunctional Landscapes Continuity and Change.Wit Press, USA, 289p.

39. I. L. Mcharg, 1967Design in Nature.John Wiley & Sons. Washington, USA, 198p.

40. V. Mervade, D. Maidment, and O. Robayo, 2009Watershed and Stream Network Delineation GIS in Water Resources. University of Texas, Center for Research in Water Resources, Printed Lecture Note, Texas, USA, 26p.

41. V. Mervade, 2012Watershed and Stream Network Delineation Using ArcHydro Tools. University of Purdue, School of Civil Engineering, Printed Lecture Note, USA, 22p.

42. J. P. Metzger, and E. Muller, 1996Characterizing the Complexity of Landscape Boundaries by Remote Sensing.Landscape Ecology, 11 (2), 6577

43. I. L. Mcharg, 1991Design With NatureWiley and Sons, New York, USA, 198p.

44. S. Mostaghimi, S. W. Park, R. A. Cooke, and S. Y. Wang, 1997Assessment of Management Alternatives On a Small Agricultural WatershedJournal of Water Resources, 31 (8), 18671997

45. P. Opdam, E. Steingröver, and S. Van Rooij, 2006Ecological Networks: A Spatial Concept for Multi-Actor Planning of Sustainable LandscapesLandscape and Urban Planning322332

46. O. Callaghan, J. R. 1996Land Use: The Interaction of Economics, Ecology and Hydrology.London: Chapman & Hall, 216p.

47. O. Keefe, T. C. Elliott, S. R. Naiman, R. J. 2012Introduction to Watershed Ecology. Watershed Academy Web Documents, Environmental Protection Agency, USA, 137

48. N. Page, K. Rood, T. Holz, P. Zandbergen, R. Horner, and M. Mcphee, 1999Proposed Watershed Classification System for Stormwater Management in The GVS & DD Area. Environmental Monitoring and Assessments Task Group, Washington, USA, 125p.

49. B. Schroder, 2006Pattern, Process, and Function in Landscape Ecology and Catchment Hydrology-How can Quantitative Landscape Ecology Support Predictions in Ungauged Basins (PUB). Hydrology and Earth System Sciences Discuss, 11851214

50. P. Selman, 2006Planning at the Landscape ScaleRoutledge Publisher, Newyork, USA, 214p.

51. F. Steiner, 1999The Living Landscape: An Ecological Approach to Landscape Planning.McGrawHill, New York, 462p.

52.   G. Strassberg, N. L. Jones, and D. R. Maidment, 2011Arc Hydro Groundwater: GIS for Hydrogeology.Esri Press, USA, 160 p.

53.   S. Sahin, 1996A Research on Determining and Evaluating the Landscape Potential of Dikmen Valley. Ph. D. Thesis. Ankara University, Graduate School of Natural and Applied Sciences, Department of Landscape Architecture, Ankara, Turkey, 160p.

54.   S. Sahin, 2007Co-Operative Approach in the Implementation of European Landscape Conventionand European Water Framework Directive in Turkey: Joined up Thinking. International Congress River Basin Management, 1218227

55.   S. Sahin, 2009aLandscape Ecology: Concepts, Methods and Applications. Public Administration Institute for Turkey and the Middle East (TODAIE) Publisher, Ankara, Turkey, 231p.

56.   . Sahin, Sustainable Landscape Assessment of River Catchments in the Example of Dikmen Brook in Ankara, Turkey. International Journal of Geosciences, 57 (2), 3346

57.   N. Tangtham, 1996Watershed Classification: The Macro Land-Use Planning for the Sustainable Development of Water Resources. Advances in Water Resources Management and Wastewater Treatment Technologies Workshop. Suranaree University of Technology, Bangkok, 24p.

58.   M. G. Turner, R. H. Gardner, And O'Neill, R.2002Landscape Ecology in Theory and Practices: Pattern and ProcessSpringer Verlag, New York, USA, 404p.

59.   O. Uzun, 2003Landscape Assessment and Development of Management Model for Düzce, Asarsuyu Watershed. Ph. D. Thesis. University of Ankara, Graduate School of Natural and Applied Sciences, Department of Landscape Architecture, Ankara, Turkey, 470p.

60.   O. Uzun, and P. Gültekin, 2012Process Analysis in Landscape Planning, The Example of Sakarya/Kocaali, TurkeyScientific Research and Essays313331

61.   O. Uzun, E. F. Ilke, G. Çetinkaya, F. Erduran, and S. Açiksöz, 2012Landscape Management: Conservation and Planning Project for Konya, Bozkir-Seydisehir-Ahirli-Yalihüyük Districts and Sugla Lake. The Project. Ministry of Environment and Forestry, General Directorate of Nature Conservation and National Parks, Division of Landscape Conservation, Ankara, Turkey, 175p.

62.   U. Waltz, 2011Landscape Structure, Landscape Metrics and Biodiversity. Living Reviews in Landscape Research, 5(3), 135

63.   A. Weekes, 2009Process Domains as a Unifying Concept to Characterize Geohydrological Linkages in Glaciated Mountain HeadwatersPh. D. Thesis. University of Washington, USA, 151 p.

64.   E. Zaimoglu, 2003A Search on Landscape Planning of Selçuk (Izmir) and It's Around. Msc Thesis. Ege University, Graduate School of Natural and Applied Sciences, Department of Landscape Architecture, Ankara, 141p.

CHAPTER 10

Energy Transition: Missed Opportunities and Emerging Challenges for Landscape Planning and Designing

Renée M. de Waal * and Sven Stremke

Landscape Architecture Group, Wageningen University, P.O. Box 47, 6700 AA Wageningen, The Netherlands

ABSTRACT

Making the shift from fossil fuels to renewable energy seems inevitable. Because energy transition poses new challenges and opportunities to the discipline of landscape architecture, the questions addressed in this paper are: (1) what landscape architects can learn from successful energy transitions in Güssing, Jühnde and Samsø; and (2) to what extent landscape architecture (or other spatial disciplines) contributed to energy transition in the aforementioned cases. An exploratory, comparative case study was conducted to identify differences and similarities among the cases, to answer the research questions, and to formulate recommendations for further research and practice. The comparison indicated that the realized renewable energy systems are context-dependent and, therefore, specifically designed to meet the respective energy demand, making use of the available potentials for renewable energy generation and efficiency. Further success factors seemed to be the presence of (local) frontrunners and a certain degree of citizen participation. The relatively smooth implementation of renewable energy technologies in Jühnde and on Samsø may indicate the importance of careful and (partly) institutionalized consideration of landscape impact, siting and design. Comparing the cases against the literature demonstrated that landscape architects were not as involved as they, theoretically, could have been. However, particularly when the aim is sustainable development, rather than "merely" renewable energy provision, the integrative concept of "sustainable energy landscapes" can be the arena where landscape architecture and other disciplines meet to pursue global sustainability goals, while empowering local communities and safeguarding landscape quality.

Keywords: renewable energy; sustainable energy landscapes; landscape architecture; operational design; strategic design; climate change mitigation; transition management; Güssing; Jühnde; Samsø

1. INTRODUCTION

Making the shift from fossil fuels to renewable energy (commonly referred to as sustainable energy transition, renewable energy transition or, simply, energy transition) seems inevitable [1]. Important drivers are the adverse effects of the use of fossil fuels on the environment, geopolitical tensions and the security and affordability of energy in the long run (see [2,3,4,5]). For the European Union, it has been agreed that, by 2020, the share of renewable energy should be 20% of the total energy provision. In 2011, the share was at 13% [6]. Energy transition has the potential to contribute to sustainable development [4,5,7] when, among other conditions, "equitable availability of energy services to all people and the preservation of the Earth for future generations" is met [4] (p. xix). Since aspiring sustainable development is worthwhile beyond fulfilling international commitments, there is broad consensus that the implementation of renewable energy requires paramount attention.

According to ECLAS, the European Council of Landscape Architecture Schools, landscape architecture is "the discipline concerned with mankind's conscious shaping of his external environment. It involves planning, design and management of the landscape to create, maintain, protect and enhance places so as to be both functional, beautiful and sustainable (in every sense of the word), and appropriate to diverse human and ecological needs." [8]. Energy transition is relevant to landscape architecture, because it is in line with the discipline's striving for sustainability and, because changes take place in the physical landscape, its material object of work. Similar to, but more intensively than conventional energy provision, renewable energy technologies occupy land and influence the environment around the world.

Landscape architects have been involved in energy transition, for instance by planning and designing renewable energy technologies in the landscape. Beyond that, increasingly, there is a belief among landscape architects that the spatial domain can (and should) contribute more strategically to energy transition. This could be done, for instance, by energy-conscious spatial organization of land use functions, enabling energy savings and facilitating renewable energy provision (see [9,10,11,12]). The new challenges that energy transition poses to landscape architects, among others, are specified by Radzi and Droege [13] (p. 238) as follows: "Globally, the ground is shifting for local planning organizations and their tools. Mapping renewable energy capacity, understanding energy flows, realizing which roof and open space assets are available for renewable electricity and thermal energy conversion: such knowledge forms the basis for achieving renewable energy independence in an efficiently structured and purposeful manner." Yet, within landscape architecture, energy transition processes and sustainable energy systems are still relatively new topics [14]. As in other fields, case study research is seen as an important way to advance the discipline [15,16]. However, studies on the interface between landscape architecture and energy transition, whether theory building or focusing on design and planning methods, tend to revolve around hypothetical projects and/or projects in the initial phase, rather than implemented cases (see [9,17,18,19]). Many of the publications about realized energy transitions take an interdisciplinary, a spatial planning or a governance perspective (see [12,13,20]). Studies that focus on what landscape architecture can learn from implemented energy transition cases seem to be absent so far.

The purpose of this paper is to add to the small, but growing, body of literature on energy transition from a landscape architecture perspective. This is done by conducting an exploratory, comparative case study (see [21]) of three successful, realized energy transitions in Europe. By describing the transitions in the municipalities of Güssing (Austria), Jühnde (Germany) and Samsø (Denmark), the paper focuses on four aspects (A–D). First, the paper discusses the transition processes (A) and the renewable energy systems that have been realized (B). Then, how landscape impact was considered in the process of siting and designing renewable energy installations (C) is described, as well as to what extent experts from the spatial domain, such as landscape architects, planners, designers and architects, were involved in the transition (D). By comparing the cases with each other and against the literature, a number of lessons can be learned.

The paper commences by accounting for the selection of cases and the methodology in Section 2, followed by a brief introduction into energy transition processes and renewable energy systems in Section 3. Thereafter, in Section 4, the literature regarding energy transition and landscape architecture is discussed, followed by a presentation of the cases in Section 5, Section 6 and Section 7. In Section 8, the cases are compared with each other and against the literature, while the final section contains the conclusion.

2. STUDYING THREE RENEWABLE ENERGY MUNICIPALITIES

Over the past century or two, a number of territories in Europe have shifted to renewable energy. For the study presented in this paper, the municipalities of Güssing, Jühnde and Samsø have been selected, because they represented realized and well-documented examples of energy transition in Europe. They were among the few examples that went beyond the scale of the neighborhood and that used two or more renewable energy sources and technologies, which means that the transitions have certain spatial dimensions and complexity. The cases differed in geographical, socio-economic and planning context, which is why it is expected that they offer a wide range of insights and experiences, which suits the explorative nature of this study. Although, due to their context dependency, transition processes can hardly be transferred to other places, it is reported that all three cases inspired other regions inside and outside of Europe [22,23,24,25].

The study was structured according to case study research in landscape architecture [15]. Francis provides a systematic format for data collection and reporting, to cover a number of aspects relevant in a case description. Because the purpose of this study is to explore realized energy transitions from a landscape architecture perspective, the literature on transition management, renewable energy and landscape architecture was used to structure and frame the study. The multiple case design allowed for systematic comparison of the three cases [26] on the four aspects (A–D) central to the study.

The research drew from scientific and professional literature about the cases and information on the cases' websites. Further, Güssing, Jühnde and Samsø were visited several times between 2010 and 2012 for data collection. During the fieldwork, guided tours were attended to study the energy installations, their location and design in the landscape. In Güssing and Samsø, three people were interviewed; and four in Jühnde. Because of the limited number of interviewees, the interview results were triangulated with the available literature and fieldwork. The interviewees were key persons in the transition process and/or work(ed) for the local authorities, for instance as a project manager, an architect, a researcher or the mayor [27]. The interviews were semi-structured, conducted face-to-face and varied in length between 35 and 120 minutes. In Austria and Germany, the interviews were conducted in German. In Denmark, they were conducted in English. All interviews were transcribed in English. The interviews were coded to structure the data according to the four central aspects, A–D, which have their origin in the literature. A grounded approach is used to explore what was said about each of the aspects. Excerpts from the interviews presented in the case descriptions refer to the interviewees as G1–G3 for Güssing, J1–J4 for Jühnde and S1–S3 for Samsø.

3. ENERGY TRANSITION AND RENEWABLE ENERGY SYSTEMS

Energy transition has been (and continues to be) a particular subject for transition research (see [28,29,30,31,32]). In this context, transitions are defined as "large-scale transformations within society or important subsystems, during which the structure of the societal system fundamentally changes" [31] (p. 295). Energy transitions are long-term processes, triggered by multiple problems, containing multiple social and technological components and concerning multiple (scale) levels, phases and stakeholders [31]. Energy transition, therefore, goes far beyond mere interventions, such as the installation of wind parks or solar panels [11]. While it is agreed that insights from transition management apply to guiding energy transition, Grin, Rotmans and Schot [30] (p. 325) pointed out that "The spatial turn in many of the social sciences, which brought a new sensitivity to the importance of locating change in specific spaces beyond the national, and to the importance of the circulation of things, people and ideas between local, national, regional and global spaces, still needs to be incorporated into transitions theory"; a critique shared by Coenen et al. [33]. Although this paper's perspective is landscape architecture and not transition theory, it may shed light on how landscape architecture can bridge this gap, by approaching energy transition from the integrative nature of planning and design and of the concept of landscape itself (see [34,35]).

For realizing energy transition, increasing both energy efficiency and renewable energy provision are the key strategies (see [4,5]). According to the 'Trias Energetica' concept by Lysen [36], energy efficiency should be addressed first, then renewables should replace fossil fuels, and if fossil fuels remain to be used, this should be done in the most environmental-friendly way. Energy efficiency is improved when more services are delivered with the same input of energy or the same services are delivered with less input of energy [37]. Typical examples are the insulation of buildings and the use of energy-saving devices. Renewable energy is defined as "energy obtained from natural and persistent flows of energy occurring in the immediate environment" [5] (p. 7). It can be harvested from renewable sources by conversion technologies, such as solar boilers, geothermal power plants, hydroelectric stations, photovoltaic cells (PV), wind turbines, biogas plants, and so on. It is expected that, in the long run, a balanced mix of renewable energy sources and technologies will be able to substitute the current energy system based on fossil resources. Beyond renewable energy generation, energy transition also implies adjusting current ways of energy distribution and storage [4,5,11]. For a more exhaustive discussion of the characteristics of (regional) renewable energy systems and the challenges of their design, see de Waal, Stremke, van Hoorn and van den Brink [14].

4. ENERGY TRANSITION AND LANDSCAPE ARCHITECTURE

With regard to the way(s) in which landscape architects discuss and take part in energy transition, the familiar distinction between operational and strategic activities is considered helpful (see [38,39]). In transition management, too, a multilevel framework, including the strategic, tactical and operational level, is used [31].

First, landscape architects work on the siting and design of renewable energy technologies in the landscape, mainly, but not exclusively, wind turbines [40,41,42,43]. In landscape architecture practice, these activities mostly concern operational projects. Operational projects take place on lower spatial scale levels within limited time frames. Designs and plans serve as the input for implementation, aiming for landscape transformation [44]; the emphasis is on the product rather than the process [38,45]. In line with this, landscape architects are involved with environmental impact assessment (EIA) studies. An environmental impact assessment is an examination of the possible environmental consequences of the implementation of projects, programs and policies [46]. With regard to renewable energy, an environmental impact assessment may be required, for instance, for the construction of wind farms and hydroelectric power plants [47]. Especially in carrying out landscape and visual (cumulative) impact assessments, as a preparation for or as a part of the environmental impact assessment, landscape architects use their expertise on (visual) landscape quality, ecology, etc. [40,48].

Second, landscape architects can contribute to energy transition by means of strategic landscape planning and design. Strategic planning and design is employed to explore possible (far) futures, addressing landscape developments on various scale levels. Multiple actors, interests and issues are at stake in strategic projects [38,49]. The typical contributions of landscape architects include problem analyses, spatially explicit scenarios, long-term visions and visual representations of the proposed changes (see [49,50,51]). When landscape architects get involved at an early stage, they can add to agenda-building and/or influence the design and planning processes [38,44,52]. At times, they become project managers in strategic planning and design processes (see [53]). In the case of energy transition, the strategic contributions of landscape architects can be illustrated by the example of the recent book Landscape and Energy, Designing Transition [1]. There, it is visualized what the spatial requirements and impacts are of generating a certain amount of energy on the basis of different renewable and non-renewable sources. Further, landscape architects focus on developing diversified energy landscapes by means of spatially explicit energy potential mapping (EPM). In EPM studies, the physical potentials and limitations for renewable energy are mapped, for instance in GIS, to identify suitable locations for renewable energy technologies [54]. Similarly, Austrian examples are provided by Stoeglehner and Narodoslawsky, who discussed the tools of energy zone mapping (EZM) and the Energetic Long-Term Analysis of Settlement Structures (ELAS) calculator for identifying energy demand and saving potentials in (urban) settlements [55]. Whereas EZM focuses on energy-saving potentials for room heating, hot water production and district heating, the ELAS-calculator is a more holistic tool. Next to providing insight into the energy demand of settlements, it aims to determine the environmental and socio-economic impacts of interventions, as well. Mapping studies, suitability studies and modelling tools such as these can spark and inform the debate on sustainable energy transition in the initial phase and precede and support the making of spatial scenarios and strategic visioning, as has been done, for instance, in Switzerland [17] and Canada [18].

Going beyond technical analyses and modeling, some landscape architects have turned to ecology, thermodynamics and system science to develop principles and concepts for so-called energy-conscious landscape planning and design, which aim to foster energy transition by spatial (re)organization and (re)design of the existing physical environment [56,57]. Stremke and Koh [57], for example, present a number of design strategies to address periodic fluctuations in energy supply, low energy densities and the limited utilization of available energy; constraints that are associated with renewables and commonly found to inhibit sustainable energy transition. To explicate, three strategies for strategic energy-conscious landscape planning and design are summarized here:

- The environment holds potentials for storing thermal, chemical or other forms of energy, e.g., in aquifers or abandoned mines. Mapping and using these storage potentials aids the use of renewable energy sources that tend to fluctuate, such as wind and solar energy.
- The low energy density of many renewable energy carriers, compared to fossil fuels, makes the transport of energy over a long distance less

favorable. A principle, such as (re)locating energy sinks and sources in proximity of each other, aids the efficient use of renewable energy.

- When energy quality is also taken into account in the process of (re)locating energy sources and sinks, energy cascades can be created. A heat cascade, for instance, makes use of residual heat from heavy industry in areas with lower quality heat demand, such as greenhouses.

These and other design strategies have been applied in strategic planning and design projects, for example in the south of the Netherlands. In order to envision sustainable energy landscapes, a methodological framework was employed by the landscape experts and other experts. This framework comprises the following five steps: analyses (of the present conditions and near future developments), scenario-making (identification of possible far futures by concretizing existing context scenario's), development of integrated spatial visions (development of desired far futures) and identification of energy-conscious spatial interventions [19,50]. The study showed, next to the description of the methodological framework, that it is possible for that region to achieve a sustainable and self-sufficient energy system, based on existing technologies.

Based on the above publications, it is safe to state that a growing number of landscape architects, both from practice and academia, focus on the transition to sustainable energy. Landscape architects already contribute to the transition by means of operational activities, such as siting, the design of technologies and environmental impact assessments. Moreover, it is outlined how landscape architects could, more strategically, contribute to the development of a built environment that makes better use of locally available, renewable energy sources by means of strategic landscape planning and design. In addition to the question about what landscape architects can learn from the successful, realized energy transitions in Güssing and Samsø, a second question emerges based on this literature discussion: whether and to what extent operational and strategic activities of landscape architects have been employed in the realization of sustainable energy transition in the three cases.

5. ENERGY SELF-SUFFICIENT GÜSSING

Güssing is a town in the Burgenland region of Austria (see Figure 1) that is well known for its historic castle. For three decades, the region suffered from close proximity to the Iron Curtain and poor connections to the other parts of Austria, which made it unattractive for industries. A lack of local employment forced many inhabitants to work elsewhere, to commute long distances or to move away. Forest is the largest land use in Güssing, followed by farmland and residential areas [58,59]. The municipality comprises 45 km^2 and has 4500 inhabitants, resulting in a population density of 100 inhabitants/ km^2 [23].

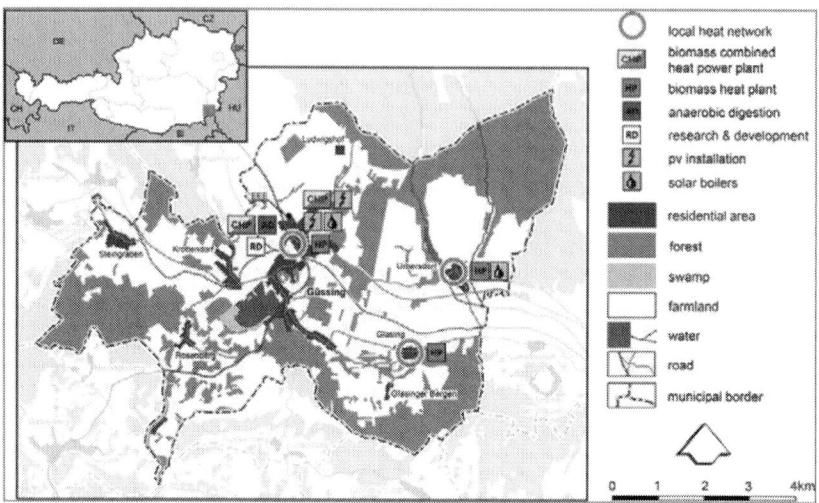

Figure 1. Map of Güssing: location in Austria, land uses, infrastructure and renewable energy technologies.

5.1. The Transition Process

The combination of a poor economy, low employment and large amounts of money that were spent for energy imports provided the context in which change was instigated in Güssing [23,58]. At the end of the 1980s, Peter Vadasz, a member of the municipal council, and Reinhard Koch, a local technical engineer, recognized the potential of local wood as a renewable energy source and energy transition as a way to improve the economy and employment in Güssing. In 1990, Vadasz, Koch and some other experts developed a strategy to provide heat, electricity and fuel for Güssing, all on the basis of local wood [58]. When the plan was presented to the municipal council, it was accepted by an absolute majority, whereby the expected spinoff for the local economy and employment was an important motivation [23].

The transition really took off in 1992, when Vadasz was elected mayor. He appointed Koch as manager to the energy transition [23]. Together, they became frontrunners [60] and succeeded to raise public support (G2, G3). Implementation started with interventions in the town of Güssing and gradually involved the larger municipality and the district [23,59]. In 1996, the European Centre for Renewable Energy (EEE) was founded, coordinating the energy transition and spin-offs, such as eco-energy tourism. The EEE also stimulates research activities and disseminates the so-called Güssing Model [61] nationally and internationally. In 2001, energy self-sufficiency was realized for the municipality. Hereafter, the transition was expanded to the district and combined with research and development on renewable transport fuels [58,62].

Typical for transition processes, the energy transition in Güssing addressed multiple issues in multiple domains. It took place at various scale levels, namely individual buildings, the town, the municipality and, currently, the district. The transition occurred in phases, in which both strategic thinking and operational implementation intertwined. The government was involved: first as the instigator and later as the consolidator, whereby the later role has been taken over by the EEE in recent years. Especially in the initial phase, it was important that the transition be supported by the inhabitants. According to interviewee G2: "A critical mass should be cooperative, and in fact, also a mix of future consumers must be interested; not only the users in winter, but also consumers that need heat in the summer. For that sake, we involved private consumers from the beginning by organizing information sessions." However, citizen participation became less important during the course of the transition and remained limited to the development of the heat network. According to interviewee G2, the current, less active role of citizens is a pity in light of continuing the transition to renewable transport fuels and other goals of the EEE, such as extending the eco-energy tourism concept.

5.2. The Renewable Energy System

Energy efficiency was addressed by insulating public buildings, resulting in a 40%–50% savings [23,24]. To provide renewable energy, a number of technologies were installed (see Figure 1). Biomass heat plants and heat networks were constructed in the villages of Glasing, Urbersdorf and in the town of Güssing, along with two combined heat and power plants (CHPs). Güssing also has a small PV plant, and its grammar school has PV panels and solar boilers on the roof. More recently, an aerobic digestion plant was erected, where poultry manure and corn silage is used to produce biogas [63]. Table 1 presents an overview of the renewable energy provision in Güssing.

Local authorities in Güssing speak of 100% energy self-sufficiency, because the renewable energy provision exceeds their energy use [58]. In reality, transportation still relies on fossil fuels. Starting in 1991, biodiesel was produced from locally-grown rapeseed, but due to a change in the EU biofuel policy, the plant was outcompeted and had to close in 2005 [23]. Currently, the generation of fuel gas, synthetic gas, petrol, diesel, methanol and hydrogen from wood is being developed, in an experimental setting near the newest CHP plant.

Another drawback is that energy provision relies heavily on local (waste) wood [24]. A more balanced mix of sources would enhance energy security (see [57]). The potential for wind energy is indeed low. The small share of solar energy, however, could be increased, especially since Güssing is located in one of the sunniest regions of Austria.

5.3. Considerations on Landscape Impact, Siting and the Design of Renewable Energy Technologies

According to interviewees G2 and G3, in Güssing, the impact of renewable energy technologies on the landscape was not considered until problems arose. Soon after the opening of the heat plant and the heat network in Güssing (see Figure 2), the chipping of the wood caused a noise and dust nuisance for the neighboring school, which led to a "massive protest" according to interviewee G2. However, according to the same interviewee: "That has been ended very quickly by the municipality. The operators of the district heat plant and the school management agreed that the chipping should take place in the forest instead of the plant."

Table 1. Renewable energy provision in Güssing [58,64].

Facility	Location	Energy source	Capacity
CHP (combined heat and power plant); steam turbine with heat network	Güssing	Saw dust	8.6 MW fuel capacity, 1.7 MW electrical capacity and 3.5 MW thermal capacity
CHP; wood gasification, R&D, heat network	Güssing	Wood chips	8 MW fuel capacity, 2 MW electrical capacity and 4.5 MW thermal capacity
Heat plant with heat network	Güssing	Wood chips, waste wood	17 MW (only heat)
PV (photovoltaic cells) installation + solar boilers	Güssing (grammar school)	Solar	10 kW peak electrical capacity and 40m² solar thermal panels
PV installation	Güssing	Solar	27.9 kW peak electrical capacity
Heat plant + solar boilers with heat network	Urbersdorf	Wood chips, solar	650 kW + 320 m² solar thermal panels (only heat)
Heat plant with heat network	Glasing	Wood chips	300 kW (only heat)

When the newest CHP, with anaerobic digestion and research and development facilities, was planned along the main road to Güssing (see Figure 3), the inhabitants of Ludwigshof (see Figure 1) protested, because they feared noise and dust nuisance (G2, G3). In spite of the protests, the plant has been built at the intended location. There, the plant also significantly affects the view from the regional road to Güssing's historic castle. Remarkably, the inhabitants did not complain about the visual impact of the installation. Interviewee G2 commented on that as follows: "Personally I regret that. It is a general thing that the aesthetics of the buildings, whatever their function, is not really taken care of in this region. In the western states, such as Tirol, Vorarlberg, that is much better; industrial buildings can be wonderful over there, but they also have significantly more money to spend. When a carpenter builds his firm over there, he has the ambition that his building should look great, and that is different over here. Here, they prioritize having the plant in the first place."

With regard to the forests, it was said by interviewee G2 that the harvesting had so far no negative impact, neither on sustainability nor on visual quality. The forest organization manages the forest in a sustainable way to safeguard the wood potential for the future, which is possible with the current and near-future energy demand. The visual quality of the wood actually improved due to the energy transition in Güssing because the forest management is now much better (G2).

Figure 2. Heat plant in Güssing.

Figure 3. The CHP (with research and development center) as seen from the regional road located at the edge of an industrial area and in the view of the historic castle.

5.4. Involvement of Landscape Architects

In Güssing, clearly, the focus was foremost on the economic, the technical and, to a limited extent, the social dimensions of energy transition. The contribution of energy transition to overall sustainable development became a motivation only later in the process (G2). Landscape planning and design were not part of the transition process, which, in some instances, resulted in opposition during the implementation of energy technologies. Interviewee G2 regretted that and considered it even problematic. When asked whether the municipality of

Güssing employs planners or designers, the same interviewee replied that this is, in general, much weaker in Austria than in Germany, where it is better institutionalized. According to him, there are also differences in this respect within Austria. In Oberösterreich and Salzburg, for instance, planning is also better institutionalized.

Beyond the implementation of the renewable energy technologies in Güssing, a landscape planner working for the Burgenland state government was involved in the design of a cycling route, as part of developing the eco-tourism concept (G3).

6. JÜHNDE BIO-ENERGY VILLAGE

Jühnde is a village in the south of Lower Saxony, Germany, and forms, together with Barlissen, the municipality of Jühnde. The population density is 44 inhabitants/ km^2. The area is characterized by large-scale farmland and forest areas (see Figure 4), and the closest city is Göttingen. Following Jühnde's successful implementation of energy transition, Barlissen adopted a similar plan. In this case, the focus of the description will be on Jühnde, because this was the first village in Germany to adopt the bio-energy village (Bioenergiedorf in German) concept [20].

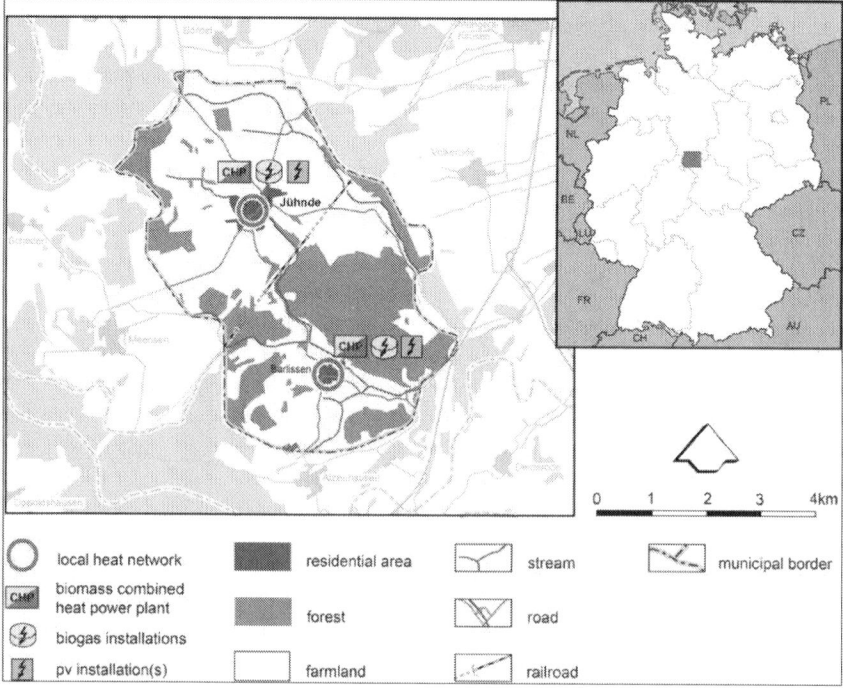

Figure 4. Map of Jühnde: location in Germany, land uses, infrastructure and renewable energy technologies.

6.1. The Transition Process

In Jühnde, the energy transition started in 2001. At that time, researchers from the Interdisziplinäre Zentrum für Nachhaltige Entwicklung (IZNE; Interdisciplinary Centre for Sustainable Development) of the University of Göttingen initiated an action research, to study energy transition as a strategy for enhancing sustainability and societal and economical welfare in rural areas [25]. Among the researchers were geoscientists, agricultural scientists, social scientists and economists. The research team selected one of initially 23 villages that wanted to become Germany's first bio-energy village by means of a feasibility study and a number of additional criteria (J3, J4). Jühnde's application for this project was carefully prepared by the village community, and its selection was enthusiastically received by the villagers [25] (J1). In Jühnde, becoming independent of fossil fuels by using local renewable sources was perceived to save money, stabilize local energy prices and support the local economy by creating employment in the rural area. Indeed, becoming a bio-energy village was supported by the farmers, because they could enter into long-term contracts to provide biomass, which meant increased income stability [25]. One full-time and five part-time jobs were created to operate the biogas installation and to deal with the 7,000 tourists that now visit Jühnde each year [25] (J1, J2). Moreover, the villagers pay significantly less money for their energy than before, when they heated with LPG (liquefied petroleum gas), oil or electrical systems (J1, J3).

From the beginning until realization in 2005, IZNE was involved in the transition process by sharing knowledge, motivating the community and progress monitoring [24,25]. Community participation is a part of the bio-energy village concept [25] and also deliberately stimulated by local frontrunners. Important in this respect were the mayor and a local physician, who later became an operational manager of the biogas installation (J1, J2). The two of them organized information meetings for the villagers, visited people at home, organized excursions to reference projects in Germany and abroad and acted as intermediaries between the researchers and the inhabitants. According to three interviewees, this was essential to the success of the project (J1–J3). Villagers were consulted, cooperated in working groups, contributed financially and were involved in the construction and managing of the heat network and the biogas installation, either unpaid or commissioned.

In 2004, the cooperative partnership, Bioenergiedorf Jühnde eG, was founded as the future operating company and owner of the biogas installation, as well as the CHP and the local heat network [25] (J2). Within the cooperative, every heat consumer is a member having a voting right. Over 70% of the households are now connected to the local heat network, allowing the system to operate effectively. Today, the project is actively disseminated in the region and (inter)nationally [20,65], inspiring other bio-energy villages and beyond. The interviewees, J2 and J3, mentioned e-mobility as the next step to enhance CO_2 reduction, to go beyond the achieved energy self-sufficiency (see also [65]).

6.2. The Renewable Energy System

As was the case in Güssing, the locally-available renewable sources and the energy demand influenced the decision on the different technologies and their capacities. The cooperative partnership in Jühnde operates a biogas installation and a CHP, complete with a heat plant running on wood chips to serve peak demands (see Figure 4). Biomass from 250–300 ha is delivered by six farmers in Jühnde, together with manure from 800 cows and 1400 pigs [24]. Yearly, 350 tons of wood chips from the regional forest are used, which is 10% of the annual growth [24]. The biogas installation generates two and half-times the electricity demand and fulfils the entire heat demand of the village. Heat is transported to about 145 households via the newly constructed, 5.5-km heat network ([24,25]. The generated electricity is transmitted via the existing grid.

Next, there are PV panels on the roof of the nursery school, the community house, individual houses and stables and at the site of the biogas installation. They are owned by another cooperative, private households and a private firm, respectively (J3). Table 2 provides an overview of the renewable energy provision in Jühnde. Because Barlissen is part of the municipality of Jühnde and adopted a similar renewable energy system, we included the information on Barlissen in Table 2. Energy efficiency was not explicitly addressed in Jühnde, and no achievements in this regard have been reported.

Table 2. Renewable energy provision in the municipality of Jühnde [25,65].

Facility	Location	Energy source	Capacity
Biogas installation (CHP) with heat network	Jühnde	Manure, energy crops	700 kW
Heat plant	Jühnde	Wood chips	550 kW
PV installations	Jühnde	Solar	(unknown)
Biogas installation (CHP) with heat network	Barlissen	Manure, energy crops	250 kW
Heat plant	Barlissen	Biomass	500 kW
PV installation	Barlissen	Solar	30 kW

6.3. Considerations on Landscape Impact, Siting and Design of Renewable Energy Technologies

Jühnde concentrated the energy installations at a site in the north of the village. In the siting process, several factors played a role (J1–J4). It had to be in proximity of the village in order to minimize the length of the heat network. Building on municipal land would be practical and economical. The installation could not be built close to the historic country estate, which is a monument. Among the villagers, the visibility of the installation was not perceived as problematic (J1–J3).

Initially, villagers did worry about odor nuisance, but there is little or no odor from the biomass that is stored before fermentation or from the biogas emerging from fermentation. The fertilizer that remains after fermentation is used on the fields instead of liquid manure; it is of outstanding quality and does not have the pungent smell [25]. Yet, the facility was located in the north of the village, so that the prevailing westerly wind would blow odor, if any, away from the village (J1–J3). To prevent noise nuisance, the heat plant is well insulated (J3). At the chosen location, north of the village, the installation is visible from the edge of the village, but not from the center nor in combination with the estate. From a walking trail and local road near Jühnde, the installation seems well embedded within the rolling landscape (see Figure 5).

Figure 5. The biogas installation is embedded in the landscape, as seen from the local road.

To get the permits for building the installation, an environmental impact assessment was conducted, and for that, a landscape maintenance plan was drawn up. This plan specified the plantations, envisioned the future landscape image and fitted the installation in the surroundings (J3). To mitigate the impact on ecology and the landscape image, the authorities required compensation. This was proposed in the form of an orchard, which is situated on the fields next to the installation (J2, J3).

Where the finances and technical requirements allowed it, the aesthetic design of the biogas installation site was addressed (J2, J3). Overall, the chosen strategy was to embed the installation within the existing landscape, rather than letting it stand out. The fire water pond and the staff building, for instance, have a natural look, and the inclination of the roof of the storage building is exactly that of the surrounding landscape (J3). Further, plantings were used to blend the installation in with its surroundings (J2).

To conclude, the university took the aesthetic value of the landscape into account when advising the farmers about energy crops, preventing monocultures from coming into existence. It was advised to vary and rotate crops and to allow for certain weeds to grow in between the crops, which enhances the attractiveness of the agricultural landscape and biodiversity [25] (see Figure 6).

Figure 6. The landscape around Jühnde with a variety of (energy) crops.

6.4. Involvement of Landscape Architects

In the case of the biogas installation in Jühnde, several formal procedures had to be followed with regard to the landscape image and ecological values. Because the village community had little experience with that, the cooperative commissioned an engineering firm for the project management and for dealing with the formal procedures (J2, J3). For creating the landscape management plan, the engineering firm hired a landscape planner, as is required in Germany (J2, J3).

Towards the implementation of the biogas installation, the landscape management plan needed to be further specified to allocate and design the buildings and green spaces on the site. A number of guidelines and legal requirements were applied to the construction of buildings and installations. For this stage, a local construction architect was commissioned. Although he had few experiences with landscape, he also did the green space design of the installation (J3).

7. SAMSØ RENEWABLE ENERGY ISLAND

The municipality of Samsø encompasses an island of 114 km² in the Danish Kattegat, east of the mainland. Samsø is linked by car ferry to Hou (Jutland) and to Kalundborg (Zealand). The largest settlement is the town of Tranebjerg with 829 inhabitants, and there are several smaller villages and parishes (see Figure 7). The landscape is varied, featuring rolling hills, forest, heathland and beaches. The predominant land use on Samsø is agriculture. The island has 4,120 inhabitants, resulting in a population density of 37 inhabitants/km². The population has been decreasing for the last two decades, primarily due to a lack of employment for young people [66].

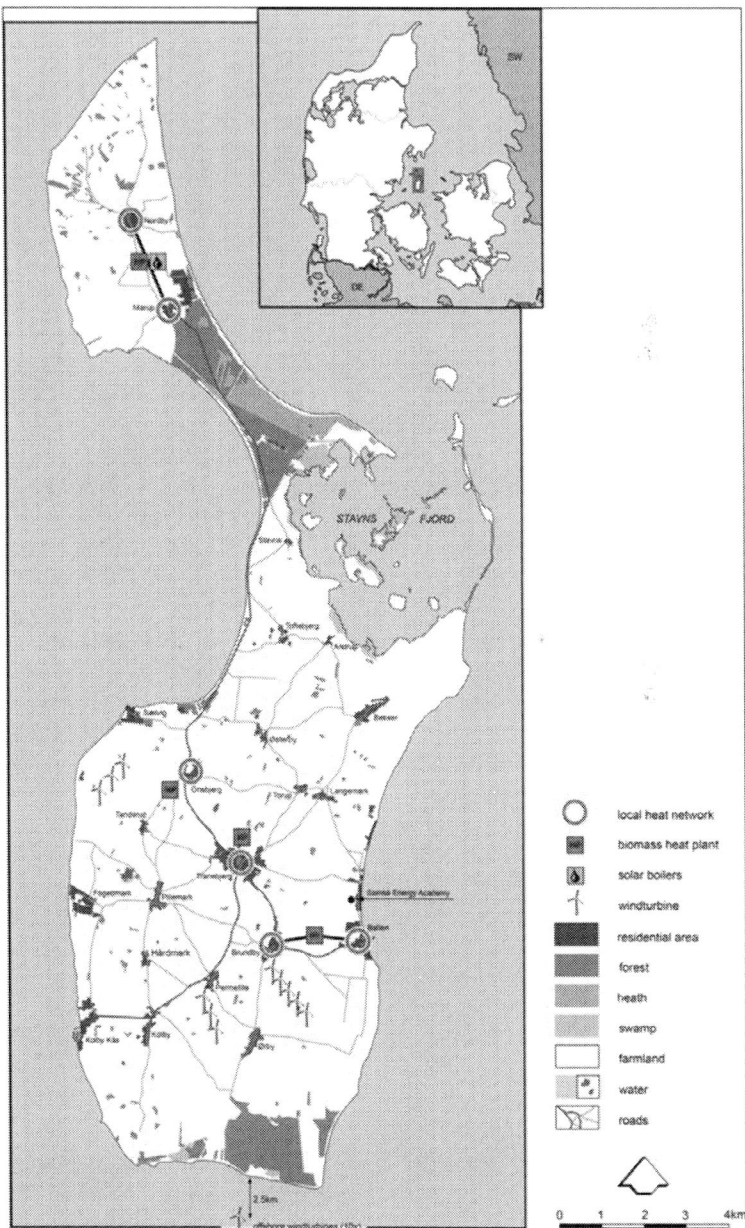

Figure 7. Map of Samsø: Location in Denmark, land uses, infrastructure and renewable energy technologies.

7.1. The Transition Process

In 1997, the Danish Ministry of Energy and Environment organized a competition for municipalities to submit the most realistic plan for energy transition. According to the announcement, the plan should be realized within ten years and without additional subsidies, while making use of local resources and proven technologies. The participation of Samsø in the competition was instigated by the engineering firm, PlanEnergi, in consultation with the municipal administration. Samsø won the competition with the plan created by PlanEnergi [66].

Whereas the competition was motivated by the sustainable development goals of the national government, the local decision to initiate the energy transition was more economically driven. As was the case in Güssing, the start of renewable energy transition on Samsø coincided with unfavorable economic conditions in the municipality. In 1998, the slaughterhouse was closed, a major employer on the island. About 100 people needed to find a new job, and energy transition was seen as a potential to create jobs and boost the economy (S1, S2). Interviewee S2 stressed that "In the making of the 10-year report ([66]), they interviewed a number of people about their motivation for entering the project. Number 1 was to 'help the local economy and independency of other sources', and Number 5 was 'CO$_2$-neutrality'. That means, this island isn't green at all!"

After revising and concretizing the initial plan, financed by the national government, the involvement of the community and other stakeholders became of crucial importance to the success of the transition. A frontrunner in the process was project leader Søren Hermansen who, being from Samsø himself, succeeded in actively involving many people. From the start of the implementation, Samsingers participated financially, in working groups and in the construction and management of local heat networks and other technologies. For the realization of heat networks, similar to Güssing, support from a large part of the inhabitants was necessary. This largely succeeded, however, for "one or two villages, it didn't work out in the end" (S3). Farmers participated by providing the heat plants with straw, investing in wind turbines and by experimenting with renewable transport fuels. The fact that inhabitants and farmers participated financially was important to the success of the transition, especially because of the fact that, as part of the competition, the goal was to achieve energy transition without subsidies, other than those normally available (S2, S3).

In the beginning three organizations were involved in realizing energy transition on the island, which later merged into the Samsø Energy Academy. Hermansen eventually became director of this institution, and the Academy is still guiding the developments today [67]. The local trade organization advocates renewable energy, because of the economic activity that the transition stimulates—among others, an increase in tourism. The municipality of Samsø was involved (but not leading) from the beginning, and in the final, crucial stage for realizing energy self-sufficiency within ten years, they provided finances for realizing the offshore wind turbines (S2).

The phasing of the transition on Samsø revolved around different interventions of increasing complexity and size; it started with smaller (domestic) renewable energy projects, followed by the heat plants and networks, the land-based wind turbines and, eventually, by the offshore wind turbines (S3). Samsø reached energy self-sufficiency within ten years [66], which is short for such a complex transition. Hereafter, the transition scope was expanded by starting a new program: Fossil Free Island. The island now aims to phase out the use of fossil fuels completely towards 2030 [67]. Similar to Güssing and Jühnde, multiple issues in multiple domains were the reason for, and the focus point of, energy transition. Again, a multi-phased and multi-scalar approach was developed, ranging from the individual households to the entire municipality in the end.

7.2. The Renewable Energy System

On Samsø, the renewable energy system is based on the abundant potential for wind energy, solar energy and available agricultural (waste) products on the island. The wind turbines on Samsø produce more than 100% of the electricity consumption, and biomass sources cover 70% of the heat demand [66]. In Tranebjerg, Onsbjerg and Ballen-Brundby, heating is provided by plants that run on straw, a waste product from wheat and rye cultivation on the island. The Nordby-Mårup plant uses wood chips from the Brattingsborg estate in the south of the island (80%) and solar energy via boilers (20%). Local heat networks distribute heat to the consumers in the towns and villages. Owners of the more than 2000 residences and summerhouses outside the settlements are supported to replace their oil-fuelled furnaces with alternative installations, such as heat pumps and solar boilers. For an overview of the renewable energy sources and technologies on Samsø, see Table 3.

Furthermore, on Samsø, some drawbacks occurred during the transition to a renewable energy system. Despite several campaigns, the efforts of energy advisers and implemented efficiency measures, the household electricity consumption is increasing; a rebound-effect that has been observed across Europe [68]. Similar to Güssing and Jühnde, the use of fossil fuels for transportation is compensated for by the export of renewable electricity. The great potential of biogas to provide electricity and heating is unused so far, but studies on the feasibility of biogas production on the island are being conducted currently.

Table 3. Renewable energy provision on Samsø [66,69].

Facility	Location	Energy source	Capacity
5 land-based wind turbines	Brundby	Wind	1 MW each (electricity)
3 land-based turbines	Permelille	Wind	1 MW each (electricity)
3 land-based turbines	Tanderup	Wind	1 MW each (electricity)
10 offshore wind turbines	South of the island Samsø	Wind	2.5 MW each (electricity)
Heat plant with heat network	Tranebjerg	Straw	3 MW (heat)
Heat plant with heat network	Onsbjerg	Straw	0.8 MW (heat)
Heat plant with heat network	Brundby-Ballen	Straw	1.6 MW (heat)
Heat plant + solar boilers with heat network	Nordby-Mårup	Wood chips and solar	1.6 MW (heat) (2500 m^2)

7.3. Considerations on Landscape Impact, Siting and Design of Renewable Energy Technologies

According to the interviewees, S1 and S2, while siting the first set of turbines on land, their height and visibility were discussed with a wide range of stakeholders. This process was initiated by the Samsø Energy Company, one of the predecessors of Samsø Energy Academy. This process was also the preparation for the formal environmental impact assessment that had to be conducted. As a result of this process, the turbines are located in three groups: three turbines near Tanderup, three near Permelille and five near Brundby (see Figure 8). It was decided that all turbines that would be visible from one location should have identical designs. Studies revealed that all three clusters of wind turbines are visible from some locations; hence, eleven identical turbines were installed. Further, it was decided to use the same tower heights (instead of custom made ones), so that the turbines would reflect the landscape contours. For energy reasons, it was determined that the turbines should have a capacity of 1 MW (50-m tower height, 54-m diameter, 77-m total height). Interestingly, interviewee S3 acknowledges that the process of wind turbine siting went relatively smooth compared to other regions in Denmark and that the small number and size of the turbines on the island contributed to that. Next to the visual and energy considerations, land ownership was vital in the process of siting the turbines and managing resistance among the inhabitants. By locating some of the turbines on private land and others on public land, the ownership of all turbines could be organized in a way that the Samsingers could agree on it (S2). Interviewee S2 explained further that, normally, for each of the three groups of turbines, a separate, environmental impact assessment should have been conducted. However, because the Samsingers considered it important that the groups were designed and developed as a unity, similar to the rationale behind cumulative impact studies, they managed to have the three groups assessed in one study. In this formal assessment procedure, landscape planners and other experts were involved (S2). The off-shore turbines were sited in one, curved line, so that they least spoil the view from the island. Fortunately, they also receive the most wind in this spatial constellation.

For siting the heat plants, visibility, as well as potential noise and dust nuisance played a role. Locations were proposed, for example, by the Samsø Energy Company and then discussed with the inhabitants and other stakeholders, such as the municipality. By organizing an open planning process, similar to the turbines, consensus was reached without having to compromise restricted areas, such as the nature area in the north of the island or areas that have many (summer) houses.

A local architect designed the heat plants and their immediate surroundings. Interviewee S2 gave the following account on the considerations around the siting and design process of the heat plants. In the case of the Ballen-Brundby plant, the excessive technical costs resulted in a merely functional design. In spite of these circumstances, the physical appearance was judged positively by inhabitants and other stakeholders. For the Nordby-Mårup heat plant, it was proposed to locate solar boilers along the road in order to make a statement. The inhabitants uniformly rejected that: they found the boilers 'ugly' and wanted them to be hidden behind the plant. Although the boilers ended up in front of the plant, they are partly hidden by shrubs. Instead of designing with the boilers, the architect gave the building a notable appearance (see Figure 9). In Onsbjerg, locals feared that the view of the church would be dominated or even blocked by the heat plant. The architect therefore placed the chimney eccentrically on one side of the building and placed the plant well between the nearby buildings.

7.4. Involvement of Landscape Architects

On Samsø, generally speaking, the impact of energy technologies on the landscape received much attention. This study has shown that, similar to Güssing, some interventions were opposed, due to landscape concerns. On Samsø, however, this was handled during the planning process rather than afterwards. Almost all interventions were sited and designed consciously, and formal planning procedures were more prominent. Landscape architects were among the experts on the environmental impact assessment committee in the county. During the preparation of the formal procedures, discussions on the siting and design of technologies took place, but without the participation of landscape architects. Occasionally, a (local) architect was involved. Upon the question of whether more involvement of experts, such as landscape architects, would have been beneficial to the transition, the interviewee, S1, stated that the process at Samsø was "one of the people" instead of experts. Participation, in his view, would enhance the commitment of inhabitants, which, in turn, was considered essential for the long-term success of the transition.

Figure 8. Land-based wind turbines near Tanderup.

Figure 9. The Nordby-Mårup heat plant, which runs on wood chips and solar boilers. The solar boilers are screened with vegetation to hide the view from the road.

Table 4. Overview of the cases of Güssing, Jühnde and Samsø.

Aspect	Güssing (Austria)	Jühnde (Germany)	Samsø (Denmark)
Transition period	1992–2001	2001–2005	1997–2007
Geographic entity	Municipality and town	Municipality and village	Municipality and island
Population density	100 inhabitants/km²	44 inhabitants/km²	37 inhabitants/km²
A. The transition process			
Context and motivations for renewable energy transition	The context and motivations in Güssing were a combination of the poor economy, low employment and the large amount of the municipal budget spent on energy imports.	The immediate cause was an action research by the University of Göttingen, to develop Germany's first bio-energy village. The concept is seen as a way to enhance sustainability and socio-economic welfare in rural areas. The community was motivated by socio-economic reasons.	The immediate cause was winning a national competition for becoming the first renewable energy municipality in Denmark. Participation was instigated by an engineering firm from outside. Continuation of this process was, at least partly, motivated by economic and demographic reasons.
Typical characteristics of transitions	Multiple issues were addressed in multiple domains. The transition took place in phases that each addressed the next scale level. Continuation focuses on renewable transport fuels and development of eco-tourism.	The village of Jühnde is a small territory for a transition. However, also here, multiple issues in multiple domains were addressed. The process was multi-phased and continues with e-mobility in the future.	Multiple issues were addressed in multiple domains. The transition took place on multiple scale levels. The process was multi-phased and continues with the Fossil Free Island program.
Frontrunners	Local frontrunners were vital for conceiving and initiating the transition. From 1996, the European Centre for Renewable Energy (EEE) took over the task of implementing and continuing the developments.	Local frontrunners were vital for communicating and mediating the bio-energy village concept between the university and the village community. The cooperative partnership Bioenergiedorf Jühnde eG has operated the biogas installation since 2004, of which one of the frontrunners is now the manager.	Local frontrunners were vital for implementing the transition. In the beginning, three organizations were important for organizing the developments, which later became one, the Samsø Energy Academy, of which, frontrunner Hermansen became director.
Government involvement	The municipality was an important stakeholder, especially in the beginning of the process, when political support was needed to start the transition.	The mayor acted as a frontrunner in the transition himself. The municipality was important as the landowner when siting the biogas installation.	The municipality was involved from the beginning, but not leading; they participated financially in the crucial, final stage of the transition.

Table 4. *Cont.*

Aspect	Güssing (Austria)	Jühnde (Germany)	Samsø (Denmark)
Citizen participation	Inhabitants were involved in the beginning, because their support and cooperation were needed for implementing the heat network; after that, inhabitants got less involved.	Inhabitants were highly involved; they were consulted, cooperated financially and participated in working groups and in the construction and management of heat networks and other renewable energy technologies. Participation is seen as an important factor for success.	Inhabitants were highly involved; they were consulted, cooperated financially and participated in working groups and in the construction and management of heat networks and other renewable energy technologies. Participation is seen as an important factor for success.
Drawbacks reported	The diminished involvement of inhabitants is considered a pity in the light of getting support for future developments.	-	-

B. The renewable energy system

Aspect	Güssing (Austria)	Jühnde (Germany)	Samsø (Denmark)
Energy efficiency	In public buildings, a 40%–50% energy savings was achieved by insulation.	-	In spite of several campaigns and implemented measures, energy consumption in households is still increasing.
Renewable energy sources	Local wood chips, saw dust and waste wood, solar energy	Manure, energy crops, wood chips, solar energy	Wind energy, straw, wood chips, solar energy
Renewable energy technologies	CHP and/or heat plants and/or solar boilers combined with heat networks, PV installations	Biogas installation (CHP) with heat network (2×), heat plant (2×), PV installations	Land-based and offshore wind turbines, heat plants combined with heat networks, heat plant and solar boilers combined with heat network
Drawbacks reported	Transportation still relies on fossil fuels in spite of attempts to provide (local) biofuels.	Transportation still relies on fossil fuels.	Transportation, including the ferry to the mainland, still relies on fossil fuels, in spite of attempts to provide (local) biofuels.

C. Considerations on landscape impact, siting and design of renewable energy technologies

Aspect	Güssing (Austria)	Jühnde (Germany)	Samsø (Denmark)
	Landscape (impact) was not pro-actively considered. Two planning-related issues arose because of noise and dust nuisance: one after implementing a heat plant and the other during the planning of a CHP. The first issue was settled, because the municipality mediated between the heat plant and the school. The second issue was not solved; the CHP has been built at the intended location.	Landscape impact was considered by the formal EIA (environmental impact assessment), for which a landscape maintenance plan was drawn up. This was followed by the detailed allocation and design of buildings and green spaces at the site. To compensate for the impact of the biogas installation on biodiversity and landscape image, an orchard needed to be realized next to the installation.	Landscape impact was pro-actively considered. For the land-based wind turbines, a formal EIA was conducted, prepared by the Samsø Energy Academy, in consultation and cooperation with the inhabitants. Similar to this process, but without formal procedures, the location and the design of the heat plants result from an open, participatory planning process.

Table 4. *Cont.*

Aspect	Güssing (Austria)	Jühnde (Germany)	Samsø (Denmark)
Drawbacks reported	The fact that opposition against the CHP was not solved in concert with the inhabitants of Ludwigshof was judged negatively by one interviewee.	-	-

D. Involvement of landscape architects

Aspect	Güssing (Austria)	Jühnde (Germany)	Samsø (Denmark)
	No landscape architects were involved, except for a planner, who created the eco-tourism cycling route. It was reported that this is in line with the limited planning and design tradition in this part of Austria.	For drawing up the landscape maintenance plan for the formal EIA, a landscape planner was hired by the engineering firm that was responsible for the project management. A local architect was involved in the detailed planning and design of the installation.	Landscape architects at the county were involved in the formal EIA procedure for the land-based wind turbines. A local architect was involved in designing the heat plants and their immediate surroundings.

8. CASE STUDY COMPARISON

Because energy transition poses new challenges and opportunities to the discipline of landscape architecture, the main question addressed in this paper was what landscape architects can learn from the transitions in Güssing, Jühnde and Samsø. After discussing the literature, a second question was raised, namely whether and to what extent the identified operational and strategic activities of landscape architects have been employed in the realization of the aforementioned cases. This study described for each case (A) the transition process, (B) the renewable energy systems, (C) the consideration of landscape impact, siting and design of renewable energy technologies and (D) the involvement of landscape architects (or other professionals from the spatial domain) in the transitions. In this section, the cases are compared with each other to demonstrate differences and similarities. Table 4 provides an overview of the three cases, structured according to the four central aspects of this study.

Güssing, Jühnde and Samsø are commonly regarded as successful examples of energy transition [22,23,24,25]. While within (academic) landscape architecture, sustainable development is a major driver to work on energy transition, the cases showed that the economic and social context motivated the transition. The case of Jühnde showed how those different motivations have been combined successfully. Sustainability was a main motivation for the university researchers to initiate the study, and one of the local frontrunners reported that for him "personally, it was one of the main reasons to bring the project forward" (J2); whereas the village community and the mayor emphasized the socio-economic side.

With respect to the processes, the cases demonstrated the main characteristics of transitions. Especially Güssing and Samsø were complex, long-term processes, triggered by multiple problems, containing social and technological components and concerning multiple (scale) levels, phases and stakeholders. In all cases, local frontrunners played a key role in the developments. In Güssing, they conceived and initiated the transition to be further developed by the European Centre for Renewable Energy. In Jühnde, they were important mediators between the university researchers who instigated the transition and the villagers who needed to implement it. On Samsø, Hermansen was a driving force as the project leader and, later, as director of the Samsø Energy Academy. Next to the presence of local frontrunners, a participatory approach appeared essential for the implementation of the renewable energy system in all three cases. Without cooperation among stakeholders, the heat networks of Güssing and Jühnde could not have been realized, because a certain number of connections is needed to make the system work effectively. On Samsø, financial participation by the inhabitants, for example in the form of the shared ownership of the wind turbines, was also crucial, because there was little external funding. The ways in which participation could be organized most effectively in energy transitions and how participation depended on the given planning contexts are beyond the scope of this study, but would be relevant issues for further research.

The renewable energy systems that were realized in these cases, acclaimed for their success, provide valuable insights for both experienced experts and 'relative newcomers' interested in energy transition. It appeared that different geographical, socio-economic and planning contexts led to different energy demands and different potentials for renewable energy generation and efficiency. Therefore, the renewable energy systems in these cases were specifically designed to meet the respective energy demand, making use of the available potentials for renewable energy generation and efficiency. As a result, the systems differed with regard to energy sources, technologies and capacities, and the transferability of the cases is limited. While they present inspiring examples, for every new case, the specific, context-dependent potentials for renewable energy generation and efficiency should be identified, as well as the energy demand.

Yet, in these cases, also, two common drawbacks could be identified regarding the energy systems, which offer opportunities to learn and hold the potential to inform and advance future plans for energy transitions. First, this study showed that all cases were not yet able to replace gasoline and diesel adequately. Instead, fossil fuel use is compensated for by a surplus of renewable electricity generation. Next, it appeared that, although Güssing succeeded in reducing energy demand by insulating public buildings, the three cases underused the potential for energy savings. Following the logic of the 'Trias Energetica' by Lysen [36], this means that the amount of energy that is to be provided by renewable sources is higher than necessary. For Güssing, Jühnde and Samsø, this has not been a problem so far, because the population density is relatively low and the limits to renewable energy generation have not yet been reached. Yet, at the global scale and in urban areas with much higher population densities in particular, reducing the energy demand deserves much more attention in the planning and design of energy-conscious environments [3,70].

Landscape impact, siting and the design of renewable energy technologies were considered more extensively in Jühnde and on Samsø than in Güssing. In Jühnde and Samsø, formal environmental impact assessments were required for realizing the biogas installation and land-based wind turbines, respectively. On Samsø, an open and participatory process was the basis for a relatively smooth transition in this respect. Inhabitants were pro-actively involved in discussing the siting and design of land-based turbines, heat plants and their surroundings. In Jühnde, preparing the environmental impact assessment and the siting and design of the biogas installation were conducted by professionals, commissioned by the cooperative partnership, in which villagers participated. The university researchers, the initiators of the project, considered the landscape image while advising farmers on energy crops. In Güssing, no environmental impact assessments were conducted for renewable energy technologies, nor were siting and design of installations considered explicitly in less formal ways. When problems about noise and dust nuisance arose, they were solved in one occasion and remained unsolved in another. How far and in which ways landscape impact and the siting and design of installations were considered seemed to depend on the planning context and the nature of the interventions; for instance, wind turbines have a much higher (visual) impact on the landscape than a heat plant. Yet, the finding that the implementation of renewable energy technologies in Jühnde and on Samsø was not hampered by structural opposition may serve as an indication for the relative importance of the careful siting and design of such technologies as part of a larger, comprehensive transition process.

The actual involvement of landscape architects was limited; only for Samsø was it reported that landscape architects contributed to the environmental impact assessment. In Jühnde other professionals from the spatial domain were involved, such as the landscape planner, who was responsible for the landscape maintenance plan and the preparations for the environmental impact assessment. Next, an architect conducted the site design of the buildings and green spaces at the biogas installation, and an engineering firm was responsible for the project management. In Güssing, where no landscape architect or similar experts contributed, the involvement of landscape architects could have positively contributed, according to one of the interviewees. Thereby, the need to pro-actively consult inhabitants to prevent opposition was stressed, as well as a well-considered location and the physical appearance of interventions. The activities in which landscape architects or similar experts were involved in these cases concern those that were framed as 'operational': they took place on lower spatial scale levels, within limited time frames; they were input in the process toward implementation and aimed for landscape transformation rather than organizing the planning and design process. The emerging approach in landscape architecture that aims to approach energy transition more strategically, and that focuses on optimizing energy efficiency and renewable energy generation by means of reorganizing the spatial arrangement of the larger physical environment, was not a reality in the cases studied. Not denying the considerable achievement of energy transition in all three cases, a theoretical example may illustrate the potential contribution by strategic, energy-conscious landscape architecture. For the three cases, it was reported that fossil fuels for transportation were compensated for by renewable electricity. Admittedly, the development of sustainable transport fuels is well beyond the expertise of landscape architects. On the basis of energy potential mapping, however, the abundance of renewable electricity would have been constituted a priori, on the basis of which energy-conscious landscape architects along with other experts could have developed strategies to change the energy sources and technologies, and the means of transportation as well (e.g., by proposing to replace fossil fuel vehicles with electric cars).

For the case of Güssing, one of the interviewees suggested some reasons for why landscape impact, siting and design of renewable energy installations were not considered explicitly, such as a weak institutionalization of planning and design. A reason for the (nearly) absence of landscape architects in the cases of Güssing and Samsø could be that the transitions started there 25 and 15 years ago. Around that time, in many countries, the first wind energy projects were taken up by landscape architects. Back then, it is important to stress that the discipline of landscape architecture was not yet ready to address energy transition in a strategic manner, as was discussed in the literature section of this paper. Yet, if landscape architecture aims to broaden its disciplinary scope and address energy transition in both operational and strategic ways, the question of why landscape architects were not involved in the cases in this study remains valid and needs to be addressed in the future. Moreover, the questions of where and how landscape architects are involved in successful cases of energy transition gain relevance for further inquiry. Some first studies on the contribution of landscape architects in realized transitions, for example in Italy, have recently been conducted, and publications are in review (e.g., [71]).

9. CONCLUSIONS

Realizing energy systems that rely entirely on renewable energy sources is a prerequisite for achieving sustainable energy transition, as was demonstrated by the cases discussed in this paper. Although these cases represented inspiring examples, it must be stressed that their renewable energy systems are hardly transferrable to other situations and that for every new case, the specific, context-dependent potentials and possibilities should be identified. Further factors for success seemed to be the presence of (local) frontrunners and a certain degree of citizen participation. Much of the literature on energy transition and landscape architecture focuses on energyefficiency and renewable energy generation, by means of energy-conscious planning and design. In future research on energy transition, the relations and possible synergies between (spatial) expertise and stakeholder participation, within the wider planning context, deserve further attention.

Landscape impact, siting and design of renewable energy installations were in none of the cases the most important aspects for realizing the transitions, yet it appeared important in the sense that resistance to the one or the other proposed intervention can be recognized and mitigated. The case of Güssing showed that (limited) opposition is not decisive for the overall success of the transition. However, Jühnde and Samsø had a relatively smooth process in this respect, demonstrating careful siting and designing, while partly institutionalizing the decision process. How far and in which ways landscape impact, siting and design of installations were considered seemed to depend on the planning context and the nature of the interventions. For some renewable energy technologies, such as wind turbines and biogas installations, environmental impact assessments may be already required. In those instances, landscape architects and similar professionals are among those that can prepare for or conduct environmental impact assessments, as was the case in Jühnde and Samsø.

Based upon the research presented in this paper, it can be concluded that in Güssing, Jühnde and Samsø, landscape architects were not as involved as they, theoretically, could have been. Some of the activities that landscape architects, according to the literature, could have conducted in the transition process were realized by other experts and, in the case of Samsø, also by non-experts. The paper illustrated that the involvement of the spatial domain could have helped to foresee and address some of the drawbacks that surfaced during the transition processes, the realization of the renewable energy system and the mitigation of landscape impacts. Provided that landscape architects continue to broaden their knowledge on the topic of energy transition, more strategic and spatially explicit approaches that have, in the past, contributed to other kinds of transitions could be introduced to energy transition. Hereby, a pro-active attitude on behalf of the discipline is essential, if only to inform the wider public, stakeholders and potential commissioners about the added value of landscape architects to energy transition.

By stating that "The energy landscape is where it happens!" [72], Søren Hermansen supported the emerging paradigm, that landscape is indeed an integrative concept in which the ecological/functional, social and aesthetic aspects of energy-related interventions can be approached together. Because of that, landscape architecture, among other disciplines, can help to integrate the multiple dimensions of energy transition. If we are to strive for long-term, sustainable development, rather than "merely" renewable energy provision, energy transition should be approached pro-actively and strategically, across disciplinary boundaries and spatial scales. The "sustainable energy landscape" concept that was put forward by Stremke and van den Dobbelsteen 10] can inform the energy-landscape discourse, where landscape architects, geographers, engineers and other experts meet to pursue global sustainability goals, while empowering local communities and safeguarding landscape quality.

ACKNOWLEDGMENTS

The study presented in this paper was funded by the Netherlands Enterprise Agency (formerly Agentschap NL) and Wageningen University. We should like to thank our interviewees for their time and enthusiasm, and we are grateful to our colleague, Adrie van`t Veer, for helping us with the figures. Last, but not least, we should like to thank our colleagues, Adri van den Brink and Ingrid Duchhart, and the anonymous referees for their useful comments.

AUTHOR CONTRIBUTIONS

Renée de Waal and Sven Stremke equally contributed to the research in terms of conception, research design, data collection and revising of the paper. Renée de Waal was the main person responsible for data analysis and drafting of the paper.

REFERENCES AND NOTES

1.   Sijmons, D. Landscape and Energy. Designing Transition; nai010 publishers: Rotterdam, The Netherlands, 2014.

2.   Intergovernmental Panel on Climate Change (IPCC). IPCC Fourth Assessment Report: Synthesis Report; IPCC: Cambridge, UK, 2007.

3.   MacKay, D.J.C. Sustainable Energy—Without the Hot Air; UIT Cambridge Ltd.: Cambridge, UK, 2009.

4. Tester, J.W.; Drake, E.M.; Driscoll, M.J.; Golay, M.W.; Peters, W.A. Sustainable Energy: Choosing among Options; MIT Press: Cambridge, MA, USA and London, England, 2005.

5. Twidell, J.; Weir, A.D. Renewable Energy Resources: Second Edition; Taylor & Francis: Abingdon, UK, 2006.

6. EEA Share of renewable energy in gross final energy consumption. Available online: http://epp.eurostat.ec.europa.eu/tgm/table.do?tab=table&init=1&plugin =1&language=en&pcode=t2020_31 (accessed on 30 January 2014).

7. Dincer, I. Renewable energy and sustainable development: A crucial review. Renew. Sustain. Energ. Rev.**2000**, 4, 157–175.

8. ECLAS About eclas/landscape architecture. Available online: http://www.eclas.org/landscape-architecture-european-dimension.php (accessed on 4 July 2014).

9. Van Timmeren, A.; Zwetsloot, J.; Brezet, H.; Silvester, S. Sustainable urban regeneration based on energy balance. Sustainability**2012**, 4, 1488–1509.

10. Stremke, S.; van den Dobbelsteen, A. Sustainable Energy Landscapes. Designing, Planning and Development; CRC Press (Taylor & Francis Group): Boca Raton, FL, USA, 2013.

11. Wächter, P.; Ornetzeder, M.; Rohracher, H.; Schreuer, A.; Knoflacher, M. Towards a sustainable spatial organization of the energy system: Backcasting experiences from Austria. Sustainability**2012**, 4, 193–209.

12. Williams, J. The role of planning in delivering low-carbon urban infrastructure. Environ. Plann. Plann. Des.**2013**, 40, 683–706.

13. Radzi, A.; Droege, P. Governance tools for local energy autonomy. In Climate Change Governance; Knieling, J., Leal Filho, W., Eds.; Springer-Verlag: Berlin Heidelberg, Germany, 2013; pp. 227–241.

14. De Waal, R.; Stremke, S.; van Hoorn, A.; van den Brink, A.; Wageningen University, Wageningen, The Netherlands. Unpublished work. 2014.

15. Francis, M. A case study method for landscape architecture. Landsc. J.**2001**, 20, 15–29.

16. Deming, E.M.; Swaffield, S. Landscape Architecture Research: Inquiry, Strategy, Design; John Wiley &Sons: New York, NY, USA, 2011.

17. Grêt-Regamey, A.; Wissen Hayek, U. Multicriteria decision analysis for the planning and design of sustainable energy landscapes. In Sustainable Energy Landscapes. Designing, Planning and Development; Stremke, S., van den Dobbelsteen, A., Eds.; CRC Press (Taylor & Francis): Boca Raton, FL, USA, 2013.

18.   Schroth, O.; Pond, E.; Tooke, R.; Flanders, D.; Sheppard, S. Spatial modeling for community renewable energy planning: Case studies in British Columbia, Canada. In Sustainable Energy Landscapes. Designing, Planning and Development; Stremke, S., van den Dobbelsteen, A., Eds.; CRC Press (Taylor & Francis Group): Boca Raton, FL, USA, 2013; pp. 311–334.

19.   Stremke, S.; Koh, J.; Neven, K.; Boekel, A. Integrated visions (part II): Envisioning sustainable energy landscapes. Eur. Plann. Stud.**2012**, 20, 609–626.

20.   Karpenstein-Machan, M.; Wüste, A.; Schmuck, P. Erfolgreiche Umsetzung von Bioenergiedörfern in Deutschland—was sind die Erfolgsfaktoren? Berichte über Landwirtschaft. Zeitschrift für Agrarpolitik und Landwirtschaft**2013**, 91, 1–25. (In German).

21.   Grix, J. The foundations of research; Palgrave MacMillan: Hampshire, UK, 2004.

22.   Nevin, J.A. The power of cooperation. Behav. Analyst**2010**, 33, 189–191.

23.   Marcelja, D. Self-sufficient community: Vision or reality? Creating a regional renewable energy supply network (Güssing, Austria). In Local Governments and Climate Change, 39th ed.; Van Staden, M., Musco, F., Eds.; 2010; pp. 217–228.

24.   Radzi, A. 100% renewable champions: International case studies. In 100% Renewable. Energy Autonomy in Action; Droege, P., Ed.; Routledge: New York, NY, USA, 2009; pp. 93–166.

25.   Ruppert, H.; Eigner-Thiel, S.; Girschner, W.; Karpenstein-Machan, M.; Roland, F.; Ruwisch, V.B.S.; Schmuck, P. Wege zum Bioenergiedorf. Leitfaden.; Fachagentur für nachwachsende Rohstoffe: Gülzow, Germany, 2008. (In German)

26.   Yin, R.K. Case study: Research, Design and Methods; Sage: Los Angeles, CA, USA, 2009.

27.   In Güssing we interviewed G1, regional guide at the EEE, G2, project manager and public relations officer of the EEE, and G3, former member of the city council and involved with the tourist association in Güssing. In Jühnde, we interviewed J1, former mayor, J2, operational manager Bioenergiedorf Jühnde eG, J3, local architect, and J4, researcher at IZNE. On Samsø, we interviewed G1, architect at the Energy Academy, G2, director of the Energy Academy, and G3, involved with the trade and tourist center of Samsø Municipality.

28.   Kemp, R. The Dutch energy transition approach. Int. Econ. Econ. Pol.**2010**, 7, 291–316.

29. Laes, E.; Gorissen, L.; Nevens, F. A comparison of energy transition governance in Germany, the Netherlands and the United Kingdom. Sustainability**2014**, 6, 1129–1152.

30. Grin, J.; Rotmans, J.; Schot, J. Transitions to Sustainable Development. New Directions in the Study of Long Term Transformative Change; Routledge: New York, NY, USA, 2010.

31. Loorbach, D.; van der Brugge, R.; Taanman, M. Governance in the energy transition: Practice of transition management in the Netherlands. Int. J. Environ. Tech. Manag.**2008**, 9, 294–315.

32. Rotmans, J.; Kemp, R.; van Asselt, M. More evolution than revolution: Transition managment in public policy. Foresight**2001**, 3, 15–31.

33. Coenen, L.; Benneworth, P.; Truffer, B. Toward a spatial perspective on sustainability transitions. Res. Pol.**2012**, 41, 968–979.

34. Motloch, J.L. Introduction to Landscape Design; Wiley: New York, NY, USA, 2001.

35. Thompson, I.H. Ecology, Community and Delight: Sources of Values in Landscape Architecture; E&FN Spon: London, UK, 2000.

36. Lysen, E.H. The Trias Energica: Solar Energy Strategies for Developing Countries. In Proceedings of Eurosun Conference Freiburg; Freiburg, Germany: 16–19 September 1996.

37. International Energy Agency. Available online: http://www.iea.org/aboutus/faqs/renewableenergy/ (accessed on 18 September 2012).

38. De Jonge, J. Landscape Architecture between Politics and Science; Blauwdruk: Wageningen, The Netherlands, 2009. [Google Scholar]

39. Van der Cammen, H.; de Klerk, L. The Selfmade Land Culture and Evolution of Urban and Regional Planning in the Netherlands; Uitgeverij Unieboek, Het Spectrum bv: Houten-Antwerpen, The Netherlands, 2012.

40. Mogen, E.A.H. The role of the landscape architect in the wind farm site selection process and best practices. In Conference of CELA and ISOMUL on Landscape Legacy: Landscape Architecture and Planning between Art and Science; Carsjens, G.J., Ed.; ISOMUL, Wageningen University: Maastricht, The Netherlands, 2010.

41. Schöbel, S. Windenergie und Landschaftsästhetik: Zur landschaftsgerechten Anordnung von Windfarmen. Jovis: Berlin, Germany, 2012. (In German)

42. Palmer, J.F. Public Acceptance Study of the Searsburg Wind Power Project: Year One Post-Construction; Clinton Solutions: Fayettevilly, New York, NY, USA, 1997.

43. Schöbel, S.; Dittrich, A.R. Renewable energies—landscapes of reconciliation? Topos**2010**, 70, 56–61.

44. Brinkhuijsen, M. Landscape 1:1: A Study of Designs for Leisure in the Dutch Countryside; Wageningen University: Wageningen, The Netherlands, 2008.

45. Duchhart, I. Designing Sustainable Landscapes. From Experience to Theory: A Process of Reflective Learning from Case-Study Projects in Kenya; Wageningen University: Wageningen, The Netherlands, 2007.

46. United Nations. Glossary of Environment Statistics; United Nations: New York, NY, USA, 1997.

47. Council Directive. On the Assessment of the Effects of Certain Public and Private Projects on the Environment (85/337/eec); Official Journal L 175; European Commission: Bruxelles, Belgium, 1985.

48. Landscape Institute. Guidelines for Landscape and Visual Impact Assessment, 3rd ed.; Routledge: London, UK, 2013.

49. De Zwart, B. A triptich of expertise. The design competition as an instrument to unite assignment, design and commissioner. In Designing for a Region; Meijsmans, N., Ed.; SUN Academia: Amsterdam, The Netherlands, 2010.

50. Stremke, S.; van Kann, F.; Koh, J. Integrated visions (part I): Methodological framework for long-term regional design. Eur. Plann. Stud.**2012**, 20, 305–319.

51. Weller, R. Planning by design. Landscape architectural scenarios for a rapidly growing city. J. Landsc. Architect.**2008**, 6–17.

52. Van Buuren, M. De erfenis van de moderne beweging. De Blauwe Kamer**2000**, 4, 46–57. (In Dutch).

53. Hajer, M.; Sijmons, D.; Feddes, F. Een plan dat werkt; NAi Uitgevers: Rotterdam, The Netherlands, 2006. (In Dutch)

54. Van den Dobbelsteen, A.; Broersma, S.; Fremouw, M. Energy potential mapping and heat mapping: Prerequisite for energy-conscious planning and design. In Sustainable Energy Landscapes. Designing, Planning and Development; Stremke, S., van den Dobbelsteen, A., Eds.; CRC Press (Taylor & Francic Group): Boca Raton, FL, USA, 2013; pp. 71–94.

55. Stoeglehner, G.; Narodoslawsky, M. Energy-conscious planning practice in Austria: Strategic planning for energy-optimized urban structures. In Sustainable Energy Landscapes. Designing, Planning, and Development; Stremke, S., van den Dobbelsteen, A., Eds.; CRC Press (Taylor & Francis Group): Boca Raton, FL, USA, 2013; pp. 355–372.

56. Stremke, S.; Koh, J. Ecological concepts and strategies with relevance to energy-conscious spatial planning and design. Environ. Plann. Plann. Des.**2010**, 37, 518–532.

57. Stremke, S.; Koh, J. Integration of ecological and thermodynamic concepts in the design of sustainable energy landscapes. Landsc. J. **2011**, 30, 2–11.

58. BMVIT Model Region Güssing. Self-sufficient Energy Supply Based on Regionally Available Renewable Resources and Sustainable Regional Development. Bundesministerium für Verkehr Innovation und Technologie: Vienna, Austria, 2007.

59. Koch, R.; Brunner, C.; Hacker, J.; Urschik, A.; Sabara, D.; Hotwagner, M.; Aichernich, C.; Hofbauer, H.W.R.; Fercher, E. Energieautarker Bezirk Güssing; Bundesministerium für Verkehr Innovation und Technologie: Vienna, Austria, 2006. (In German)

60. The term frontrunner is used in transition management, e.g., by Rotmans and Loorbach, to indicate "agents with peculiar competencies and qualities: Creative minds, strategists, and visionaries. If a new regime is to be created effectively, agents are needed at a certain distance from that regime." (Rotmans, J.; Loorbach, D. Complexity and transition management. J. Ind. Ecol. **2009**, 13, 184–196.

61. The Güssing Model for realizing energy transition was also called 'Güssing energy self-sufficient city'; it is now known as 'Güssing energy self-sufficient district', because the process is being scaled up (Koch et al., 2006). In this paper, we focus on the town of Güssing and its direct surroundings (as outlined in Figure 1), which is the municipality of Güssing, and call it Güssing Renewable Energy Municipality.

62. EEE Home-das EEE. Available online: http://www.eee-info.net/cms/ (accessed on 24 January 2014).

63. This plant did not exist at the time of our site visits and is therefore not further discussed in the paper.

64. Keglovits, C. Renewable energy system Güssing. In E-Mail. 2011. [Google Scholar]

65. Fangmeier, E. Bioenergiedorf Jühnde. Available online: http://www.bioenergiedorf.de/con/cms/front_content.php?idcat=13 (accessed on 4 July 2014).

66. Jørgensen, P.J.; Hermansen, S.; Johnsen, A.; Nielsen, J.P.; Jantzen, J.; Lundén, M. Samsø—A Renewable Energy Island. 10 Years of Development and Evaluation; PlanEnergi, Samsø Energy Academy: Samsø, Denmark, 2007.

67. The Energy Academy Front page. Available online: http://energiakademiet.dk/en/ (accessed on 22 January 2014).

68. Herring, H. Energy efficiency, the rebound, and the steady state society. In Proceedings of Green Economic Conference, Oxford, UK, 20 July 2012.

69.   The   Energy   Academy   RE-island.   Available   online:
       http://energiakademiet.dk/en/vedvarende-energi-o/ (accessed on 30
       January 2014).

70.   Sijmons, D.; Hugtenburg, J.; Hofland, A.; De Weerd, T.; van Rooijen,
       J.; Wijnakker, R. De kleine Energieatlas; Ministerie van
       Volkshuisvesting, Ruimtelijke Ordening en Milieu: Den Haag, The
       Netherlands, 2008. (In Dutch)

71.   Minichino, S.; Stremke, S. Landscape planning and design for
       sustainable energy transition. A comparison between dutch and italian
       practices. Landsc. Res.**2014**. submitted.

72.   This quote origins from the (Dutch) television broadcast 'Tegenlicht'
       from 8 October 2012, published by VPRO.

CHAPTER 11

Assessing Landscape Ecological Risk in a Mining City: A Case Study in Liaoyuan City, China

Jian Peng [1,*], Minli Zong [2,3], Yi'na Hu [1], Yanxu Liu [1] and Jiansheng Wu [2]

[1] *Laboratory for Earth Surface Processes, Ministry of Education, College of Urban and Environmental Sciences, Peking University, Beijing 100871, China*
[2] *Key Laboratory for Environmental and Urban Sciences, School of Urban Planning and Design, Shenzhen Graduate School, Peking University, Shenzhen 518055, China*
[3] *Shanghai Urban Planning and Design Research Institute, Shanghai 200040, China*

ABSTRACT

Landscape ecological risk assessment can effectively identify key elements for landscape sustainability, which directly improves human wellbeing. However, previous research has tended to apply risk probability, measured by overlaying landscape metrics to evaluate risk, generally lacking a quantitative assessment of loss and uncertainty of risk. This study, taking Liaoyuan City as a case area, explores landscape ecological risk assessment associated with mining cities, based on probability of risk and potential ecological loss. The assessment results show landscape ecological risk is lower in highly urbanized areas than those rural areas, suggesting that not only cities but also natural and semi-natural areas contribute to overall landscape-scale ecological risk. Our comparison of potential ecological risk in 58 watersheds in the region shows that ecological loss are moderate or high in the 10 high-risk watersheds. The 35 moderate-risk watersheds contain a large proportion of farmland, and the 13 low-risk watersheds are mainly distributed in flat terrain areas. Our uncertainty analyses result in a close range between simulated and calculated values, suggesting that our model is generally applicable. Our analysis has good potential in the fields of resource development, landscape planning and ecological restoration, and provides a quantitative method for achieving landscape sustainability in a mining city.

Keywords: ecological disturbance; ecological vulnerability; ecological importance; uncertainty analysis; Liaoyuan City; China

1. INTRODUCTION

Landscape sustainability is defined as the capacity of a landscape to consistently provide long-term, landscape-specific ecosystem services, which is essential for maintaining and improving human wellbeing [1]. Sustainability is an object that not only meets the demands of humans nowadays, but also ensures future benefits. Thus, the forecast for future benefit and security is critical in sustainability research. Risk assessment, which focuses on future damages, is appropriate for evaluating sustainability as both of the concepts concern the future benefits and security of system. In other words, ecological security framework for ecosystem health protection and ecological risk control are essential for the improvement of ecological sustainability in regions and landscapes [2]. Therefore, ecological risk assessment is one effective way to determine regional and landscape-scale ecological sustainability [3]. In 1989, Hunsaker defined regional ecological risk assessment as the evaluation of regional-scale risk faced by environmental resources, or the risk caused by regional-scale pollution and natural disturbance [4], and he proposed to apply ecological risk assessment to regional and landscape scales in 1990 [5]. In the decade that followed, a series of ecological risk assessments were conducted in watersheds and other large areas [6,7,8,9,10]. However, growing recognition of the interplay of factors that might influence risks faced by ecosystems, such as global urbanization, land use change, and climate change, has led to the realization that unilateral risk management is unlikely to be useful in the management of complex systems. The need for a multilateral approach that incorporates the roles of various landscape factors in influencing risk and, therefore, sustainability was addressed with the first landscape ecological risk assessments that were conducted in the 21st century [11].

Landscape ecological risk refers to the possibility of harm to the structure and function of ecosystems from disturbances, such as human activities and natural disasters [12]. Landscapes are typically impacted by multiple disturbances that operate at different spatial and temporal scales [13,14]. Therefore, in contrast to general ecological risk assessment, landscape ecological risk assessment comprehensively assesses various types of potential ecological impacts, and their cumulative effects. It explores the effects of a variety of hazards for large-scale units, and is the complement and expansion of general ecological risk assessment. Landscape ecological risk assessment also accounts for differences in ecological characteristics and risks between different landscapes and assessment units via spatial heterogeneity, and time-series analyses. For example, Graham et al. used contagion index to evaluate the regional ecological risk considering the terrestrial and aquatic linkages, and the landscape pattern [15]; Kapustka et al. suggested risk management countermeasures grounded in landscape ecology [16]; Liu et al. drew attention to potential ecological risks caused by the intensification of soil erosion, and ecological vulnerability [17].

Within the last decade, landscape ecology work in China has produced relatively independent quantitative evaluation models which took landscape patterns and function into consideration [18]. One such model provides a landscape ecological risk index developed from spatial patterns and several indices, such as disturbance and vulnerability indices [19,20], landscape exposure, stability, and external pressure indices [21], as well as threat and intensity indices [22]. However, models of these type tend to calculate risk directly from probability values by the superimposition of landscape pattern indices, ultimately predicting only the probability of the occurrence of an adverse ecological event, and paying little attention to the probability of ecological losses following the event. Models based on threat and intensity indices focus on ecological risk caused by external threats, and ignore the inherent features of landscapes, including their vulnerability, resilience, stability, and the value of certain landscape features. Some other models, that characterize risk loss (the loss of risk), by calculating ecosystem services, are often limited by the direct conversion of land use, and fail to consider the effects of landscape patterns on it [23]. In addition, uncertainty analysis methods, highlighted in ecological risk assessment, is essential for the reliability of the results, and it has drawn increasing attention from scholars [24].

Mining area is a special man-land system, where people are engaged in complex interactions with land, for the exploitation of mineral resources, which has caused serious impacts on the environment [25,26]. Ecological risk assessment plays an important role in both theoretical support, and practical guidance for the implementation of regional sustainable development, and ecological restoration in mining cities [27]. Environmental problems in mining cities result primarily from mining itself, and ecological risk assessment in these cities should be built on the specific social, economic, and natural environments of the mining city in question. When combined with traditional ecological problems and risk type analysis, landscape ecology can form an important component of ecological risk assessment. Current work on ecological risk assessment in mining cities in China largely address the ecological effects of mining on soils, such as heavy metal pollution [28], land use, vegetation and landscape patterns [29,30]. However, quantitative assessments of integrated ecological risks stemming from a variety of risk sources, targeting a number of risk receptors, and driving varied ecological effects, are scarce.

Mining landscape ecological risk assessment entails the analysis of the direct and indirect risks posed by mining on a macro-scale in view of overall regional sustainability. It will be an important warning, as well as a practical guide for the planning and sustainable development of mining landscapes, including mining cities. In this study, using Liaoyuan known as the coal capital of Jilin Province as a case study, a mining landscape ecological risk assessment was quantitatively explored. Liaoyuan belongs to the first batch of China's resource exhausted cities, and urban development and environmental protection have been significantly influenced by mining activities. In the transformation of economic development, the urban ecological problem and regional ecological risk can probably be stimulated. As a result, ecological risk assessment at landscape scale is in great need to provide a spatial approach for ecological security and sustainability. Specifically, the goals of this study were to (1) quantify the disturbance and vulnerability degrees of the landscape using multiple indicators developed on the basis of traditional landscape pattern indices for mining cities; (2) calculate the risk of loss based on the landscape pattern index, in order to explore a new model for the landscape ecological risk assessment of mining cities; (3) construct landscape ecological risk zoning based on risk assessment results to provide direction for the sustainable development of mining cities; and (4) conduct landscape ecological risk uncertainty analysis to verify the reliability of our risk evaluation results.

2. MATERIALS AND METHODS

2.1. Study Area and Data Source

Liaoyuan Prefecture City (5140 km²; 42°17′40″N to 43°13′40″N, 124°51′22″E to 125°49′52″E), which contains 33 towns, lies in the south-central part of Jilin Province, China. It is upstream of the Dongliao and Huifa Rivers, within the transition zone between the Changbai Mountains and the Songliao Plain, and across Liaohe River and Songhua River (Figure 1). Liaoyuan enjoys a semi-humid, temperate, and continental monsoon climate, with abundant water, and forest resources. Liaoyuan is a coal resource-based city. By the end of 2007, 33 types of minerals had been discovered in Liaoyuan, of which coal and building stone comprise a large proportion. There are 152 mineral ore fields in Liaoyuan, most of which are small, and their distribution is concentrated. Mining has boosted economic development in the region, but has also resulted in severe damage to regional eco-environment. The environmental problems such as ground subsidence, air pollution, and excessive heavy metal waste production, have had serious impacts on local people's life and social development [31]. In March 2008, the National Development and Reform Commission (NDRC, China) listed Liaoyuan in the first batch of resource-exhausted cities that were in urgent need of transformation in economic development.

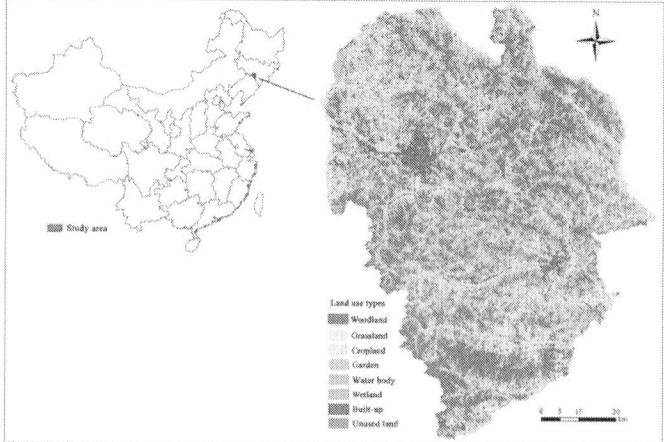

Figure 1. Location of Liaoyuan City, China.

The land use data is from the second national land survey of Liaoyuan. According to land use classification and planning criteria published by China Land Resource Bureau, the landscape of Liaoyuan may be divided into 8 types, i.e., woodland, grassland, cropland, garden, water body (including rivers, surface water ponds and reservoirs), wetland (including inland beaches and marshes), built-up, and unused land (including bare, sandy and saline land; Figure 1). Among these types, wetlands and unused land refer to natural reserves referring to the secondary indicators of classification system.

In this study, cities, rural settlements, mining sites, and other construction sites are treated as ecological risk sources, and natural and semi-natural landscapes are treated as risk receptors, whose risk is assessed. The location of mines is provided by Liaoyuan City Land Bureau, which is used for evaluating the disturbance of mines. Data on gross regional product, population, industry, and other socio-economic data is taken from the "Liaoyuan Statistical Yearbook (2009)" and "Liaoyuan Yearbook (2010)" to measure the development of Liaoyuan [32,33]. Based on a 30 m × 30 m Digital Elevation Model (DEM) image, and 1:200,000 topographic maps of Liaoyuan, in 2009, the study area is divided into 58 small watershed basins (Figure 2). Watershed, which in relation to hydrology process and topography, can determine ecosystem processes in a relatively integrated region without being subjectively sliced. Thus, since watershed unit contains more ecological meanings than town unit, it is set as assessing units for landscape ecological risk assessment. The hydrological analysis module (Hydrology) of ArcGIS10.0 is used for zoning the study area.

2.2. Calculation of Landscape Ecological Risk

The formula for risk measurement is the methodology basis of regional landscape ecological risk assessment [34]. In this paper, overall landscape ecological risk (R) is calculated as a function of the probability of ecological risk (P), and risk loss (D), wherein P is the product of ecological disturbance (E) and ecological vulnerability (V), and risk loss represents ecological importance (S). However, the final probability (between 0 and 1), sometimes, may be too small to have a meaningful impact on the partitioning of ecological risk. In order to avoid such insignificant values, the cube root result of proposed function is extracted to arrive at final landscape ecological risk.

$$R = \sqrt[3]{P \times D} = \sqrt[3]{E \times S \times V}$$

(1)

In order to clarify the management approach most appropriate to a specific landscape ecological risk, such as risk control guidelines, the calculated risk is "zoned" after the risk assessment. First, landscape ecological risk is divided into three grades (high, moderate and low), and then, using the "Natural Break" function in ArcGIS10.0, the risk is stacked onto ecological risk probability, together with risk loss, to produce a map of landscape ecological risk zones.

Figure 2. Extracted watersheds based on DEM in Liaoyuan City, China. (Note: R-River, U-Upstream, M-Midstream, D-Downstream).

2.2.1. Ecological Disturbance

Ecological disturbances are relatively discrete events that alter ecosystems, communities, or demographic structure, and result in changes in resources, substrates, or the physical environment [35]. It is the external cause of regional ecological risk, and one of the main sources of landscape heterogeneity [36]. Land use degree is a representation of how broad and deep landscapes are utilized under the influence of a combination of human activities and social development factors [37]. It reflects not only the natural attributes of the land, but also the comprehensive effect of human activity on the land [38]. However, land use degree represents only current patterns of disturbance, and does not incorporate, in its measure, the probability of future human disturbance. Thus, on the basis of previous researches on land use classification, in this study, a model of landscape ecological disturbance (E), which attempts to incorporate settlement, mining, and road disturbances into the overall measure of human disturbance, is constructed,

$$E = aU + bM + cR + dT$$

(2)

where U is landscape disturbance; M, R, and T are mining, settlement and road disturbances, respectively, and a, b, c, d are weights (of 0.4, 0.3, 0.2, and 0.1, respectively) assigned to each disturbance parameter such that a + b + c + d = 1.

Mining, settlement, and road disturbances (M, R, and T) are calculated using the "Buffer Analysis" modules in ArcGIS 10.0. Based on the characteristics of the disturbance and the decreasing relationship between the magnitude of influence of the disturbance and distance from its source, where different buffers are set to measure the influence of different disturbances (Table 1). On the other hand, the influence of mines and settlements are assigned maximum values in area where they overlapped, and the influence of roads is assigned summation due to its additive effects. Depending on the intensity of human activities, different landscapes are assigned different levels to quantify the influence of human activities on landscape disturbance. These levels are as follows: unused (0.2), natural renewable (woodland, grassland, water body, and wetland; 0.4), half natural renewable (cropland, and garden; 0.6), and artificial non-renewable (built-up; 0.8).

Table 1. Influence values of various disturbances based on the size of the disturbance and the distance from the disturbance source.

Disturbance	Mining				Settlement			Road			
Level	Large mining (>10 ha)	Medium mining (1–10 ha)	Small mining (<1 ha)	City	Designated Town	Rural settlement	National road	Provincial road	County road	Highways & railways
Number of Buffer	4	3	2	2	2	2	3	3	2	2
Distance/meter (influence value)	300(0.8) 600(0.6) 1000(0.4) 2000(0.2)	300(0.6) 600(0.4) 1000(0.2)	300(0.4) 600(0.2)	600(0.8) 1200(0.4)	400(0.6) 1000(0.3)	200(0.4) 500(0.1)	50(0.8) 250(0.5) 500(0.1)	50(0.7) 100(0.5) 500(0.1)	50(0.6) 100(0.1)	30(0.5) 50(0.1)

2.2.2. Ecological Vulnerability

Ecological vulnerability (V), an integral part of ecological risk assessment, refers mainly to the vulnerability of ecosystems to the strong external disturbance [39], and is a function of vulnerability of landscape type, and vulnerability of landscape structure. Ecological vulnerability of landscape type denotes the probability of the landscapes to deviate from their steady states, or suffer enormous damage from outside interference. Landscapes are classified into seven kinds and given different weights: unused land (7), wetland (6), water body (5), cropland (4), garden (3), grassland (2) and woodland (1), then normalized, and finally multiplied by the area ratio of that type of landscape to the surrounding landscape, to obtain the ecological vulnerability of landscape types.

Based on the pattern-process feedback mechanism, in landscape ecology, landscape patterns can also affect ecological vulnerability, and thereby, landscape ecological risk. As an intrinsic attribute of ecosystems, ecological vulnerability is closely related to ecosystem sensitivity, resilience, and stability [40]. Ecosystem sensitivity refers to the ecosystem's internal adaptive capacity to external pressure or external interference; ecosystem resilience is its ability to recover from these interferences; and ecosystem stability represent the ability to maintain the normal dynamic ecological system. Different landscape types, with different sensitivity and resilience, may play different roles in maintaining biodiversity, protecting species, improving landscape structure, and promoting the overall functioning of the landscape [41]. In this study, such two indicators as landscape fragmentation and area ratio of various landscapes whose slope is greater than 15°, are used to measure ecological sensitivity. The fragmentation of a landscape by natural or human factors is a consequence of changes in landscape patterns, from a continuous structure to patches, and more fragmented landscapes with greater slopes are assumed to be more ecologically sensitive. Ecological resilience, assumed to be positively related to landscape connectivity (the degree to which the landscape facilitates or impedes movement among resource patches), and landscape dominance (the degree to which one or a few land cover types predominate the landscape in terms of area proportion), is calculated as a function of these terms based on the formula described by Wu et al. [31]; weights were assigned using the "Analytic Hierarchy Process" in Matlab R2010a (Table 2).

Table 2. Indicators system for structural vulnerability evaluation.

		Woodland	Grassland	Cropland	Garden	Water Body	Wetland	Unused Land
Ecological sensitivity	Landscape fragmentation	0.9001	0.7286	0.9821	0.4864	0.9518	0.8936	0.4563
	Area ratio (slope > 15°)	0.0999	0.2714	0.0179	0.5136	0.0482	0.1064	0.5437
Ecological resilience	Landscape connectivity	0.9359	0.2309	0.9768	0.3165	0.2951	0.2897	0.2500
	Landscape dominancy	0.0641	0.7691	0.0232	0.6835	0.7049	0.7103	0.7500

Finally, ecological vulnerability (V) is calculated based on the following model:

$$V_j = \sum_{i=1}^{7} k_{ij} \times L_j = \sum_{i=1}^{7} k_{ij} \times \frac{F_{ij}}{C_{ij}}$$

(3)

where i is landscape type, j is basin unit, k is ecological vulnerability of landscape type, L is structural vulnerability, F is ecological sensitivity, and C is ecological resilience.

2.2.3. Ecological Importance

Ecological importance refers, fundamentally, to the intrinsic value of an ecosystem. Thus, the greater the ecological importance of a landscape (as measured by the importance of various constituent landscapes or units in the region), the greater the ecological loss associated with adverse impacts on it. In this study, the value of ecosystem services is used as a measure of ecological importance, as suggested by Costanza et al. [42] and Xie et al. [43].

Ecosystem services are vital functions of the life-support system [44]. They contribute to human welfare, both directly and indirectly, and therefore represent a portion of the total economic value of the planet [42]. Ecosystem services value is the product of interactions between nature and humans, and is directly affected by human activities via changes in landscape patterns [45]. It is also closely related to the spatial distribution of landscape, whose impact may be negative. Landscape fragmentation is a driver of biodiversity loss [46,47]. Increase in fragmentation, implying corresponding increases in patches, leads to reduction in ecosystem services, and therefore decrease in ecosystem services value. In this study, landscape fragmentation is incorporated into the calculation of the ecosystem service value of each watershed, and ecological importance (S) is modeled as follows:

$$S_j = \sum_{i=1}^{7} V_i' \times A_{ij}/A_j = \sum_{i=1}^{7} \frac{V_i}{F_{ij}/F_i} \times A_{ij}/A_j \tag{4}$$

where i is type of landscape, j is the assessing unit, V_i' is the ecosystem services value of the original ecosystems, V_i is the ecosystem services value after fragmentation disturbance, F_{ij} is the landscape fragmentation of type i in watershed j, F_i is the fragmentation of landscape type i in the whole study area, A_{ij} is the area of type i in watershed j, and A_j is the total area of watershed j.

2.3. Monte Carlo Analysis of Assessment Uncertainty

Uncertainty, in the form of incomplete information and data, and diversity of risk sources, is inevitable in ecological risk assessment. As a part of results of probabilistic uncertainty analysis, sensitive analysis reflects dynamic disturbance of the evaluation results of each parameter [48]. It judges the impact of each parameter by its correlation with evaluation results. If the correlation is high, the parameter has great influence on the result and it is sensitive. Especially for a multi parameter model, sensitive analysis can identify the parameters with higher influence on the evaluation results, so that in the further analysis, we can take effective measures to reduce the uncertainty.

In order to make the results more robust, it is necessary to carry out uncertainty analyses using Monte Carlo methods, which explores the uncertainty associated with the ecological risk assessment process, and its possible impact on results. Based on a Monte Carlo analysis, uncertainty arising from land use classification assignments, and their associated ecological vulnerability, are explored using the program Crystal ball 11.1.2.2.000. Specifically, the simulation computing sets two possible uncertainty distributions (low 20%, high

40%), and carries out 10,000 simulative iterations to calculate the simulated value (Table 3 and Table 4). Outliers are excluded by working within a 95% confidence interval; i.e., simulation results above the maximum 2.5%, and minimum 2.5% were excluded before the final simulated value is calculated.

Table 3. Probability distribution of land use degree under low (20%)/high (40%) uncertainty scenarios.

Land USE degree	Possible Land Use Degree			
	Level-1 (0.2)	Level-2 (0.4)	Level-3 (0.6)	Level-4 (0.8)
Level-1 (0.2)	0.8/0.6	0.2/0.4	0/0	0/0
Level-2 (0.4)	0.1/0.2	0.8/0.6	0.1/0.2	0/0
Level-3 (0.6)	0/0	0.1/0.2	0.8/0.6	0.1/0.2
Level-4 (0.8)	0/0	0/0	0.2/0.4	0.8/0.6

Table 4. Probability distribution of landscape ecological vulnerability under low (20%)/high (40%) uncertainty scenarios.

Vulnerability of Landscape Types	Possible Vulnerability of Landscape Types						
	Level-1 (0.0357)	Level-2 (0.0714)	Level-3 (0.1429)	Level-4 (0.1071)	Level-5 (0.1786)	Level-6 (0.2143)	Level-7 (0.2500)
Level-1 (0.0357)	0.8/0.6	0.2/0.4	0/0	0/0	0/0	0/0	0/0
Level-2 (0.0714)	0.1/0.2	0.8/0.6	0.1/0.2	0/0	0/0	0/0	0/0
Level-3 (0.1429)	0/0	0.1/0.2	0.8/0.6	0.1/0.2	0/0	0/0	0/0
Level-4 (0.1071)	0/0	0/0	0.1/0.2	0.8/0.6	0.1/0.2	0/0	0/0
Level-5 (0.1786)	0/0	0/0	0/0	0.1/0.2	0.8/0.6	0.1/0.2	0/0
Level-6 (0.2143)	0/0	0/0	0/0	0/0	0.1/0.2	0.8/0.6	0.1/0.2
Level-7 (0.2500)	0/0	0/0	0/0	0/0	0/0	0.2/0.4	0.8/0.6

Setting Table 3 as an example, we define that the uncertain values should be attributed to the adjacent class. On one hand, the low (20%) uncertainty means 80% of the values is unchanged, while the other 20% should change the class. Namely in Level-1, 80% is still in Level-1, 20% will change to the adjacent Level-2. On the other hand, the high (40%) uncertainty means 60% of the values is unchanged, while the other 40% should change the class. Namely in Level-2, 60% is still in Level-2, and 40% will change to the adjacent class, where the proportion of adjacent Level-1 and Level-3 should either be 20%, respectively.

3. RESULTS

3.1. Probability of Landscape Ecological Risk

Based on the data of Liaoyuan City, maps showing the spatial distribution of the four types of disturbances—landscape, mining, settlement, and road—are generated (Figure 3). The distribution of ecological disturbances in each watershed (Figure 4), shows that, of the 58 watersheds included in the study, the Banjie River, Donglishu River, and Xiaoliushu River Downstream watersheds have the highest ecological disturbance, and the Dahengdao River Upstream, Dasha River Midstream, and Dasha River Upstream watersheds, have the lowest ecological disturbance. Whereas, in general, ecological disturbance is strongly positively correlated with urbanization and human activities, and in those areas with high ecological disturbances, mines, settlements and roads all strongly influence the landscape types and land uses. In addition, ecological disturbance is significantly correlated with population density (R = 0.76); watersheds with high population densities, have high ecological disturbances. For example, in the Banjie River Watershed, where population density is 2657 person/km^2, ecological disturbance is 0.4608; whereas in the Muchang River Watershed, where population density is 78 person/km^2, ecological disturbance is 0.2573.

Ecological vulnerability across watersheds (Figure 4) reveals that the Muchang River Watershed has the highest vulnerability, with Erdao River Upstream Watershed for the lowest. Located in the north-central part of the Liaoyuan City, the Muchang River Watershed has the largest reservoir in the study area and is surrounded by mountains, and other landscapes, which isolate it from the surrounding landscapes. Its high landscape topographic index and low landscape connectivity, are likely responsible for its high ecological vulnerability. On the contrary, Erdao River Upstream Watershed, located in northwest Liaoyuan, is a low-lying area with unbroken rivers, large and centrally located croplands, and woodland cover along the river, mainly adjoined by croplands. Spatially, ecological vulnerability increases substantially from the middle to the edges, where the peak appears in the middle, and the higher areas concentrated in the southeast. It is worth noting that regions in relatively flat terrain have high ecological vulnerability, those located in mountainous areas have intermediate levels of vulnerability, and regions located in transitional zones between mountains and plains have the lowest ecological vulnerability.

Ecological risk probability, which is calculated based on ecological disturbance and ecological vulnerability, regardless of ecological losses, is mapped at three levels (low, moderate, and high), for the 58 watersheds (Figure 5). The Banjie River and Xiaoliushu River Downstream watersheds have the highest ecological risk probability, while the low-probability regions, such as the Mei River and Xiaosha River watersheds, mostly concentrated in the southwest, west, northeast and north of Liaoyuan. The overall ecological vulnerability of mountains falls within the moderate and low levels. In summary, the ecological risk probability of mountains (low mountains, hills, and terraces) is lower than that of relatively flat areas.

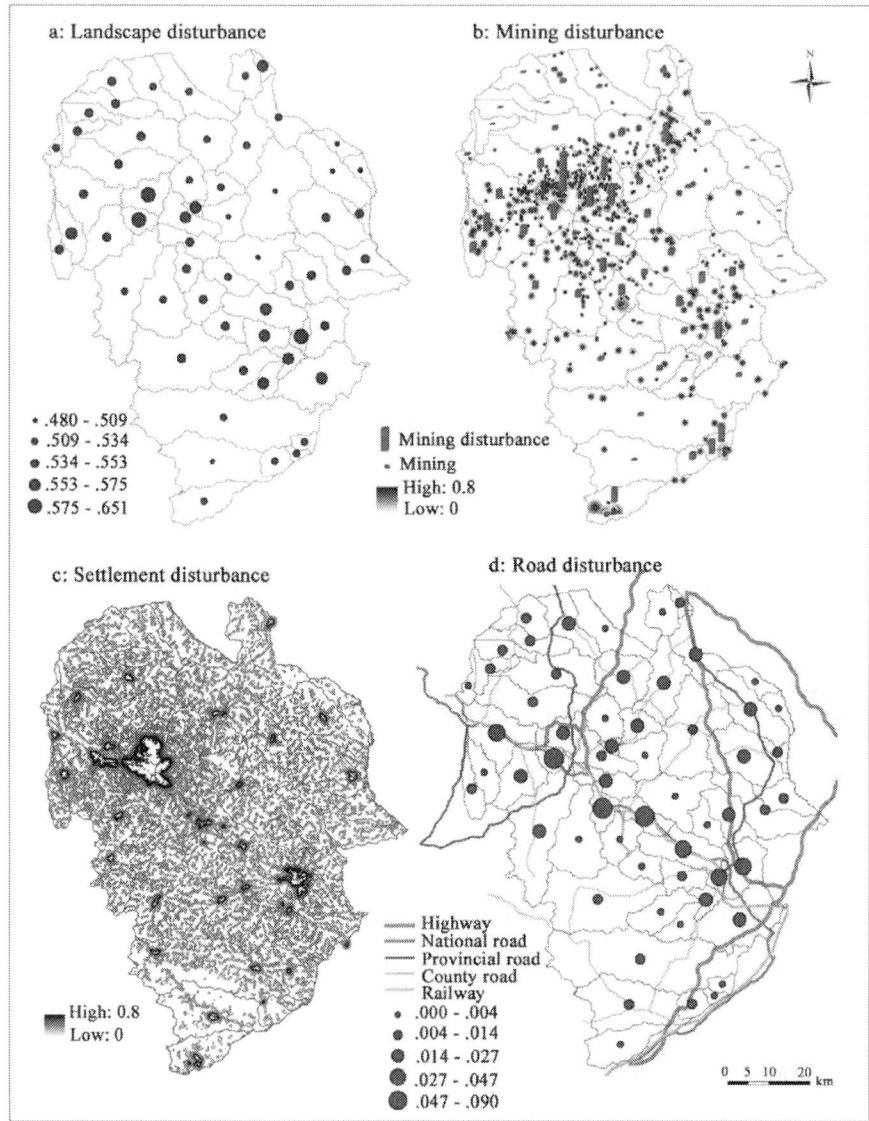

Figure 3. Spatial distribution of four types of disturbance: (**a**) Landscape disturbance; (**b**) Mining disturbance; (**c**) Settlement disturbance; (**d**) Road disturbance.

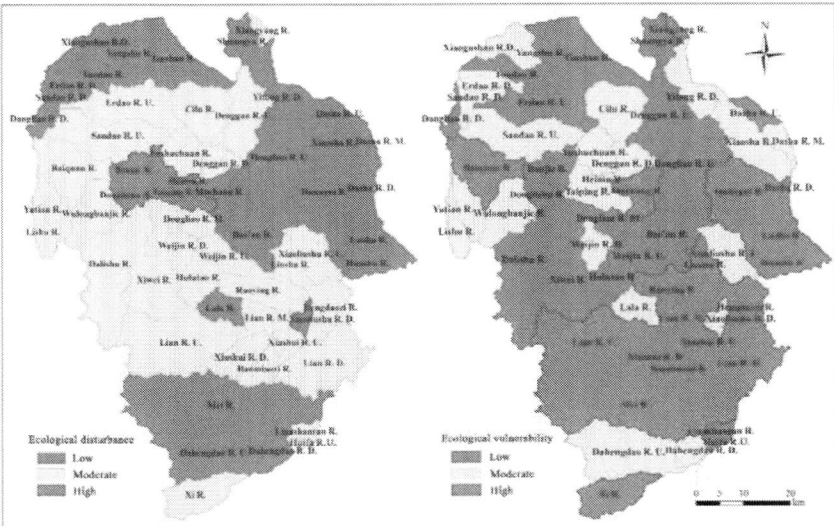

Figure 4. Ecological disturbance and vulnerability of watersheds in Liaoyuan City, China.

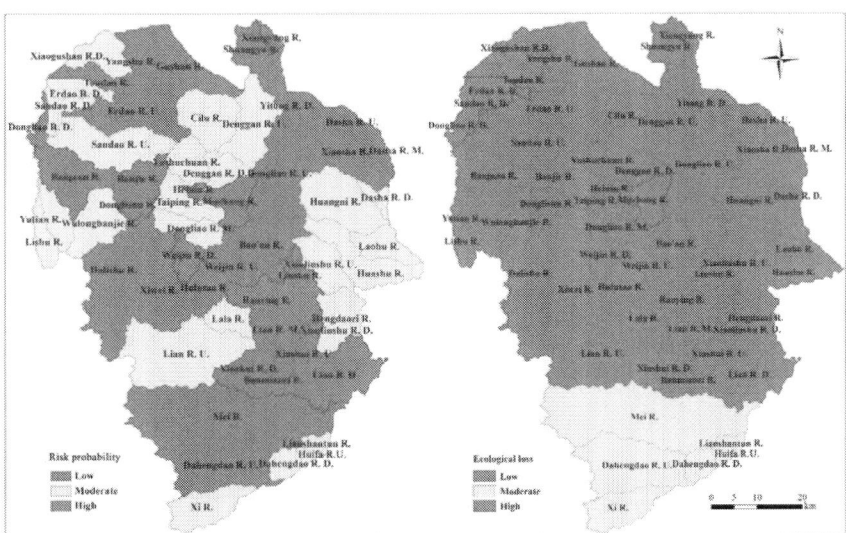

Figure 5. Landscape ecological risk probability and ecological loss in Liaoyuan City, China.

3.2. Ecological Loss and Landscape Ecological Risk

A map of the ecological importance and thus the ecological loss due to landscape ecological risk for 58 watersheds, divided into three levels (high, moderate, low; Figure 5), reveals significant spatial differences across the

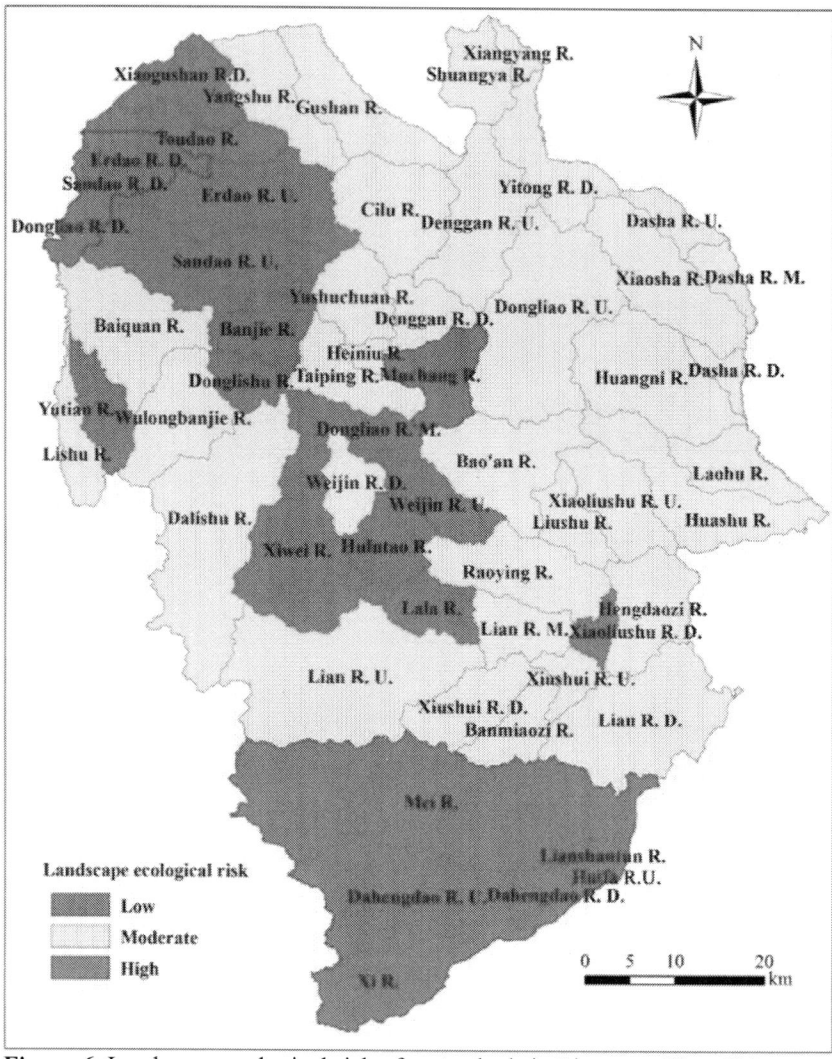

Figure 6. Landscape ecological risk of watersheds in Liaoyuan City, China.

Liaoyuan City. Sandao River Downstream, Dongliao River Downstream, and Erdao River Downstream watersheds have the highest ecological importance, and Xiaoliushu River Downstream, Banjie River and Donglishu River watersheds, have the lowest ecological importance. The top three most important watersheds are adjacent to one another, and located in the northwest of Liaoyuan, and have a high proportion of little fragmented water bodies, which play an important role in ecosystem services in the watershed. These watersheds also have low proportions of built-up land and cropland, and large proportions of woodland resulting in over 95% of the area of these watersheds consisting of natural and semi-natural landscapes.

The three "lowest loss" watersheds all have experienced high urbanization. Due to urban sprawl, natural and semi- natural landscapes are being continuously converted to built-up land. Consequently, the proportion of construction land in these watersheds is higher than that in others, and resulting in the decline of ecosystem services, and thus their ecological importance. Overall, the ecological importance of watersheds with superior natural resources endowment is higher than that of the watersheds with construction land expansion. That is to say, human activities have a negative impact on the ecological importance of watersheds.

Ecological risk, calculated based on landscape ecological risk probability, and ecological loss, mapped for the watersheds of Liaoyuan City (Figure 6), shows that, except for the Muchang River Watershed, regions with high ecological risk (e.g., Dahengdao River Upstream, Mei River, and Xi River watersheds) are concentrated in northwestern and southern Liaoyuan, whereas regions with low ecological risk (e.g., Lala River, Xiwei River, and Banjie River watersheds) are along the southeast-northwest direction. What is more, cities and counties almost belong to low ecological risk areas, which is due to the natural/semi-natural landscapes, as the object of study, have low ecological loss in these areas. Regions with medium ecological risk constitute the majority of the study area, and are concentrated in the northeast, east, and the west.

3.3. Zoning of Landscape Ecological Risk

The results of our zoning (Figure 7) show that 10 watersheds face high ecological risk, 35 watersheds face moderate risk, and 13 watersheds are at low risk. Each of these risk classes is addressed as follows: (1) High-risk areas can be divided into five types, accounting for 18.39% of the total area. These areas are mainly concentrated in watersheds where there are large water bodies, high coverage of grassland and woodland (Figure 7), such as Muchang River, Mei River and Dahengdao River watersheds. The ecosystem services offered by water, woodland and grassland and other landscapes, such as water conservation, windbreak and sand-fixation, and biodiversity conservation, have important significance for regional eco-environment. Thus, it is important to keep maintaining and strengthening the protection of woodland, grassland through reforestation; (2) Moderate-risk area includes three types, accounting for 62.02% of the total area. In these areas, the area ratio of cropland is larger than that in high-risk areas. Cropland is a semi-natural and man-made landscape which is vulnerable to human activities. During the long winter in the Northeast China, as the fallow period is long, cropland may be frequently destroyed by human activities and weather disasters. Therefore, it is necessary to strengthen the management of cropland and prevent soil erosion; (3) Low-risk area comprises three types, mainly in relatively flat terrain in the northwestern, accounting for 19.59% of the total area. The area ratio of built-ups in these areas is higher than the other two, while the area ratio of natural and semi-natural landscape is the lowest, as it is deeply influenced by human activities.

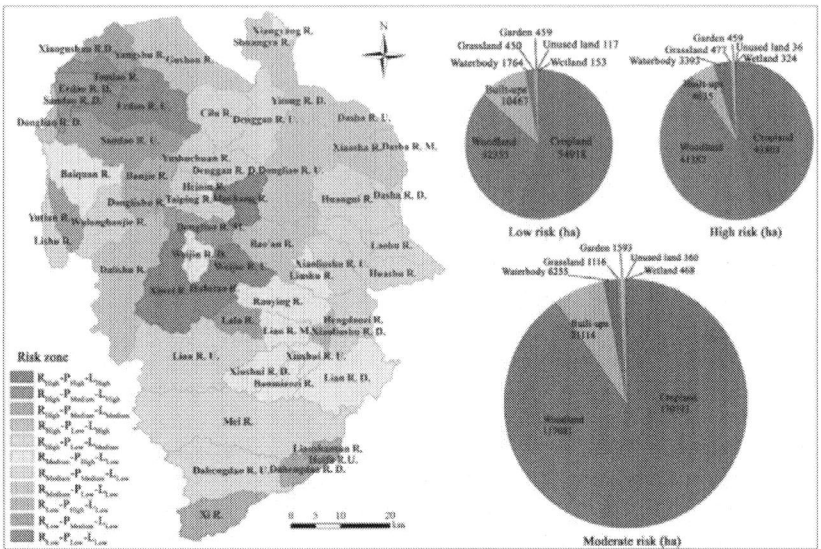

Figure 7. Landscape ecological risk zoning and total area of land use types in each grade of landscape ecological risk in Liaoyuan City, China (Note: R means ecological risk, P means ecological risk probability, and L means ecological loss).

4. DISCUSSION

4.1. Implications of Assessment Results of Landscape Ecological Risk

In the view of ecological risk probability, the high probability may be accounted for by the fact that these watersheds are relatively flat, and adjoin the city and county centers (which are suitable for human habitation, and have higher human disturbances than other areas). The low-probability regions are exposed to the lowest amount of human activity. However, although human interference in mountainous regions is lower, the ecological sensitivity of these regions is high due to the effects of terrain and slope. Nevertheless, the high or complete vegetation coverage of mountainous lands results in high ecological resilience, such that the overall ecological vulnerability of mountains falls within the moderate or low levels. This phenomenon results in a relatively high ecological risk probability in flat areas.

In the view of landscape ecological risk, the areas with high urbanization are faced with lower risk than rural areas with good ecological conditions, a pattern that is consistent with the findings of Guo et al. for Beijing [49]. Despite this, we recommend that that watersheds with low ecological risk should not be damaged, but rather their protection must be strengthened, and attention should be paid to the reduction in their extant, albeit low, levels of risk. In the zoning of landscape ecological risk, the internal city and the surroundings are of high possibility to be transformed into built-ups due to urbanization. Therefore, the planning and construction of these areas should pay more attention to protecting the integrity and connectivity of natural and semi-natural landscapes, and changing the growth mode to the approach of sustainable development.

4.2. Uncertainty of Landscape Ecological Risk Assessment

The results of the Monte Carlo analysis (Figure 8) show that the simulated and calculated risk values are similar under both high uncertainty (40%) as well as low uncertainty (20%). This implies that the graded assignment of land use degree and vulnerability of landscape types, is reasonable, and the model is generally applicable in the study area. Whereas under low uncertainty, the ratio of simulated risk to calculated risk is concentrated between 0.99 and 1.05, while under high uncertainty it is concentrated between 0.99 and 1.1; exceptions are the Dasha River Downstream and Xiangyang River watersheds. Results under low uncertainty are slightly better than those under high uncertainty, suggesting that the reliability of calculation values is higher under low uncertainty, and the possibility of bias increases as uncertainty increases [50].

Moreover, under both uncertainties, the spatial distribution of simulated and calculated risk values, in the 58 watersheds, are essentially the same. Under low uncertainty, simulated values, in the east and north of the study area, are closer to calculated values, than they are in the west and southwest. Under high uncertainty, although results are more scattered, the difference between simulated and calculated risk values of the west and southwest is larger than that of the east and north. In general, simulated values under low uncertainty are more geographically concentrated than those under high uncertainty.

Sensitivity analysis, which is important for making recommendations for future monitoring and research, is conducted for two cases. The land use degree is divided into four levels (U-level), and the vulnerability of landscape type is divided into seven levels (V-level). The intervals are equidistant, and a higher level represents to a higher risk value. A variance contribution rate of 1% is set as the threshold in the sensitivity analysis, and under low uncertainty there are five major factors, with four under high uncertainty (Figure 9). Two additional factors, the sixth factor under low uncertainty (V-level 2; level 2 in vulnerability of landscape type), and the fifth factor under high uncertainty (U-level 4; level 4 in land use degree), have a combined variance contribution rate of 0.76%, less than the threshold.

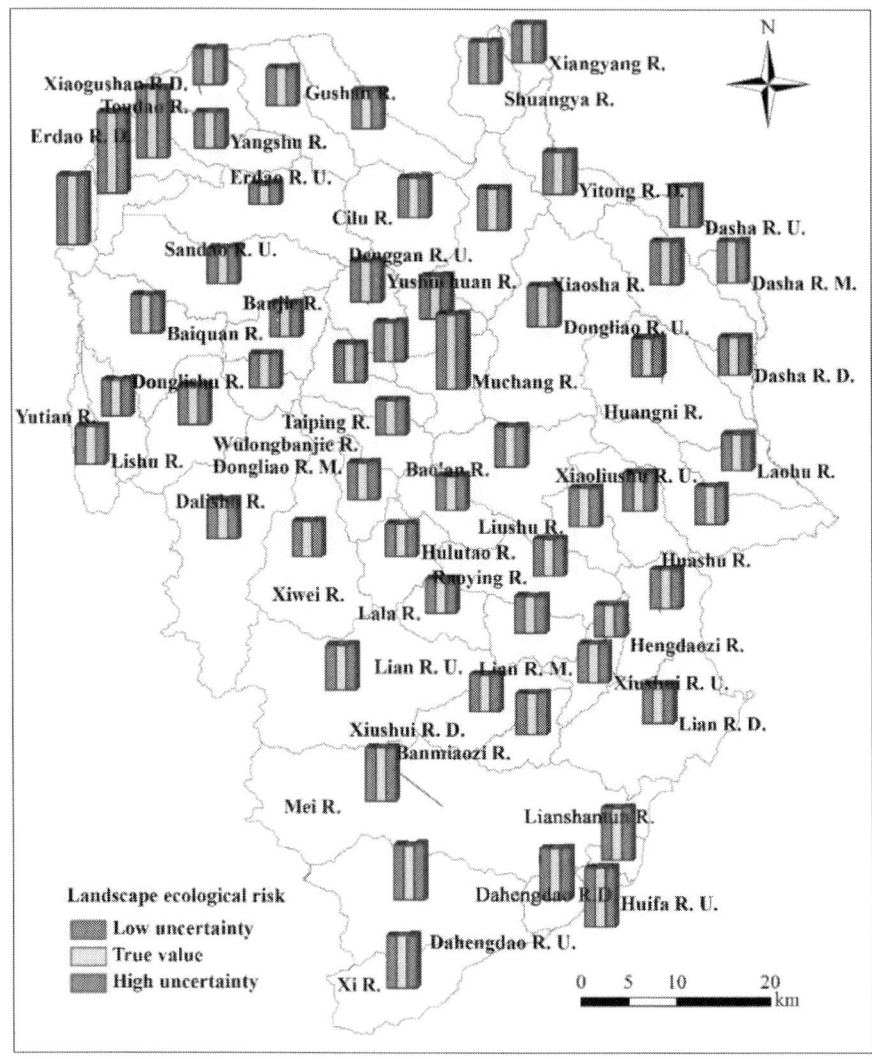

Figure 8. Simulated landscape ecological risk under low or high uncertainty of watersheds in Liaoyuan City, China.

A variance contribution threshold of 85% results in four major factors (with the highest sensitivity) under low uncertainty, and three major factors under high uncertainty. This suggests that the most sensitive factors are similar, showing only slight differences between low and high uncertainty. For example, under high uncertainty, the variance contribution rate of V-level 4 is 43.17%, nearly 5% different from that under low uncertainty; other factors show only approximately 2% difference in variance contribution rates between low and high uncertainties. Under both uncertainty scenarios, U-level 3 has a lower average variance contribution rate (24.69% for low uncertainty, and 26.40% for high uncertainty) than V-level 4 (Figure 9). In addition, V-level 1, U-level 2, and U-level 3 all affect the simulated values to varying degrees. Overall, the impact of vulnerability of landscape type on simulated value is greater than that of land use degree. Considering that landscape types corresponding with V-level 4, V-level 1, and U-level 2 account for high area ratio in each watershed, it testifies that when grading assignments, ensuring which landscape types are most dominated in the watershed is essential for the accuracy of the simulated results.

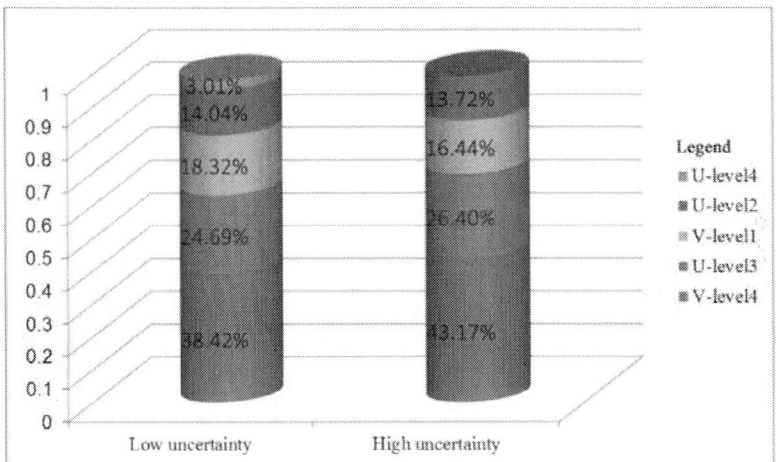

Figure 9. Variance contribution rate of sensitive factors under low or high uncertainty.

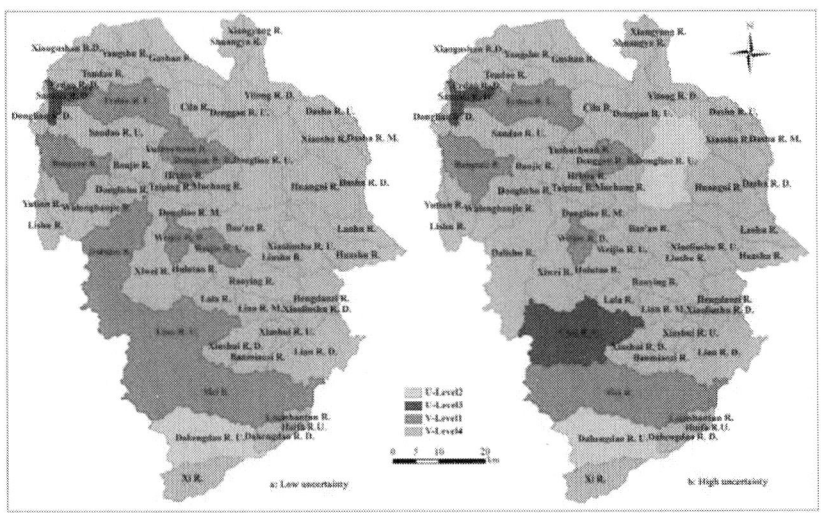

Figure 10. Most sensitive factors in assessing landscape ecological risk under low or high uncertainty of watersheds in Liaoyuan City, China.

Sensitivity analyses also shows that under high or low uncertainty, the most sensitive factors change in only five watersheds, i.e., Yushuchuan River, Dongliao River Upstream, Weijin River Upstream, Lian River Upstream, and Dalishu River watersheds (Figure 10). The most sensitive factors of over 90% of the watersheds are stable, indicating that the assignments of land use degree, and vulnerability of landscape types are reliable. From a regional perspective, in both uncertainty scenarios, the most sensitive factors change the most in the western region, such as Dalishu River and Lian River Upstream watersheds, whereas the factors are stable in the southern, northwestern, and eastern regions, implying that the assignments in these areas is the most reliable.

5. CONCLUSIONS

Landscape ecological risk assessment is useful in representing landscape ecological security, which is a critical domain of landscape sustainability [50,51,52]. In the assessment process of this study, the higher landscape ecological risk means the more vulnerability and disturbance on the important landscape units relate to less benefit in the future. The result with high risk may suffer more damages in the future which threatens the sustainability of the landscape. Finally, a zoning to identify the spatial risk should be useful to prevent risks and support a sustainable landscape planning. In the former landscape ecological risk assessment studies, ecological disturbance and vulnerability have attracted much attention [53]. However, the concern for ecological importance is less than those two domains. In the study, on one hand, fragmentation of landscape has been introduced to modify the ecosystem services in the assessment of ecological importance, which is an improvement in the evaluation; on the other hand, the uncertainty of risk was often ignored in the former assessment, and the sensitivity analysis in this study may be a novelty in regional ecological risk study.

Mining areas are susceptible to damage through the economic development of mining cities, and the ecological risk posed by mining has gradually drawn increasing attention. Liaoyuan City, once the coal city of Jilin, is a typical mining city in that it has become a resource-exhausted city because of serious damage to its environment, caused by mining. In the mining process, the sustainability of landscapes is disturbed, and the specific ecological risks and uncertainties associated with this disturbance needs to be clarified. Our study of landscape ecological risk in Liaoyuan has found that areas with high probability of risk are mainly concentrated in the relatively flat terrain of the landscape. Not only cities, but also natural and semi-natural landscapes, contribute to ecological risk. Woodland and water dominated watersheds are high-risk areas, cropland-dominated watersheds are moderate-risk areas, and the watersheds with high proportion of built-up land are low-risk areas. Although the similarity of simulated and calculated values is higher under the low uncertainty scenario, in both scenarios, the spatial distribution of both sets of values among the watersheds are the same. Overall, vulnerability of landscape type has a higher impact on the simulated values than land use degree.

Landscape ecological risk assessment, based on the risk probability and ecological loss, directly guides the planning and restoration of landscapes by recognizing spatial structure, by contributing to regional risk management. One of the gaps in our study is that it is quantified through the relative value of ecological services for several landscape types within a watershed, rather than the calculation of an overall absolute value which may be compared among various regions. Another gap is the absence of information on how landscape patterns affect the per unit area value of ecosystem services. Thus, future research in the field of landscape sustainability needs to focus on quantifying landscape ecological risk in ecologically sensitive and fragile areas. It also needs to incorporate further robust analyses of uncertainty, which, although common in regional ecological risk assessment, is still relatively rare in the field of landscape ecological risk assessment. In addition, the directions for improving uncertainty analyses in landscape ecological risk assessment include: (1) analyzing multiple sources of uncertainty; (2) putting forward new methods that are particularly applicable for landscape ecological risk assessment [54]; (3) using a grid system to carry out evaluation at grid scale, in order to improve the accuracy of assessing results [55]; and (4) investigating more specifically the sensitivity of each parameter in the assessment model [56].

ACKNOWLEDGMENTS

This research is financially supported by National Natural Science Foundation of China (41271195).

AUTHOR CONTRIBUTIONS

Jian Peng, Minli Zong, and Jiansheng Wu conceived and designed the study. Jian Peng, Minli Zong, Yi'na Hu and Yanxu Liu made substantial contributions to acquisition, analysis, and interpretation of the data. Jian Peng and Minli Zong wrote the first draft of the article. Yi'na Hu and Yanxu Liu reviewed and edited the first draft. All authors read and approved the submitted manuscript, agreed to be listed, and accepted the version for publication.

REFERENCES

1. Wu, J.G. Landscape sustainability science: Ecosystem services and human well-being in changing landscapes. Landsc. Ecol. **2013**, 28, 999–1023.
2. Peng, J.; Wu, J.S.; Pan, Y.J.; Han, Y.N. Evaluation for regional ecological sustainability based on PSR model: Conceptual framework. Prog. Geogr. **2012**, 31, 933–940. (In Chinese).

3.   Xu, E.G.B.; Leung, K.M.Y.; Morton, B.; Lee, J.H.W. An integrated environmental risk assessment and management framework for enhancing the sustainability of marine protected areas: The Cape d'Aguilar Marine Reserve case study in Hong Kong. Sci. Total Environ. **2015**, 505, 269–281. [Google Scholar] [CrossRef] [PubMed]

4.   Hunsaker, C.T.; Graham, R.L.; Suter, G.W.; O'Neill, B.L.; Jackson, B.L.; Barnthouse, L.W. Regional Ecological Risk Assessment: Theory and Demonstration; Oak Ridge National Lab: Oak Ridge, TN, USA, 1989.

5.   Hunsaker, C.; Graham, R.; Suter, G.; O'Neill, R.; Barnthouse, L.; Gardner, R. Assessing ecological risk on regional scale. Environ. Manag. **1990**, 14, 325–332.

6.   Adams, S.M.; Bevelhimer, M.S.; Greeley, M.S.; Levine, D.A.; Teh, S.J. Ecological risk assessment in a large river-reservoir: 6. Bioindicators of fish population health. Environ. Toxicol. Chem. **1999**, 18, 628–640.

7.   Baron, L.A.; Sample, B.E.; Suter, G.W. Ecological risk assessment in a large river-reservoir: 5. Aerial insectivorous wildlife. Environ. Toxicol. Chem. **1999**, 18, 621–627.

8.   Cook, R.B.; Suter, G.W.; Sain, E.R. Ecological risk assessment in a large river-reservoir: 1. Introduction and background. Environ. Toxicol. Chem. **1999**, 18, 581–588.

9.   Jones, D.S.; Barnthouse, L.W.; Suter, G.W.; Efroymson, B.A.; Field, J.M.; Beauchamp, J.J. Ecological risk assessment in a large river-reservoir. 3. Benthic invertebrates. Environ. Toxicol. Chem. **1999**, 18, 599–609.

10.   Sample, B.E.; Suter, G. Ecological risk assessment in a large river-reservoir: 4. Piscivorous wildlife. Environ. Toxicol. Chem. **1999**, 18, 610–620.

11.   Landis, W.G. Twenty years before and hence: Ecological risk assessment at multiple scales with multiple stressors and multiple endpoints. Hum. Ecol. Risk Assess. **2003**, 9, 1317–1326.

12.   Peng, J.; Liu, Y.X.; Pan, Y.J.; Zhao, Z.Q.; Song, Z.Q.; Wang, Y.L. Study on the correlation between ecological risk due to natural disaster and landscape pattern-process: Review and prospect. Adv. Earth Sci. **2014**, 29, 1186–1196. (In Chinese).

13.   Hemstrom, M.; Mezenich, J.; Reger, A.; Wales, B. Integrated analysis of landscape management scenarios using state and transition models in the upper Grande Ronde River Subbasin, Oregon, USA. Landsc. Urban Plan. **2007**, 80, 198–211.

14. Lange, A.; Siebert, R.; Barkmann, T. Sustainability in land management: An analysis of stakeholder perceptions in rural northern Germany. Sustainability **2015**, 7, 683–704.

15. Graham, R.L.; Hunsaker, C.T.; O'Neil, R.V. Ecological risk assessment at the regional scale. Ecol. Appl. **1991**, 1, 196–206.

16. Kapustka, L.A.; Galbraith, H.; Luxon, M. Using landscape ecology to focus ecological risk assessment and guide risk management decision-making. Toxicol. Ind. Health **2001**, 17, 236–246.

17. Liu, S.L.; Cui, B.S.; Dong, S.K.; Yang, Z.F.; Yang, M.; Holt, K. Evaluating the influence of road networks on landscape and regional ecological risk a case study in Lancang River Valley of Southwest China. Ecol. Eng. **2008**, 34, 91–99.

18. Yin, H.; Wang, Y.L.; Cai, J.L.; Lv, X.F.; Liu, X.Q. Regional ecological risk assessment: Its research progress and prospect. Chin. J. Ecol. **2009**, 28, 969–975. (In Chinese).

19. Xie, H.L. Regional eco-risk analysis of based on landscape structure and spatial statistics. Acta Ecol. Sin. **2008**, 28, 5020–5026. (In Chinese).

20. Jing, Y.P.; Zhang, S.W.; Li, Y. Ecological risk analysis of rural-urban ecotone based on landscape structure. Chin. J. Ecol. **2008**, 27, 229–234. (In Chinese).

21. Li, J.G.; He, C.Y.; Li, X.B. Landscape ecological risk assessment of natural/semi-natural landscape in fast urbanization regions—A case study in Beijing. China. J. Nat. Resour. **2008**, 23, 33–47. (In Chinese).

22. Wang, C.Y.; Zhang, J.; Xi, H.M.; Fu, J. Ecological risk assessment of island exploitation based on landscape pattern. Acta Ecol. Sin. **2008**, 28, 2811–2817. (In Chinese).

23. Li, Y.F.; Luo, Y.C.; Liu, G.; Ouyang, Z.Y.; Zheng, H. Effects of land use change on ecosystem services: A case study in Miyun reservoir watershed. Acta Ecol. Sin. **2013**, 33, 726–736. (In Chinese).

24. Lanis, W.G.; Wiegers, J.K. Ten years of the relative risk model and regional scale ecological risk assessment. Hum. Ecol. Risk Assess. **2007**, 13, 25–38.

25. Liu, L.; Liu, J.; Zhang, Z.G. Environmental justice and sustainability impact assessment: In search of solutions to ethnic conflicts caused by coal mining in Inner Mongolia, China. Sustainability **2014**, 6, 8756–8774.

26. Himley, M. Global mining and the uneasy neoliberalization of sustainable development. Sustainability **2010**, 2, 3270–3290.

27. McLellan, B.C.; Corder, G.D. Risk reduction through early assessment and integration of sustainability in design in the minerals industry. J. Clean. Prod. **2013**, 53, 37–46.

28.   Chang, Q.; Qiu, Y.; Xie, M.M.; Peng, J. Theory and method of ecological risk assessment for mining areas based on the land destruction. Acta Ecol. Sin. **2012**, 32, 5164–5174. (In Chinese).

29.   Li, Z.Y.; Zhang, N.; Tang, J.; Ji, Y.; Liu, J.L. Analysis on the landscape ecological risk of Jilin coal mining area. J. Jilin Univ. (Earth Sci. Ed.) **2011**, 41, 207–214. (In Chinese).

30.   Li, L.X.; Wang, B.; Zhou, L.B.; Yu, X.B. Ecological landscape risk evaluation in mineral resources exploitation. Conserv. Util. Miner. Resour. **2011**, 2, 1–5. (In Chinese).

31.   Wu, J.S.; Zong, M.L.; Peng, J. Assessment of mining area's ecological vulnerability based on landscape pattern: A case study of Liaoyuan, Jilin Province of Northeast China. Chin. J. Ecol. **2012**, 31, 3213–3220. (In Chinese).

32.   Wang, E.H. Liaoyuan Statistical Yearbook (2009); Liaoyuan Bureau of Statistics: Liaoyuan, China, 2010. (In Chinese)

33.   Chang, Y.J. Liaoyuan Yearbook (2010); Jilin People's Publishing House: Liaoyuan, China, 2010. (In Chinese)

34.   Chen, Q.Y.; Liu, J.L. Development process and perspective on ecological risk assessment. Acta Ecol. Sin. **2014**, 34, 239–245. (In Chinese).

35.   Pickett, S.T.A.; White, P.S. Chapter 21—Patch Dynamics: A Synthesis. In The Ecology of Natural Disturbance and Patch Dynamics; White, S.T.A., Pickettp, S., Eds.; Academic Press: San Diego, CA, USA, 1985; pp. 371–384.

36.   Chen, P.; Pan, X.L. Ecological risk analysis of regional landscape in inland river watershed of arid area—A case study of Sangong River Basin in Fukang. Chin. J. Ecol. **2003**, 22, 116–120. (In Chinese).

37.   Deal, B.; Pallathucheril, V. Sustainability and urban dynamics: Assessing future impacts on ecosystem services. Sustainability **2009**, 1, 346–362.

38.   Scott, C.; Lake, P.; Sergi, S.; John, M.; John, S. The effects of land use changes on streams and rivers in Mediterranean climates. Hydrobiologia **2013**, 719, 383–425.

39.   Song, G.B.; Li, Z.; Yang, Y.G.; Semakula, H.M.; Zhang, S.S. Assessment of ecological vulnerability and decision-making application for prioritizing roadside ecological restoration: A method combining geographic information system, Delphi survey and Monte Carlo simulation. Ecol. Indic. **2015**, 52, 57–65.

40.   De Lange, H.J.; Sala, S.; Vighi, M.; Faber, J.H. Ecological vulnerability in risk assessment—A review and perspectives. Sci. Total Environ. **2010**, 408, 3871–3879.

41. Wayne, C.Z.; Robert, L.B.; Ralph, D.N. Patterns of deforestation and reforestation in different landscape types in central New York. For. Ecol. Manag. **1990**, 36, 103–117.

42. Costanza, R.; D'Arge, R.; Groot, R.; Farber, S.; Grasso, M.; Hannon, B.; Limburg, K.; Naeem, S.; O'Neill, R.V.; Paruelo, J.; et al. The value of the world's ecosystem services and natural capital. Nature **1997**, 386, 253–260.

43. Xie, G.D.; Lu, C.X.; Leng, Y.F.; Zheng, D.; Li, S.C. Ecological assets valuation of the Tibetan Plateau. J. Nat. Resour. **2003**, 18, 189–196. (In Chinese).

44. Banzhaf, H.S.; Boyd, J. The architecture and measurement of an ecosystem services index. Sustainability **2012**, 4, 430–461.

45. Hessburg, P.F.; Reynolds, K.M.; Salter, R.B.; Dickinson, J.D.; Gaines, W.L.; Harrod, R.J. Landscape evaluation for restoration planning on the Okanogan-Wenatchee national forest, USA. Sustainability **2013**, 5, 805–840.

46. Manuela, D.G.; Rolf, H.; Silvia, T. Effects of habitat and landscape fragmentation on humans and biodiversity in densely populated landscapes. J. Environ. Manag. **2009**, 90, 2959–2968. [Google Scholar]

47. Matteo, M.; Duccio, R.; Francesco, G.; Giovanni, B.; Valerio, A. Biodiversity, roads, & landscape fragmentation: Two Mediterranean cases. Appl. Geogr. **2013**, 46, 63–72.

48. Campolongo, F.; Saltelli, A.; Tarantola, S. Sensitivity analysis as an ingredient of modeling. Stat. Sci. **2000**, 15, 377–395.

49. Guo, X.R.; Hu, H.W.; Chen, D.S.; Wei, W.; Cheng, S.Y. Urban ecological risk assessment based on land utilization analysis. J. Beijing Univ. Technol. **2012**, 38, 1114–1120. (In Chinese).

50. Gao, Y.; Wu, Z.F.; Lou, Q.S.; Huang, H.M.; Cheng, J.; Chen, Z.L. Landscape ecological security assessment based on projection pursuit in Pearl River Delta. Environ. Monit. Assess. **2012**, 184, 2307–2319.

51. Zhou, K.H.; Liu, Y.L.; Tan, R.H.; Song, Y. Urban dynamics, landscape ecological security, and policy implications: A case study from the Wuhan area of central China. Cities **2014**, 41, 141–153.

52. Cen, X.T.; Wu, C.F.; Xing, X.S.; Fang, M.; Garang, Z.M.; Wu, Y.Z. Coupling Intensive Land Use and Landscape Ecological Security for Urban Sustainability: An Integrated Socioeconomic Data and Spatial Metrics Analysis in Hangzhou City. Sustainability **2015**, 7, 1459–1482.

53. Peng, J.; Dang, W.X.; Liu, Y.X.; Zong, M.L.; Hu, X.X. Review on landscape ecological risk assessment. Acta Geogr. Sin. **2015**, 70, 664–677. (In Chinese).

54. Li, Y.F.; Sun, X.; Zhu, X.D.; Cao, H.H. An early warning method of landscape ecological security in rapid urbanizing coastal areas and its application in Xiamen, China. Ecol. Model. **2010**, 221, 2251–2260.

55. Ma, L.Y.; Xu, X.G.; Xu, L.F. Uncertainty Analysis of integrated ecological risk assessment of China. Acta Sci. Nat. Univ. Pekin. **2011**, 47, 893–900. (In Chinese).

56. Li, Y.F.; Shi, Y.L.; Qureshi, S.; Bruns, A.; Zhu, X.D. Applying the concept of spatial resilience to socio-ecological systems in the urban wetland interface. Ecol. Indic. **2014**, 42, 135–146.

CHAPTRT 12

Urban Planning for a Renewable Energy Future: Methodological Challenges and Opportunities from a Design Perspective

Han Vandevyvere [1,*] and Sven Stremke [2]

[1] Department of Architecture, Urbanism and Planning, KU Leuven, Kasteelpark Arenberg 1, B-3001 Leuven, Belgium
[2] Landscape Architecture Group, Environmental Sciences Department, Wageningen University and Research, Droevendaalsesteeg 3, 6708PB, Wageningen, The Netherlands;

ABSTRACT

Urban planning for a renewable energy future requires the collaboration of different disciplines both in research and practice. In the present article, the planning of a renewable energy future is approached from a designer's perspective. A framework for analysis of the planning questions at hand is first proposed. The framework considers two levels of inquiry: the technical environmental aspect, and its wider embedding in sustainable development. Furthermore, life cycle analysis and exergy studies are discussed for their application potential in design. An altered trias energetica as proposed in earlier publications appears to remain a robust concept for low exergy, renewable energy based urban design. When considering sustainable development, environmental assessments shall be completed by an inquiry of the socio-cultural, economical, juridical, aesthetical and ethical aspects characterizing the planning or decision process. The article then presents a number of practical design principles that can help envisioning a built environment that can be sustained on the basis of renewable energy sources. In accordance with the altered trias energetica concept, elements of passive urban energy design, exergetic optimization of energy provision systems and the sourcing of renewable energy are identified, and their respective potentials assessed.

Keywords: energy transition; built environment; renewable energy; sustainable urban development; energy potential mapping; regional planning; sustainable energy landscape; energy-conscious design

1. INTRODUCTION

The way by which the transition towards a renewable energy future and the transition towards sustainable cities can be geared to one another presents a major challenge for research and practice.

Sustainable building measures taken at the level of the individual building (i.e., micro-scale), such as increased insulation levels or solar-based energy provision, have become common practice but these must not be considered in isolation from questions related to the larger scale energy infrastructure (i.e., macro-scale). An additional and increasingly important level of intervention, the meso-scale, has to be considered next to the micro and the macro-levels [1,2]. This holds true for both energy provision and sustainable urban development issues. In terms of energy provision one may think of CHP or boiler plants for district heating and cooling, geothermal applications serving groups of buildings or grouped solar arrays. In terms of urban planning, questions regarding urban compactness, morphology and building orientation, related mobility effects, or the exchange of waste heat are examples of how urban planning may influence the overall energy consumption figures. Because of the many interrelations between 'energy' and the 'built environment', it is essential to coordinate the sustainability efforts in both fields of application. As a matter of fact, the urban fabric with its energy behavior in terms of demand and supply must be regarded as a 'whole system' in exchange with its larger environment. In this paper we will focus on the energy aspects of such whole systems approach, however without neglecting the overarching perspective of sustainable development.

The intended optimizations—we may hypothesize at the outset of this paper—will be scale- and context dependent. It may, for example, make more sense to renovate an existing compact urban district with related advantages in terms of combined energy consumption for buildings and mobility, than to aim at developing a new passive housing district on a greenfield outside the city where inhabitants may primarily depend on car traffic. The better performance of each single passive house in terms of energy consumption may indeed be overruled by the amount of energy needed for the increased traffic demand of its inhabitants. Another critical factor that has been studied by Stöglehner and Narodoslawsky is the energy consumed by the construction and maintenance of new infrastructure networks in low-density settlements [3].

Similar observations can be made regarding the integration of energy technologies at different scale levels within the overall energy infrastructure, and in particular within smart grids. As such, it could be more efficient to construct a solar farm outside an urban district rather than to install PV-panels or solar boilers on each individual building of the same district. Location and land use pressure, spatial quality, orientation and shading, investments, maintenance, emissions and nuisance control are factors that require a proper trade-off for the given context.

The quantitative, technical optimization of building related energy scenarios can develop along two axes, by assessing environmental impacts on the one hand, and by considering costs and benefits on the other hand. A combination of both parameters is however possible as well. This can, for example, be achieved by including environmental costs within an economic assessment [4].

Apart from environmental and economical evaluations however, socio-cultural, juridical, aesthetical and ethical considerations should come to complete the assessment when the goal is to achieve integrated sustainable development [5,6,7,8]. A typical example within the built environment is the trade-off between heritage value and improvement of the energy performance: it is unlikely that packing an iconic art nouveau building with an external insulation layer would be considered a good option. If modeling such decisions is desired, a multi-criteria decision analysis will have to be brought in. It becomes clear that at this level, both quantitative and qualitative factors will have to be considered. Several urban sustainability tools account of this fact, but respond to it in different ways [6,9,10].

The present paper will discuss this problem through two key questions. Firstly, which types of information and assessment method are available to select the most sustainable urban energy scenarios? And secondly, what can be the constituent parts of a built environment that supports renewable energy based functioning?

The resulting overview is constituted through a literature review that targets different research fields: life cycle assessment studies, energy and exergy research, urban morphological studies, and integrated urban sustainability evaluation. Combining output from these different disciplinary fields allows drawing up a systematic overview of the current challenges and opportunities for designing sustainable cities that make use of renewable energy sources.

2. THEORETICAL FRAMEWORK AND LITERATURE DISCUSSION

For the present analysis we consider two major levels of assessment. On the one hand, the environmental performance of a given urban (energy) system, and on the other hand its integrated sustainability. By the latter we mean the quantitative-qualitative evaluation that considers socio-cultural, economic, juridical, aesthetical and ethical aspects as a complement to the set of environmental indicators.

2.1. Environmental Assessments

2.1.1. Life Cycle Assessment

Life Cycle Assessment (LCA) is a method to analyze the impacts on the environment resulting from a given process, e.g., the construction, use and demolition of a building. It is a well-established and regulated method [11] that is now finding its way into the evaluation of the built environment, from single construction components over buildings to entire urban fragments. It typically considers a series of impact categories like climate change, resource depletion, eco-toxicity or human health risks. The aggregation of all impacts into a single 'score' or 'environmental cost' is however optional and remains contested, e.g., [12]. This is because aggregation requires weighting of the different environmental effects, and as such involves normative decisions (for example, considering human health as more important than ecosystem health or vice versa; methodological difficulties of monetizing environmental impacts in case of expression as an environmental cost), e.g., [13]. For the present discussion, the LCA of energy systems is the most important factor to be considered, e.g., [12,14].

Despite its possible critics, the LCA of energy systems permits to reasonably identify technologies with less environmental impact, independent of their energetic or exergetic efficiency (see 2.1.2). Therefore, an LCA is of paramount importance to assess the environmental sustainability of energy sources and conversion technologies. An example is the evaluation of bio-fuels through LCA. Whereas bio-fuels may seem very interesting on the basis of their potential carbon-neutrality, an LCA of the entire process chain may reveal severe impacts that result in a score that is worse than that of fossil fuels [15].

At the demand side, the LCA of buildings or larger parts of the urban tissue (such as entire neighborhoods) permits to reduce aversive environmental impacts of such built structures. The method allows, for example, us to assess if the surplus impact associated with the use of additional insulation materials is compensated by the reduced energy consumption figures of the building under consideration, e.g., [16]. At the higher scale levels of urban tissues or larger energy infrastructures, a major challenge still resides in collecting the data and building the models required for this type of evaluation, with reasonable levels of reliability [12,17].

2.1.2. Primary Energy versus Exergy

A second important sustainability aspect of the energy chain is its efficiency. More efficient energy use leads to reduced primary energy consumption and hence better sustainability scores. Although such effects should become manifest through LCA, it is advisable to conduct specific energy studies about primary energy use and exergetic efficiency for reasons of the particular insights they provide.

Expressing energy use in terms of primary energy consumption, on the one hand, provides a correct image of the real quantities of energy needed for a given application. For example, a passive house requiring 15 kWh/m² per year for space heating and cooling based on typical current electricity mixes may require up to 40 kWh/m² per year of primary energy through fossil fuel based electricity generation. Therefore, expression of consumption patterns in primary energy units deserves priority over current practices where this differentiation is not being made.

Exergy studies, on the other hand, provide a qualitative assessment of energy use. They allow for a complementary type of optimization, taking into account the intrinsic quality of an energy type in relation to its use [18]. Such an approach leads to fundamental insights, in particular with regard to the planning and design of the built environment [19]. In brief, exergy stands for the ability of a given quantity of energy to produce work. This quantity of work is not only dependent on the energy source, but as well on the surrounding environment. Therefore exergy (or work capacity) of a particular energy carrier must always be expressed with reference to the given environment.

A common (exergy related) problem of today's energy technologies is that they systematically demand high quality energy sources. One example is the combustion of high quality fossil fuels to provide for low quality energy applications such as space heating or cooling [18,20,21]. Or expressed in numbers, a typical home boiler burns natural gas at temperatures of over 1000 °C to provide space heating of around 20 °C. The latter could also be realized by waste heat at 45 °C. Because of its novel character, the application of exergy considerations in the planning and design of the built environment is considered in more detail in the following paragraphs.

First of all, it should be remarked that a major share of the energy required in buildings is for heating and cooling, in a typical temperate climate around 70–80% for heating alone, e.g., [22,23]. Both applications aim at heat transfer processes that only require low exergy inputs, for example heating a space up to 22 °C or cooling it to 18 °C starting from temperatures that are in a close range. A similar remark can be made about domestic hot water, although the required temperature levels are higher in this case and so the exergetic quality of the energy source should increase accordingly. High quality sources like electricity are, by contrast, only strictly required for a specific range of building appliances, including household devices and lighting. Therefore, both combustion processes and electricity use should be avoided for space heating and cooling, and in a somewhat lesser degree, for domestic hot water production. Alternative, low-exergy heat or cold sources must be searched for. In practice, this could imply waste heat recovery, solar heat production and heat extraction from underground.

A group of experts working for IEA Annex 49 has researched this issue and recently brought together its findings and recommendations in a report on low exergy ('lowex') design strategies for buildings and communities [18]. Within the same project, software tools have been developed to assist designers in exergetic optimization. In the context of the discussion in this paper, it is important to note that the researchers of this project emphasize that exergy cannot be considered as a sustainability indicator ([18], p.114). Herein they differ from other scientists that will be discussed below. Some of the practical conclusions of the above study deserve particular attention because they help to define the constituents of a sustainable urban energy paradigm as discussed in the third section of this paper. Concerning renewable energy, both a large share of renewable sources and a high exergetic efficiency must be strived for simultaneously. In practice, trade-offs will have to be made. Solar based systems that take the heat immediately from the sun are favorable for two reasons: they help reducing primary (fossil fuel based) energy demand and optimize exergetic performance. The re-use of residual heat, especially from CHP, is another solid option. Heat pumps can be a good choice for low exergy systems, as far as their electricity use does not wipe out the advantages of using waste or ambient heat. This means in particular that the bridged temperature differences must remain rather small. As lowex energy systems make use of waste and ambient heat, considering the district, community or even regional scale becomes an important factor of optimization. This fits very well with the approach of 'energy potential mapping' or 'energy landscapes' as brought forward in several recent studies [24,25].

2.1.3. A Unification Theory?

Within the research field of exergy studies, another school exists where the concept of exergy is taken further than the strictly energetic optimization of energy flows. Hereby, the exergy accumulated into or extracted/dissipated from a natural or human system is taken into consideration in order to assess its overall resource efficiency. This can go as far as considering the exergy embedded in informational structures such as DNA. Following the same lines of thought, exergy could then be used as a sustainability indicator. In this context, Jørgensen uses the concept of 'eco-exergy' [26].

In order to assess these new research directions, a group of researchers has mapped the current state of affairs in an overview article [27]. They distinguish four paths of development: ecosystem analysis, industrial system analysis, (thermo-)economic analysis based on extended exergy accounting (EEA) and environmental impact assessment, for example through the Cumulative Exergy Extraction from the Natural Environment (CEENE) approach. The latter category aims at reconciling LCA, which is mainly focused on harmful emissions, with exergy by considering those negative environmental impacts in terms of their related exergy effects or through the exergetic cost of abating the emissions in case ([27], pp. 2228–2229). The authors argue that industrial system analysis and environmental impact assessment based on exergy have indeed a mature potential, while ecosystem analysis and thermo-economics remain controversial. Techniques like EEA or CEENE would moreover avoid the application of (subjective) weight factors, as occurs in LCA, through a unified comparison basis [27,28].

From a practical point of view however, case studies in this area illustrate the complexity of transferring extended exergy concepts to the daily practice of urban design, planning and construction [26,28,29]. Therefore we propose not to include them in the implementation framework proposed here, at least for the time being.

2.1.4. The Altered 'Trias Energetica' or 'New Stepped Strategy' as an Analytical Work Tool

In energy-related engineering, a common approach is to apply a principle known as the trias energetica [30]. This is a three-step method for increasing the sustainability of energy systems by respectively 1° reducing the demand for energy, 2° applying renewable energy sources wherever possible and 3° filling in the remaining need as efficiently and cleanly as possible with fossil fuels. However, thinking of a renewable energy future and taking into account the exergy aspects mentioned above, a new version of the trias has meanwhile been proposed by van den Dobbelsteen and worked out in several research projects [20,31].

This altered trias or 'New Stepped Strategy' maintains the first step of the original sequence, inserts a new one and converts the fossil fuel part into a temporary, transitional step as follows ([20], p. 12):

1.   Reduce consumption (using intelligent and bioclimatic design)
2.   Reuse waste energy streams
3.   Use renewable energy sources and ensure that waste is reused as food
4.   (Supply the remaining demand cleanly and efficiently)

It is to be noted that the use of materials is now included in the strategy, but for the present discussion we will solely focus on the energy part. The second step targets the exergetic optimization discussed in Section 2.1.2, however without being exhaustive in its description. We will, for the sake of completeness, formulate a solidified trias or expanded 'stepped strategy' in Section 3 of this paper.

When comparing van den Dobbelsteen's proposal with the work of IEA Annex 49 [18], step 1 appears to remain largely valid from an exergetic point of view as well: reducing the energy demand is mostly effective for limiting the exergetic losses at the same time. By contrast, step 3 addresses the use of renewable energies which is, in the strict sense, not being considered through exergy analysis. As a conclusion, the new stepped strategy approach may be regarded as a solid strategy from both an energy and exergy point of view, as well as from an impact reduction/LCA point of view. It coherently addresses every side of the problem. Nevertheless, detailed energy/exergy and LCA studies will always allow fine-tuning the potential design options, and therefore remain desirable in many cases.

2.2. Environmental Assessment versus Integrated Sustainability Evaluation

Urban development questions present complex decision contexts concerning ill-defined problems, in particular when integrated sustainable functioning is pursued [32,33]. Therefore, environmental assessments should in this context be complemented by an evaluation in the social, economic and policy/process realms, e.g., [5,6,7,8].

For the present discussion, we will group these dimensions together into a second step that aims at putting sharp the picture of integrated sustainable functioning. As a detailed analysis of sustainable urban development evaluation is beyond the scope of this article, we summarize a methodological approach based on multimodal system analysis to frame the present discussion [7,8,34,35,36]. The multimodal system theory states that reality manifests itself in human experience through independent modal aspects or law-spheres. In fact, these modal aspects can roughly be considered as knowledge fields. Herman Dooyeweerd, founder of the theory, identified 15 such spheres that logically succeed each other as follows: the numerical, the spatial, the kinetic, the physical, the biological, the sensitive-psychical, the analytical-logical, the historical-cultural, the linguistic-communicative, the social, the economic, the aesthetic, the jural, the ethical and the credal law-sphere [34]. Sphere sovereignty on the one hand and functional links between the law-spheres on the other hand, are two important characteristics of the modal aspects. For example, a law-sphere cannot exist without its 'substrate' of preceding spheres, but at the same time it is fundamentally different from the latter. Hereby the character of the modal aspects evolves from determinative (ruled by the laws of mathematics, physics or biology) to normative (regulated by human convention). Dooyeweerd's theory has been successfully applied in, among others, system sciences and sustainable urban development studies [7,35,36]. For the present analysis we thus propose an aspectual lecture of reality that has been interpreted in a way to formulate a series of principles for sustainable (urban) development (SD) [8]:

- The range of modal aspects concerning SD reaches until the ethic dimension (similar to the field of Human Rights), as we can derive from expressed sustainability descriptions such as the Brundtland definition [37]. Therefore SD requires more than a strict Planet, People, Prosperity approach, and should explicitly address the value-related trade-offs that characterize decision-making, in particular in the contexts of urban and spatial planning. These trade-offs are indeed deeply influenced by the social and ethic norms held by the actors involved in the decision process;

- Sustainability is thus a human normative concept. However, it remains founded in the biophysical world characterized by its own, deterministic laws. This implies that normative conventions describing sustainability should account of related biophysical effects. As an example, this is reflected in the structure of the Kyoto protocol where a normative convention on preventing climate change defines how much $CO2$ can be emitted. The amount of climate change (and therefore of carbon emissions) we judge to be acceptable is however based on a normative appreciation;

- In accordance with the former principles, no social or economic sustainability are possible without environmental sustainability for reasons of retrocipative foundation [34]. Weak sustainability (e.g., [38]) is therefore not feasible, or in other words, sustainable development is integrated, or it is not at all;

- From an analytical point of view, a multidisciplinary framework is needed for the proper assessment of all the independent modal spheres concerned by SD;

- And thereby scientific disciplines deliver constitutive, although partial assessments through their specific idiom. Determinative (quantitative) idioms, as well as strictly social or economic idioms make up the building stones of an integrated assessment, but do not deliver an exclusive or autonomous ordeal by themselves. This observation holds in particular for monetizing techniques such as cost benefit analysis.

Based on the research mentioned higher, we opinion that a method considering these principles of extended sustainability assessment is beneficial for setting up a well-balanced design or planning process. In particular, the method allows situating questions of energy provision or intervention in the built environment in a wider landscape of mutual interferences. It facilitates structuring the question at hand through in-depth analysis, and thus helps to pave the way for informed assessment, design, and decision making.

3. CONSTITUENTS OF THE SUSTAINABLE BUILT ENVIRONMENT versus CONSTITUENTS OF THE RENEWABLE ENERGY INFRASTRUCTURE

In this section we aim at composing a methodologically coherent outline of the constituents of the transition towards sustainable urban energy futures, based on the concepts discussed in Section 2.

At this moment, and still for many years to come, renewable energy sources will not be able to replace the current volume of fossil energy consumption entirely. Therefore, when considering a transition towards 100% renewable sources, reducing energy demand will remain a prime objective, coming before any other consideration and thus reinforcing the first step of both the trias and the 'new stepped strategy' approach.

As discussed in Section 2.1.4, we propose to upscale the 'new stepped strategy' to fully account for the urban factors that come into play when considering the built environment as a whole. We thus arrive at the following principles (leaving out the fourth, transitional step):

1. **Reduce energy demand** by passive architectural and urban control measures: location choice, response to local climate and relief conditions, architectural and urban compactness, orientation and daylighting features, building insulation levels, use of thermal capacity and buffer spaces, insertion of natural elements for shading, wind braking and urban heat island reduction, etc. It is interesting to observe that many buildings and cities constructed prior to the industrial revolution heavily relied on this type of climate control because energy sources were scarce and costly (see e.g., [39]). Effects of urban morphology on traffic demand make up a second important contributing factor, and related savings in transport energy should as a principle also be considered as a passive benefit;

2. **Optimize energy streams** from an exergetic point of view. For the built environment, low temperature (and thus low exergy) heat sources are of particular importance. They however require proper building installations (e.g., floor or wall heating and heat storage facilities). A similar remark could be made about cooling. Moreover, in an urban context, cooling and heating demands exist simultaneously or can at least be time-buffered to do so. Therefore, two major constituents of exergy optimization emerge: direct sourcing of low exergy heat sources (solar heat, ambient heat, recovery of waste heat, etc.), and heat exchange and storage between different building programs.

3. **Provide renewable energy** to fill in the remaining demand. This energy may be provided at the following three scale levels: single building (intra-urban micro-scale), building group/district (intra-urban meso-scale), or city/region (extra-urban macro-scale). The question which energy provision is most appropriate at which scale level shall basically be answered through an assessment as discussed in Section 2. It thus concerns an efficiency matter, but needs embedding in a wider framework of SD goals.

Some further observations can be made regarding the constitutive elements of this expanded stepped strategy.

3.1. Reducing Urban Energy Demand through Passive Measures

When considering the impact of urban energy use in buildings, several contributing factors may be distinguished as represented in Figure 1, based on [40,41]. The impact is determined by the quantities of the different energy uses multiplied by the environmental impact of every used energy type. When looking at the typical ranges of difference between good and bad performance, the factors displayed in Figure 1 appear. It is to be noted that simply multiplying the separate limit values for obtaining the overall variation range is not correct, since the contributing factors influence each other. Rebounds, for example low energy bills, may cause inhabitants of a passive house to waste more energy through lighting and household appliances.

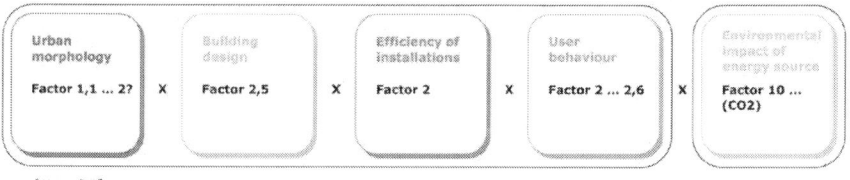

Amount of energy

Figure 1. Contributing factors to the amount of energy needed, and corresponding environmental impact, of intra-building energy use. Urban morphology and building design indicate the potential of passive design measures. Scheme based on [40,41]. The location aspect (relief and climate) is not considered here.

Among the above factors, urban morphology and building design should be considered as the pre-eminent passive design variables that can make a difference with regard to sustainable performance.

Estimations on the role of urban morphology in terms of building textures vary considerably. Ratti, Baker and Steemers arrived at variations in the order of magnitude of 10%, considering heating and cooling, lighting and ventilation [41]; whereas Salat identified 80% differences for heating energy and concludes, more generally, that urban morphology can make up for a factor 2 in energy use variations [40,42].

In addition, this analysis does not yet include higher order design parameters concerning location choice and the related effects of site, relief, vegetation and local climate. This is in fact a parameter that has been historically important, but later ignored, especially in the era of technological innovation and urban expansion that has come with the industrial revolution. If any, the focus in the past was on energy conservation through careful site planning, climate-responsive building techniques and proper use of vegetation and water structures [43,44,45] (Figure 2). A renewed interest in these aspects can now be found in studies inquiring the influence of vegetation on the urban climate, and in particular the urban heat island [46,47,48]. In a similar way, vegetation and water structures are gaining wider interest for their potential contribution to climate adaptation strategies within cities.

Figure 2. Climate-sensitive and passive design in traditional architectures: (**a**) urban palace in Damascus, with inner patio, an iwan as a half-outdoor living space and a central pond for evaporative cooling; (**b**) house in Romania with veranda for shading (open part) and as a buffer space (closed part); (**c**) double façade houses in La Coruña, Spain, where the galerias create an intermediate zone protecting and regulating the climate in the house behind; (**d**) rural house in the wind-exposed polders of the Belgian coastal region, with an arrangement of protective greenery. Image source: authors.

Increasing urban compactness must hereby not become an absolute principle. There are quantitative as well as qualitative reasons for restricting the compaction to certain limits. First, energy consumption in compact and deep structures will increase by the rising need for artificial lighting and mechanical ventilation ([41], pp. 772–773). Second, qualitative requirements like the access to daylight in living and working spaces or the sufficient availability of open, green areas within the city put equally a limit on the compactness of the urban texture. Consequently, defining optimum morphology and compactness is an extremely difficult and location-specific task, regardless of the energy or sustainability parameters under consideration.

While urban morphology is an aspect that has until recently remained undervalued, the influence of the individual building design is better known. Orientation, glazing ratios, building envelope characteristics and thermal capacity define the major passive impacts on energy use [23,49].

When considering the influence of urban morphology on energy needs for transport, general propositions such as the hyperbolic correlation between urban density and personal transport energy derived by Newman and Kenworthy [50] may be challenged in particular local situations, not least because transport user choices and behavior are fairly unpredictable. Urban compactness appears to facilitate, but not necessarily to guarantee, reduced transport energy demand. The estimation of related transport impacts, for example at the neighborhood scale, thus becomes a daunting task. Surveys have proven to be a way of monitoring the real outcomes of the many mutually interfering aspects of urban transport. Examples hereof can be found in some flagship sustainable neighborhood projects, e.g., Hammarby Sjöstad (Stockholm) ([51], pp. 6–7) and BedZED (London) ([52], pp. 26–28). For Hammarby, a compact brownfield redevelopment with good public transport connections to Stockholm's city center, CO_2-emissions related to private car use were found to be 50% lower than for reference mobility patterns. In a similar way, private car mileage in BedZED was estimated to drop some 65% under the British average.

From these observations, we may conclude that the general principles of passive urban design appear as relatively robust, including transport related issues, but that detailed studies are essential when intervening in particular contexts. These detailed studies will depend upon proper methodologies and data sets at the planning stage of an urban project. Apart from environmental and cost efficiencies, a number of qualitative parameters have to be taken into account, as discussed in Section 2.2.

It should be noted that compact urban and building structures also facilitate savings in the required amounts of construction materials (e.g., [53]). Research on building insulation has indicated that the energetic pay-back times of insulating materials are relatively short [16]. In conclusion, these types of passive energy-saving measures can concur well with the overall goal of reducing total environmental impact, and therefore the first action of the stepped strategy remains principally uncontested.

3.2. Exergetic Optimization of Building Related Energy Streams

As argued above, it is essential to differentiate between energy qualities when considering building related energy streams. Taking into account the observations made in Section 2, a fairly new branch of research called energy potential mapping investigates the opportunities for optimized exergetic use of available energy sources at the urban and regional scale levels [20,24,31,54]. Hereby the spatial aspect of energy potentials becomes critical for two reasons.

First, electricity and heat (or cold) are characterized by different transportation and storage possibilities. In basic terms and considering the present state of affairs, this may be resumed as electricity being easy to transport and difficult to store, while the opposite applies to heat and cold [19]. As a result, local heat and cold applications are feasible within the built environment, while the generation of electricity can be delocalized at larger distances from the consumers where appropriate. Some nuances must be made, however. In the future, electricity storage may become more feasible, for example by using car batteries as a storage medium within smart grids (e.g., [55]). Exploiting beneficial interferences between electricity and heat/cold applications is another aspect that gains importance (e.g., [56]). The upscaled storage of heat and cold can be achieved with buffer tanks of increasing size, e.g., concrete or polymer tanks containing up to several thousands of m^3 of water (e.g., [57], Figure 3), but also by using the underground as a free reservoir through the different types of geothermal storage (aquifer-, soil-, or cavity-based).

Figure 3. Solar boiler field with storage tank on the island of Samsø, Denmark. Directly sourcing solar energy is favorable from an exergetic point of view. If the heat can be efficiently stored for matching supply and demand, the disadvantage of the variability of the renewable source can be partially overcome. Image source: authors.

This leads to a second aspect illustrating the importance of spatial planning in the energy equation: all of the building related energy infrastructures must find their place either on or below the ground. Spatial arrangements and distances between applications thus become an important consideration. Moreover, as it was the case for the reduction of the energy demand, an optimization in the environmental sense alone will not be sufficient to arrive at sustainable solutions. Spatial planning involves many other social, economic, juridical as well as historic, cultural and aesthetic aspects having a decisive influence on the feasibility of any proposed intervention. This complex exercise leads to urban or regional energy plans such as worked out, for example, in the Dutch SREX and Lowex research projects (e.g., [31], Figure 4).

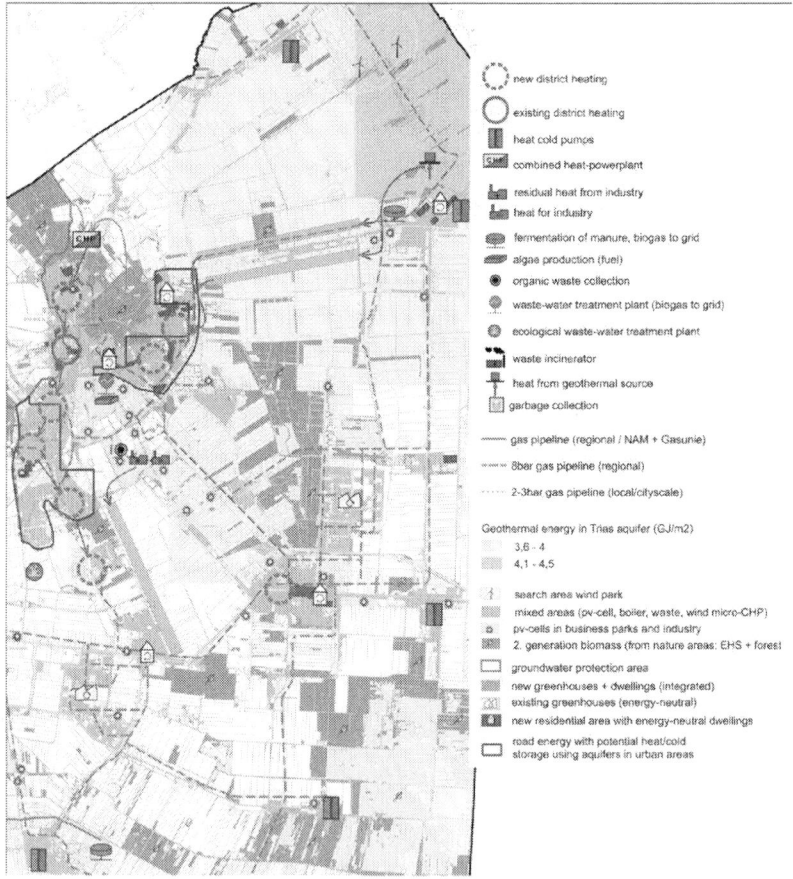

Figure 4. Extract of a spatially explicit energy vision for the region of Southeast Drenthe, The Netherlands. The map shows how renewable and residual energy sources can be integrated in a regional energy landscape with its urban, industrial and rural functions (composed image based on [31]).

Two types of application, waste heat recovery and exchange of heat and cold, deserve particular attention. Waste heat recovery is most feasible from high temperature industrial and electricity generation processes, and enables the cascading of heat through low temperature applications in the built environment. Heat exchange between building programs, on the other hand, is possible between net suppliers and demanders. Another optional but related strategy is to store excess heat or cold until it can be used. Here, the more known shallow geothermal applications come in view and perhaps gain more momentum in the near future. One example is to manage the demand and supply of both heat and cold on the neighborhood or district level.

District heating and cooling networks are defined by similar operational conditions. The heat or cold demand should be sufficiently high and spatially concentrated to allow for satisfactory efficiency, both in environmental and in economic terms. Regarding the important question if district heating remains interesting in combination with low energy building districts, a Danish case study points out that this can be the case if transmission losses in the network are reduced to the maximum, combined with the use of heat storage tanks in the serviced buildings [18]. The opposite situation appears in historic town centers, where it is difficult to upgrade all of the building patrimony to low energy or passive standards. Here it may be efficient to invest in district heating and cooling, while at the same time topping the efforts in building renovation to levels that are environmentally responsible, yet culturally acceptable and economically feasible.

Therefore, first of all, an environmental evaluation should allow formulating adequate and sufficiently precise trade-offs between the available applications. Which scenario delivers the best energy/exergy/LCA-scores? Economic, social and other qualitative assessments should then build further on the outcome of this environmental evaluation.

In practice, a deeper level of restructuring and life cycles must be addressed as well. Since the built environment has a high permanence, questions regarding the very long term (more than 100 years) should be considered. This is linked to the fact that the rotation in the built environment is typically around one or a few per cent per year (e.g., [2], p.59). How shall the building stock evolve to realize increasing sustainable performance? This means that, for example, the following type of problem needs an address: is renovation the best option; is it demolition and rebuilding; or is it demolition and building at another location? A similar reflection should be made concerning energy infrastructures whose life cycle may span across even longer periods of time [25]. Each of these options may deliver other preferable energy systems.

The fact that both the energy and exergy efficiency of the current transport sector are poor (e.g., [28]), further emphasizes the need to address mobility as well, and this at all levels (demand, modal switch, technology).

3.3. Renewable Energy Provision

Renewable energy provision may take place at three scale levels, which can be conceptualized as follows:

- Macro-scale: large hydropower, wind farms, concentrated solar power plants (CSP), large-scale solar parks, large biomass plants, deep geothermal heat and/or electricity generation. This scale is somewhat similar to current electricity production patterns;
- Meso-scale: district systems including CHP plants and related networks, mid-scale solar applications such as asphalt collectors and small to medium solar fields (PV or heat), geothermal applications at the district scale (open heat/cold systems);
- Micro-scale: renewable production at the scale of the individual building: mainly PV, solar boilers, micro CHP and closed heat/cold systems.

In theory, a purely technical trade-off could define which applications are most favorable in a given design context. As mentioned for the other constituents, there are however many more factors to be considered. This results in a complex decision making process, especially for urban planning concerned with long term development. In how far may one, for example, expect that a smart grid or super grid will be available within a few decades?

At the supranational scale, research has indicated that Europe and its neighbors can have electricity based for 100% on renewable sources under the condition that a super grid is created throughout Europe and its neighboring territories (in particular North Africa and parts of Russia) [58]. Therefore, technically spoken, this option delivers an important potential for filling in urban demands that cannot be met locally or within the proper regional context.

In a case study from Austria, Wächter et al. stress that "Critical issues are region-specific production of energy and its use, settlement and regional structures and values and role models, which all have a determining influence on energy demand." ([59], p. 193). Energy-intensive land use patterns (urban sprawl,...) are a hard nut to crack when it comes to rendering these structures more sustainable: both settlement structures with their mobility patterns, and energy infrastructures, should be profoundly transformed over the coming decades.

In another case study for a renewable energy system in the region of the Peat Colonies in the Netherlands, a strategy was proposed where areas with high renewable energy potential will have to invest less in energy savings whereas areas with small renewable energy potentials will have to focus on savings even more [60]. This proposal—to determine ambition targets for both renewable energy provision and energy savings based on the local energy potentials—once again stresses the need to map energy potentials prior to any decision with regards to alternative energy systems.

Diversity is another relevant criterion to be added for robust energy systems at all scales [61]. It implies that sources and technologies replace each other when failures appear or shortages occur. In addition, a diversified energy system can deal with intermittencies between energy supply and demand.

From this discussion it becomes clear that the third step—renewable energy provision—goes beyond individual interventions at the building or urban scale. Urban energy problems should, at all times, be considered within their regional energy landscapes. Amongst other benefits, this accommodates for the low density (or high demand for land) of most renewable energy sources. Cities should, however, also be enabled to connect to future super grids. This is especially important as macro-scale electricity generation presents itself as an environmentally efficient means of energy provision that can compensate for the lack of resources within cities and towns.

4. DISCUSSION AND CONCLUSIONS

In the previous sections we have adopted a bird's eye perspective on the challenges for renewable energy based urban planning, all from a designer's standpoint. In Section 2 we proposed an analytical framework for addressing these challenges. We distinguished two levels of analysis: (a) environmental, and (b) integrated with a view on sustainable development. Taking into account recent insights on the importance of low exergy design, we have come to the conclusion that an altered triasenergetica is a robust and useful concept for addressing environmental aspects in urban design. Although exergy remains a controversial concept, especially when it is taken out of its strict energetic context, we are convinced that it deserves particular attention in future planning practice and research. The somewhat limited capacity to serve as a more holistic sustainability indicator implies that other sustainability assessments remain important.

In Section 3 we zoomed in on the practical building stones of urban networks sustained by renewable energy. To our opinion, inquiring the role of urban morphology in relation to energy needs and exploring energy-conscious spatial planning are two main directions for further research and application in real world experiments.

A wide range of research horizons has to be considered in order to arrive at a synthetic view of the problems considered in this article. Hence, inter- and transdisciplinary research is essential for tackling problems of sustainable urban development. As argued in Section 2.2, environmentally sustainable schemes can thereby not be realized without linking to the social, cultural, economic, juridical, aesthetical, political and ethical spheres as well. For long, these spheres have been the 'battleground' for urban planners and designers and, we expect, will remain so also with regard to the energy transition. A systems perspective questioning the all too common business as usual scenarios is inevitable. In their discussion of a sustainable energy transition in Austria, Wächter et al. come to a similar conclusion when stating that the "… societal transformations towards a sustainable energy system are explicitly normative and value-laden and driven by the need to break with a business-as-usual path" ([59], p. 199). At the same time, many opportunities lie ahead.

Transiting to a renewable energy supply can result in secondary benefits such as healthy and convivial city centers, green jobs and increased welfare in more general terms. For the built environment, an important part of the transition effort can thereby be realized at a net profit from the economical point of view [62,63,64]. The renovation/transformation of the existing building stock is indeed the most promising field for climate-related intervention, under the condition that a life cycle costing approach is maintained and that certain environmental burdens, in particular CO_2-emissions, are properly taxed. From a designer's perspective, we must stress the need to include energy-conscious strategies at every stage of the planning process.

Finally, we consider this article as an invitation to open the debate on matters of 'hands on' energy-conscious urban planning. Today, we are confronted with a situation of urgently needed societal change while many of the tools needed to realize the energy transition are already at hand. What remains to be done is to creatively connect the missing links, both in research and application. The integration of new knowledge, changing societal values and innovative technologies in the sustainable transformation of the built environment remains a great challenge for urban planners and designers in many contexts.

REFERENCES

1. van Kann, F.; Leduc, W. Synergy between Regional Planning and Energy as a Contribution to a Carbon Neutral Society—Energy Cascading as a New Principle for Mixed Land-use. In Proceedings of the Scupad Conference: Planning for the Carbon Neutral World: Challenges for Cities and Regions, Salzburg, Austria, 15–18 May 2008.

2. Vandevyvere, H. Strategieën voor een Verhoogde Implementatie van Duurzaam Bouwen in Vlaanderen. Toepassing op het Schaalniveau van het Stadsfragment (Strategies Towards Increased Sustainable Building in Flanders. Application on the Scale of the Urban Fragment). Ph.D. Dissertation, K.U. Leuven, Leuven, Belgium, 2010.

3. Stöglehner, G.; Narodoslawsky, M. Energy-conscious Planning Practice in Austria: Strategic Planning for Energy-optimized Urban Structures. In Sustainable Energy Landscapes: Designing, Planning and Development; Stremke, S., van den Dobbelsteen, A., Eds.; Taylor & Francis: Boca Raton, FL, USA, 2012.

4. Allacker, K. Sustainable Building: The Development of an Evaluation Method; K.U.Leuven: Leuven, Belgium, 2010.

5. French, S.; Geldermann, J. The varied contexts of environmental decision problems and their implications for decision support. Environ. Sci. Policy**2005**, 8, 378–391.

6. Deakin, M.; Mitchell, G.; Nijkamp, P.; Vreeker, R. Introduction. In Sustainable Urban Development Volume 2: The Environmental Assessment Methods; Deakin, M., Mitchell, G., Nijkamp, P., Vreeker, R., Eds.; Routledge: London, UK, 2007.

7. Lombardi, P.; Brandon, P. The Multimodal System Approach to Sustainability Planning Evaluation. In Sustainable Urban Development Volume 2: The Environmental Assessment Methods; Deakin, M.,

Mitchell, G., Nijkamp, P., Vreeker, R., Eds.; Routledge: London, UK, 2007.

8. Vandevyvere, H. How to cut across the catch-all? A philosophical-cultural framework for assessing sustainability. Int. J. Innov. Sustain. Dev. **2011**, 5, 403–424.

9. U.S. Green Building Council. LEED 2009 for Neighborhood Development Rating System (Updated November 2011); Green Building Council: Washington, DC, USA, 2011.

10. BRE BREEAM Communities. SD5065B. Technical Guidance Manual; BRE Global Ltd: Watford, UK, 2009.

11. ISO. ISO 14040:2006 Environmental Management—Life Cycle Assessment—Principles and Framework; ISO: Geneva, Switzerland, 2006.

12. Sørensen, B. Life-Cycle Analysis of Energy Systems: From Methodology to Applications; Royal Society of Chemistry Publishing: Cambridge, UK, 2011.

13. Goedkoop, M.; Spriensma, R. The Eco-indicator 99 Methodology Report: A Damage Oriented LCIA Method; Ministerie van Volkshuisvesting, Ruimtelijke Ordening en Milieubeheer: Den Haag, The Netherlands, 1999.

14. Dones, R.; Heck, T. LCA-based Evaluation of Ecological Impacts and External Costs of Current and New Electricity and Heating Systems. In Proceedings of the Material Research Society Fall Meeting 2005, Symposium G: Life Cycle Analysis Tool for 'Green' Materials and Process Selection, Boston, MA, USA, 28–30 November 2005.

15. Kampman, B.; Bergsma, G.; Schepers, B.; Croezen, H.; Fritsche, U.R.; Henneberg, K.; Huenecke, K.; Molenaar, J.W.; Kessler, J.J.; Slingerland, S.; van der Linde, C. BUBE: Better Use of Biomass for Energy (Background Report to the Position Paper of IEA RETD and IEA Bioenergy); CE Delft/Öko-Institut: Delft, The Netherlands and Darmstadt, Germany, 2010.

16. Verbeeck, G. Optimisation of Extremely Low Energy Residential Buildings. Ph.D. Dissertation, K.U. Leuven, Leuven, Belgium, 2007.

17. Kohler, N. Life Cycle Analysis of Buildings, Groups of Buildings and Urban Fragments. In Sustainable Urban Development Volume 2: The Environmental Assessment Methods; Deakin, M., Mitchell, G., Nijkamp, P., Vreeker, R., Eds.; Routledge: London, UK, 2007.

18. Torio, H.; Schmidt, D. ECBCS Annex 49—Low Exergy Systems for High Performance Buildings and Communities (Annex 49 Final Report); Fraunhofer IBP/IEA: München, Germany, 2011.

19.   Stremke, S.; van den Dobbelsteen, A.; Koh, J. Exergy landscapes: Exploration of second-law thinking towards sustainable landscape design. Int. J. Exergy**2011**, 8, 148–174.

20.   Tillie, N.; van den Dobbelsteen, A.; Doepel, D.; de Jager, W.; Joubert, M.; Mayenburg, D. REAP Rotterdam Energy Approach and Planning: Towards CO2- Neutral Urban Development; Pieter Kers: Rotterdam, The Netherlands, 2009.

21.   Schmidt, D. Low exergy systems for high-performance buildings and communities. Energy Build.**2009**, 41, 331–336.

22.   Van Steertegem, M.E. MIRA-T Milieurapport Vlaanderen, Indicatorrapport '07; Vlaamse Milieumaatschappij: Aalst, Belgium, 2007.

23.   Hens, H. Duurzaam Bouwen; Francqui Leerstoel, Vrije Universiteit Brussel: Brussel, Belgium, 2006.

24.   van den Dobbelsteen, A.; Broersma, S.; Stremke, S. Energy potential mapping for energy-producing neighborhoods. Int. J. Sustain. Build. Technol. Urban Dev.**2011**, 2, 170–176.

25.   Stremke, S.; Kann, F.V.; Koh, J. Integrated visions (Part I): Methodological framework. Eur. Plan. Stud.**2012**, 20, 305–320.

26.   Jørgensen, S.E. Eco-Exergy as Sustainability; WIT Press: Southhampton, UK, 2006.

27.   Dewulf, J.; Van Langenhove, H.; Muys, B.; Bruers, S.; Bakshi, B.R.; Grubb, G.F.; Paulus, D.M.; Sciubba, E. Exergy: Its potential and limitations in environmental science and technology. Environ. Sci. Technol.**2008**, 42, 2221–2232.

28.   Sciubba, E.; Bastianoni, S.; Tiezzib, E. Exergy and extended exergy accounting of very large complex systems with an application to the province of Siena, Italy. J. Environ. Manag.**2008**, 86, 372–382.

29.   Jørgensen, S.E. Employing Exergy and Carbon Models to Determine the Sustainability of Alternative Energy Landscapes. In Sustainable Energy Landscapes: Designing, Planning and Development; Stremke, S., van den Dobbelsteen, A., Eds.; Taylor & Francis: Boca Raton, FL, USA, 2012.

30.   Lysen, E.H. The Trias Energetica: Solar Strategies for Developing Countries. In Proceedings of the Eurosun Conference; Freiburg, Germany: 16–19 September 1996.

31.   Broersma, S.; van den Dobbelsteen, A. Synergie Tussen Regionale Planning en Exergie: SREX; Publikatiebureau Bouwkunde: Delft, The Netherlands, 2011.

32.   Bauler, T. Indicators for Sustainable Development: A Discussion of Their Usability. Ph.D. Dissertation, Université Libre de Bruxelles, Brussels, Belgium, 2007.

33. Mondini, G.; Valle, M. Environmental Assessments Within the EU. In Sustainable Urban Development Volume 2: The Environmental Assessment Methods; Deakin, M., Mitchell, G., Nijkamp, P., Vreeker, R., Eds.; Routledge: London, UK, 2007.

34. Dooyeweerd, H. A New Critique of Theoretical Thought, Vol. 2: The General Theory of the Modal Spheres; The Presbyterian and Reformed Publisher Company: Phillipsburg, NJ, USA, 1955.

35. De Raadt, J.D.R. A sketch for humane operational research in a technological society. Syst. Pract.**1997**, 10, 421–441.

36. Basden, A. The critical theory of Herman Dooyeweerd? J. Inf. Technol.**2002**, 17, 257–269.

37. WCED. Our Common Future (Brundtland Report). United Nations— World Commission on Environment and Development: Geneva, Switzerland, 1987.

38. Cabeza-Gutés, M. The concept of weak sustainability. Ecol. Econ.**1996**, 17, 147–156.

39. Knowles, R.L. Energy and Form: An Ecological Approach to Urban Growth; MIT Press: Cambridge, MA, USA, 1974.

40. Salat, S. Energy loads, CO2 emissions and building stocks: Morphologies, typologies, energy systems and behaviour. Build. Res. Inf.**2009**, 37, 598–609.

41. Ratti, C.; Baker, N.; Steemers, K. Energy consumption and urban texture. Energy Build.**2005**, 37, 762–776.

42. Salat, S.; Nowacki, C. De l'importance de la morphologie dans l'efficience énergétique des villes. Energ. Territ. Liaison Energ. Francoph.**2010**, 86, 141–146.

43. Brown, R.D.; Gillespie, T.J. Microclimatic Landscape Design: Creating Thermal Comfort and Energy Efficiency; John Wiley &Sons: New York, NY, USA, 1995.

44. Robinette, G.O.; McClenon, C. Landscape Planning for Energy Conservation; Van Nostrand Reinhold: New York, NY, USA, 1983.

45. Thompson, J.W.; Sorvig, K. Sustainable Landscape Construction: A Guide to Green Building Outdoors; Island Press: Washington, DC, USA, 2000.

46. Shashua-Bar, L.; Hoffman, M.E.; Tzamir, Y. Integrated thermal effects of generic built forms and vegetation on the UCL microclimate. Build. Environ.**2006**, 41, 343–354.

47. Gill, S.E.; Handley, J.F.; Ennos, A.R.; Pauleit, S. Adapting cities for climate change: The role of the green infrastructure. Built Environ.**2007**, 33, 115–133.

48. Alexandri, E.; Jones, P. Temperature decreases in an urban canyon due to green walls and green roofs in diverse climates. Build. Environ.**2008**, 43, 480–493.

49. Santamouris, M. Environmental Design of Urban Buildings: An Integrated Approach; Earthscan Publications: London, UK, 2006.

50. Newman, P.; Kenworthy, J.R. Cities and Automobile Dependence: An International Sourcebook; Gower: Aldershot, UK, 1989.

51. Brick, K. Report Summary—Follow Up of Environmental Impact in Hammarby Sjöstad by Sickla Udde, Sickla Kaj, Lugnet and Proppen; Grontmij AB: Stockholm, Sweden, 2008.

52. Hodge, J.; Haltrecht, J. BedZED Seven Years on: The Impact of the UK's Best Known Ecovillage and Its Residents. BioRegional: Hackbridge, UK, 2009.

53. van den Dobbelsteen, A.; de Wilde, S. Space use optimisation and sustainability—Environmental assessment of space use concepts. J. Environ. Manag.**2004**, 73, 81–89.

54. Ramachandra, T.V.; Shruthi, B.V. Spatial mapping of renewable energy potential. Renew. Sustain. Energy Rev.**2007**, 11, 1460–1480.

55. Clement-Nyns, K.; Haesen, E.; Driesen, J. The impact of vehicle-to-grid on the distribution grid. Electr. Power Syst. Res.**2011**, 81, 185–192.

56. Saelens, D. Optimale Inzet van Thermische Opslag en Actieve Energieconcepten in Gebouwen om Maximaal in te Spelen op de Elektriciteitsmarkt en Behoeften van het net. 2011. Available online: http://www.kuleuven.be/onderzoek/onderzoeksdatabank/project/3E11/3 E110062.htm (accessed on 18 June 2012).

57. Bühl, J. Solarenergienutzung und Effizienzsteigerung in und an Gebäuden. In Klimawandel und ökologischer Umbau der Industriegesellschaft; Europäischen Informations-Zentrum in der Thüringer Staatskanzlei: Erfurt, Germany, 2010.

58. Czisch, G. Scenarios for a Future Electricity Supply—Cost-Optimized Approaches to Supplying Europe and its Neighbours with Electricity from Renewable Energies. Ph.D. Dissertation, Universität Kassel, Kassel, Germany, 2006.

59. Wächter, P.; Ornetzeder, M.; Rohracher, H.; Schreuer, A.; Knoflacher, M. Towards a sustainable spatial organization of the energy system: Backcasting experiences from Austria. Sustainability**2012**, 4, 193–209.

60. Roggema, R. INCREASE II Conference Proceedings; Province of Groningen: Groningen, The Netherlands, 2009.

61. Stremke, S.; Koh, J. Integration of ecological and thermodynamic concepts in the design of sustainable energy landscapes. Landsc. J.**2011**, 30, 194–213.

62. Barker, T.; Bashmakov, I.; Bernstein, L.; Bogner, J.E.; Bosch, P.R.; Dave, R.; Davidson, O.R.; Fisher, B.S.; Gupta, S.; Halsnæs, K.; et al. Technical Summary. In Climate Change 2007: Mitigation. Contribution of Working Group III to the Fourth Assessment Report of the Intergovernmental Panel on Climate Change; Metz, B., Davidson, O.R., Bosch, P.R., Dave, R., Meyer, L.A., Eds.; Cambridge University Press: Cambridge, UK and New York, NY, USA, 2007.

63. Jaeger, C.; Paroussos, L.; Mangalagiu, D.; Kupers, R.; Mandel, A.; Tàbara, J. A New Growth Path for Europe: Generating Prosperity and Jobs in the Low-Carbon Economy; Synthesis Report. European Climate Forum e.V.: Potsdam, Germany, 2011.

64. McKinsey-and-Company Pathways to World-Class Energy Efficiency in Belgium; McKinsey & Company: Brussels, Belgium, 2009.

Index